Y0-BQN-546

# DYNAMICS AND CONTROL OF WASTEWATER SYSTEMS

## HOW TO ORDER THIS BOOK

BY PHONE: 800-233-9936 or 717-291-5609, 8AM–5PM Eastern Time

BY FAX: 717-295-4538

BY MAIL: Order Department
Technomic Publishing Company, Inc.
851 New Holland Avenue, Box 3535
Lancaster, PA 17604, U.S.A.

BY CREDIT CARD: American Express, VISA, MasterCard

BY WWW SITE: http://www.techpub.com

WATER QUALITY MANAGEMENT LIBRARY

**VOLUME 6**

SECOND EDITION

# DYNAMICS AND CONTROL OF WASTEWATER SYSTEMS

EDITED BY


Michael W. Barnett, Ph.D.

Michael K. Stenstrom, Ph.D., P.E.

John F. Andrews, Ph.D., P.E.

LANCASTER · BASEL

**Water Quality Management Library—Volume 6**
a **TECHNOMIC** publication

Technomic Publishing Company, Inc.
851 New Holland Avenue, Box 3535
Lancaster, Pennsylvania 17604 U.S.A.



Printed in the United States of America
10 9 8 7 6 5 4 3 2 1

Main entry under title:
Water Quality Management Library—Volume 6 / Dynamics and Control of Wastewater Systems, Second Edition

A Technomic Publishing Company book
Bibliography: p.
Includes index p. 335

Library of Congress Catalog Card No. 98-85170
ISBN No. 1-56676-672-9 (Volume 6)
ISBN No. 1-56676-660-5 (11-Volume Set)

# Table of Contents

*Foreword* xiii

**Part I: Technology and Tools**

1. **DYNAMICS AND CONTROL OF WASTEWATER TREATMENT SYSTEMS: AN OVERVIEW** . . . . . . . . . . . . . . . . . . . . . . . . . . . . . . . . . . . . . . . **3**
JOHN F. ANDREWS

Benefits and Constraints 4

Modeling 6
*Model Classification* 7
*Mathematical Models* 9
*Development of Mathematical Models* 12

Computer Simulation 17
*History of Computer Simulation* 17
*Advantages and Disadvantages of Computer Simulation* 18

Control Systems 19
*Information Handling* 19
*Basic Questions* 21
*Feedback Control* 21
*Feedforward Control* 23
*Feedforward–Feedback* 24
*Use All Available Information* 24

Summary 26

Literature Sources 27

References 27

2. **MATHEMATICAL MODELING AND COMPUTER SIMULATION** ... **31**
MICHAEL K. STENSTROM, HSUEH-HWA LEE, JINN-SHIU MA and MICHAEL W. BARNETT

An Approach to Process Modeling 32
*Identify the Significant Mass Transfer/Reaction Terms* 32
*Establish the Stoichiometry* 33
*Determine the Kinetics* 34
*Specify or Characterize the Reactor* 34
*Develop Model Equations* 37

Reactor Modeling and Simulation Using Matlab 39
*Water Tank Modeling and Simulation* 40

Summary 51
References 54

3. **PROCESS CONTROL IN BIOLOGICAL WASTEWATER TREATMENT** ... **55**
GUSTAF OLSSON

Overview 55
Goals and Incentives for Control 56
*Criteria* 56

Plant Dynamics 58
*Time Scale of the Process* 58
*Disturbances* 59

Models for Control 60
*Who Needs the Model?* 61
*Dynamical Models* 62
*Model Limitations* 63
*Particular Difficulties* 64

Goals of Operation 65
Measurement and Control Variables 67
*Measurement Variables and Sensor Technology* 67
*Manipulated Variables* 68

Simple Controllers 69
*Basic Feedback Control Structure* 70
*Examples of Local Feedback Control* 70
*Open Loop Sequencing* 71
*On-Off Control* 71
*The PID Controller* 73

Plant Quality Control 75

*Control Goals for the Activated Sludge Anoxic Zone* 76
*Dissolved Oxygen Control* 79
*Choice of Set-Points for Dissolved Oxygen Control* 80
*Sludge Inventory Control* 81

Controller Implementation 83
*Sampled Control* 83
*Discretization of the PID Controller* 84
*Practical Considerations for Controllers* 85
*Sampling Rate in Control Systems* 85
*Control Signal Limitations* 86

When are Simple Controllers Not Sufficient? 86
*Time Delays* 87
*Parameter Variations* 87

Programmable Controllers 89
*A Block Language Implementation* 89
*Sequencing and Logical Circuits* 90
*Analog Controllers* 91
*Communication* 92

References 94

4. **RESPIROMETRY IN MODELING AND CONTROL OF WASTEWATER TREATMENT** ........ **99**
H. SPANJERS

Introduction 99

Fundamentals of Respiration 100
*Biochemical Background* 100
*Oxygen Demanding Processes in Activated Sludge* 101
*Modeling Respiration* 102

Measurement of Respiration Rate 105
*Techniques Based on Oxygen Measurement in the Liquid Phase* 105
*Techniques Based on Oxygen Measurement in the Gas Phase* 107
*Measurement of Related Variables* 109
*General Remarks to the Above Techniques* 110

Respirometry in Modeling 110
*Direct Methods* 111
*Model Optimization* 119

Respirometry in Control 124
*Respiration Rate as a Controlled Variable* 124

*Controlled Variable Deduced from Respiration Rate* 125
Epilogue 131
References 131

5. **INTELLIGENT SYSTEMS FOR CONTROL AND DECISION SUPPORT IN WASTEWATER ENGINEERING** .................. **133**
MICHAEL W. BARNETT

Introduction 133
Representation 135
*Objects* 136
*Rules* 139
*Methods* 141
*Certainty* 143
Modeling 149
*Mechanistic* 150
*Empirical* 152
Optimization 156
*Genetic Algorithms* 157
*Agents* 163
Control 167
Summary 168
References 170

**Part II: Applications**

6. **DYNAMICS and CONTROL OF URBAN DRAINAGE SYSTEMS** .. **175**
Z. CELLO VITASOVIC and SIPING ZHOU

Introduction to Control of Sewer Networks 175
*Background* 175
*Components of Urban Drainage Systems* 176
Levels (Modes) of RTC 177
Automatic Control 181
Modeling of Urban Sewer Networks 184
*Hydraulic Models for Sewer Systems* 186
Organizational Aspects of RTC Implementation 187
Applications of RTC to Urban Collection Systems 188
*Optimization of Sewer Networks* 189
*Combining Optimization with Hydraulic Model* 191
*Case Study 1: RTC in Seattle, Washington* 192

*Case Study 2: RTC in Hamilton, Ontario* 201

References 215

7. **THE INTEGRATED COMPUTER CONTROL SYSTEM: A COMPREHENSIVE, MODEL-BASED CONTROL TECHNOLOGY** ... **219**

IMRE TAKÁCS, GILLES PATRY, BRUCE GALL, and JASMIN PATRY

Introduction 219

*Background* 219

*$IC^2S$ Layers* 220

*The Building Blocks of $IC^2S$* 221

*Functional Modules Inside $IC^2S$* 222

The Building Blocks of $IC^2S$ 223

*Adaptive Data Filter (ADF)* 223

*Signal Tracking and Alarm (SIGTRACK)* 227

*Respirogram Evaluator (RespEval)* 228

*Dynamic Parameter Estimator (DPE)* 231

*Advanced Control Design (GMI)* 238

*Simulation Model (GPS-X)* 255

*Operator Interface (Scenario Manager)* 256

*Bridge* 257

$IC^2S$ Functional Modules 258

*Auto-Calibration Module* 259

*Advanced Fault Detection Module* 260

*Process Optimization Module* 260

*Continuous Forecasting Module* 261

$IC^2S$ Demonstration Example 262

*The Wastewater Treatment Plant Used for Demonstration* 262

*The Model Layouts Used in the On-Line System* 263

*The SCADA Link* 267

*Auto-Calibration Module* 269

*Advanced Fault Detection Module* 275

Summary 278

Acknowledgements 278

References 279

8. **APPLICATION OF INTELLIGENT CONTROL IN WASTEWATER TREATMENT** ... **281**

TAKAYUKI OHTSUKI, TETSUYA KAWAZOE, and TAKAAKI MASUI

Introduction 281

*Intrinsic Unsteadiness* 281
*Nonlinearity* 281
*Complexity of the Process* 282
*Poor Process Understanding and Dependence on Empirical Knowledge* 282
*Lack of Kinetic Information of the Process* 282
*Control Objective Change* 283

Model Representations for Wastewater Treatment System 283
*Theoretical Dynamic Models* 283
*Empirically Extracted Models* 284
*Linguistic Models* 285
*Fuzzy Models* 286

Intelligent Control System Based on Blackboard Architecture 286

Shared Data Types on Blackboard 291
*Fuzzy Number* 291
*Judgement Data* 291
*Fuzzy Membership* 293
*Trend Data* 293

Implemented Expert Modules 293
*Fuzzy Expert EM* 293
*Fuzzy Control EM* 296
*Theoretical Model (IAWQ No. 1) EM* 296
*Respirogram Evaluation EM* 296

Application Examples of the Intelligent Control System to Wastewater Treatment System Control 297
*High-Rate Activated Sludge Process* 297
*Case 1. Aeration and Methanol Control Example* 300
*Case 2. Influent Load Control Based on Nitrification Activity Forecast* 305

Discussions 312

Conclusion 313

References 313

9. **DYNAMICS AND CONTROL OF THE HIGH PURITY OXYGEN (HPO) ACTIVATED SLUDGE PROCESS** .............. **315**
MICHAEL K. STENSTROM

Introduction 315

Process Description 316

Conventional Control Strategies 319
*Reactor 4 Oxygen Purity and Total System Pressure Control Systems* 320

*Reactor Dissolved Oxygen Control* 323
*Reactor Sizes* 324
Knowledge Based Control Strategies 327
Conclusions 332
References 332

*Index* 335

# Foreword

IN 1992 the United States National Committee of IAWQ (International Association on Water Quality) organized eight specialty courses offered in conjunction with the 1992 IAWQ Biennial Conference in Washington, D.C. Designed for the practicing engineer, the specialty courses covered critical topics in environmental quality management water pollution control, wastewater treatment, toxicity reduction, and residuals management. These courses were compiled in an eight-volume series as the Water Quality Management Library. Experts from the United States and many countries contributed their expertise and experience to the preparation of these state-of-the-art texts.

The success of this series prompted the editors to expand the series to include volumes on such timely topics as water reuse, non-point source control and aeration and oxygen transfer. Additional volumes are presently being considered. In addition to the new topics, in order to keep pace with this rapidly developing field, many of the original volumes have been updated to reflect current advances in the field. In addition to providing an up-to-date technical reference for the practicing engineer and scientist as in the first series, these volumes will provide a text for continuing education courses and workshops.

The Water Quality Management Library should provide a unique reference source for professional and education libraries.

W. WESLEY ECKENFELDER
JOSEPH F. MALINA, JR.
JAMES W. PATTERSON

# Foreword

# PART I: TECHNOLOGY AND TOOLS

CHAPTER 1

# Dynamics and Control of Wastewater Treatment Systems: An Overview

WASTEWATER treatment plants are inherently dynamic because of temporal variations in influent wastewater flow rate, concentration, and composition. Quantitative descriptions of the dynamic behavior of these plants and the use of control systems to convert unsatisfactory dynamic behavior to satisfactory behavior have significant potential for improving plant performance and reducing treatment costs. Yet most mathematical models currently in use for plant design are steady state instead of dynamic which means that changes with respect to time cannot be predicted. Why is this so? One major reason is that many dynamic models are composed of differential equations for which there are no known analytical solutions. The assumption of steady state can simplify such models by reducing differential equations to algebraic equations for which solutions are available. Another major reason is that much more data is required for the validation and calibration of dynamic models than for steady state models.

The computational bottleneck for the solution of dynamic models has been essentially eliminated by the ready availability of simulation languages for digital computers. Moreover, rapid advances in both hardware and software for personal computers have resulted in inexpensive, user friendly systems for computer simulation. However, there are still problems with respect to the availability of reliable on-line sensors for data collection as well as for implementation of on-line control.

The primary objective of this paper is to introduce the reader to dynamic

John F. Andrews, Professor Emeritus, Environmental Science & Engineering Dept., Rice University, Houston, Texas. Address correspondence to John F. Andrews, 1719 Rayview, Fayetteville, AR 72703-2625.

modeling, computer simulation, and control systems for wastewater treatment plants. Although the application of these tools to process operation will be emphasized herein, there are strong interactions between process design and control system design which must be integrated to obtain the best overall performance at reasonable cost. In order to obtain this integration, process design and control system engineers must learn to speak each other's technical language. It is hoped that the information contained in this book will contribute to this goal.

## BENEFITS AND CONSTRAINTS

In a 1974 paper [1] the author presented a listing and discussion of the potential benefits from the use of dynamic models and control systems in wastewater treatment. These are briefly reviewed in Table 1.1. The reader can use his own judgment in estimating how far we have come since 1974 in realizing these benefits.

Most of the items listed are self explanatory; however, the purposeful operation of treatment plants at variable efficiency to match the dynamics of the plant effluent with those of the receiving stream is worthy of further attention. Both the quantity and quality of water in most streams vary with time, yet permit requirements for most plants are based on 10 year, 7-day low flow conditions. As a result, the wastewater may be overtreated much of the time thus wasting money. A first step in recognizing the dynamic characteristics of streams is the use of seasonal effluent requirements. However, how about using real time control based on dynamic models of the plant and receiving stream? Rossman [2] has addressed this interface area between treatment plants and receiving streams with emphasis on the types of stream models and changes in regulatory agency requirements that might be needed.

Olsson [3] has summarized the constraints on the use of instrumentation, automation, and control (ICA) in water and wastewater treatment and transport systems. This summary, given in Table 1.2, is based on brief "country status reports" by representatives from 16 countries who attended the 6th International Association on Water Quality workshop on ICA in Banff/Hamilton, Canada, June, 1993. The first constraint, legislation/regulation, is obviously pertinent to the potential for variable efficiency operation of treatment plants.

The last item given in Table 1.2, integrated control systems, needs much more attention in the future. For example, in past years many models of the activated sludge process have considered only the aeration basin unit of the process without adequate consideration of the interactions between the aeration basin, secondary settler, and air supply units. All

TABLE 1.1. **Potential Benefits from the Use of Dynamic Models and Control Systems in Wastewater Treatment Plants.**

**Performance**—Reducing effluent variability while keeping effluent quality nearer the maximum can result in significant decreases in the discharge of pollutants.

**Productivity**—Improved control can increase the amount of wastewater processed per unit of plant capacity. This can prolong the design life of existing plants or reduce required sizes of new plants.

**Stability**—Dynamic models can be used to compare the stability of different processes and control systems can be used to improve process stability.

**Reliability**—Computer monitoring and control can reduce gross process and equipment failures as well as decrease permit violations.

**Operating Personnel**—Computer monitoring and control can minimize the number of operating personnel as well as improve their productivity.

**Operating Costs**—These costs can be reduced by considering dynamic behavior and using control systems to regulate power and chemical additions.

**Start-up Procedures**—Conditions are obviously not steady state during process start-up. Dynamic models can be of value in obtaining more reliable and faster start-ups.

**Dynamic Operation**—This is the purposeful operation of processes as dynamic systems. Steady state operation may not always be the best operational mode. An example is the sequencing batch reactor.

**Variable Efficiency Operation**—Matching the dynamics of the plant effluent with those of the receiving stream should be explored as a technique for reducing operational costs.

three must function adequately for the process as a whole to perform satisfactorily. Another example of an important interaction between plant units is that between the secondary settler and influent pumping station. The on-off control of large pumps can sometimes lead to excessive discharge of solids in the plant effluent due to the generation of turbulence in the settler.

The need to integrate the dynamics of the plant with those of the receiving stream has already been discussed. However, there is also a need to integrate the plant dynamics with those of the collection system and sludge transport and disposal system. As an example for sludge transport, disposal on land may be delayed during inclement weather and provision must therefore be made for sludge storage during these periods. The chapter by Vitasovic and Zhou discusses the modeling and control of collection systems and their interface with the treatment plant.

TABLE 1.2. **Constraints on the Use of Instrumentation, Control and Automation in Wastewater Treatment.**

| |
|---|
| **Legislation/Regulation**—Many regulatory standards are based on steady state "worst case" conditions in receiving streams and do not consider dynamic variations. |
| **Education and Training**—Operators are not always well trained in instrumentation and control. Most environmental engineers need more education in process dynamics and control. |
| **Economics**—ICA is often not adequately considered in the initial plant design. Instead, it is introduced after completion of the design. Consequently, a proper balance between process costs and ICA costs is not achieved. |
| **Measuring Devices**—Easy-to-use, low maintenance, reliable on-line sensors are not available for some important measurements. |
| **Plant Constraints**—Plants are often designed to accommodate fluctuations in inputs by provision of additional capacity instead of providing flexibility for exertion of control. This can penalize those plants where skilled operators are in charge. |
| **Models for Operation**—Many of the dynamic models which have been developed are too complex for use in plant operations and include terms which are difficult to measure in the field. |
| **Software**—Many software systems are proprietary and often inflexible and "user unfriendly." Users are asking for open systems (those which can handle a mixture of software and hardware from different manufacturers) to address these problems. |
| **Integrated Control Systems**—There are strong interactions between the components of treatment plants as well as between the plant, the collection system, the receiving body of water, and the sludge transport and disposal system. Integrated models and control systems are needed to take these interactions into account. |

## MODELING

Mathematical modeling is a technique frequently used (and sometimes abused!) in both scientific and engineering research and practice. However, modeling itself is not new since scaled-down physical models have long been used in such diverse areas as astronomy (planetariums), hydraulic engineering (river models), architecture (building models), and chemical engineering (pilot plants). Biologists, chemists, and sociologists have studied model organisms (*E. coli* for example), model compounds, and model communities, respectively. Even the hypothesis which is formulated in applying the scientific method can be considered as a verbal model and simulation may be used to test the validity of a hypothesis when it is not feasible to test the real system.

## MODEL CLASSIFICATION

There are many types of models with the type selected being primarily dependent upon the purpose for which it is to be used. One possible classification is given in Table 1.3.

Each of these types of models can be further broken down into different categories. For example, visual models may consist of schematic diagrams (pictorial models) such as that presented in Figure 1.1, or a flow diagram for a wastewater treatment plant. These diagrams are adequate for illustrating system components and qualitative description of some of the interactions between components. However, they are not adequate for quantitative descriptions and a following step for design engineers is the preparation of plans (another type of pictorial model) to give dimensions of the components and distances between them. Geographical information systems (GIS) generate still another form of pictorial model.

There are also several different categories of linguistic models. An example is a procedural model consisting of a written and ordered list of tasks to be accomplished for a construction project. The addition of time to a procedural model results in a schedule (dynamic model) for completing the project. More recent examples of linguistic models are rule based expert systems. These are a class of artificial intelligence (AI) computer programs usually intended to serve as computer assistants for decision making or sometimes for direct use in automatic control. In the context of this paper, they are usually written in the form of If-Then rules for describing dynamic behavior and specifying control actions. They overlap with mathematical models since they are based on the laws of logic. An example of a rule for a pumping station might be "**IF** the water level in the wet well is above 10 ft. **THEN** turn on pump No. 2." A chapter by Barnett discusses rule based systems in more detail.

A type of model which must precede either visual or linguistic models is a mental model. A common example of the interactions between mental, linguistic, and visual models in engineering is when an engineer converts a mental model of an idea into a linguistic model in discussions with

TABLE 1.3. Classification of Models.

| |
|---|
| Visual |
| Linguistic |
| Mental |
| Physical |
| Mathematical |
| Fuzzy |

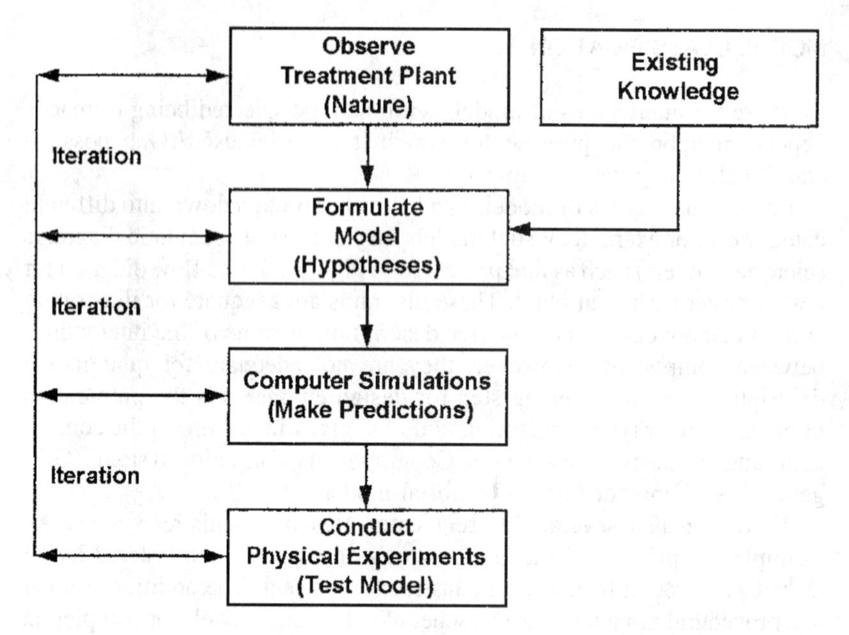


**Figure 1.1** Steps in the development of mathematical models.

coworkers. During the discussions, the key points of the model are often illustrated by sketches or visual models. A more recent example is the expert system computer program in which the mental models of skilled operating engineers are converted to If-Then rules.

Physical models are well known and often used by engineers and scientists. In wastewater treatment plant engineering, three types of physical models may be used in taking an idea (mental model) to application in the field. These are bench, pilot, and prototype. Bench scale experiments are usually designed to give ''Yes or No'' (qualitative) answers while pilot scale experiments give more quantitative answers. However, the ultimate test is whether the idea will work under field conditions so prototype testing is also an important step in physical modeling.

Mathematical models are used for more quantitative descriptions and consist of one or more equations relating the important inputs, outputs, and characteristics of a system. To those not familiar with systems engineering terminology, the term mathematical model is sometimes frightening since it may bring to mind large sets of complex equations and the use of sophisticated mathematical techniques. However, this need not be the case since most common engineering design formulae may also be called mathematical models. For example, as simple an expression as $y = mx + b$ can be considered as a mathematical model where, in systems engineering terms, the system output ($y$) is related to the system input ($x$) by the system parameters ($m,b$).

Fuzzy models are a compromise between the vague statements which we humans often use and the strict logic of rule based expert systems (Yes or No or True or False) or quantitative answers provided by mathematical models. They permit improved communication between the computer and humans by converting (translating) human statements such as high, normal, and low into numbers which the computer can understand (defuzzification) and numbers generated by the computer into statements for humans (fuzzification). A common example of the use of a mental fuzzy model is the ordinary traffic signal in which a green light means *Go* and a red light means *Stop.* However, what does a yellow light mean? If we are very close to an intersection when the light turns from green to yellow we may *Go* on yellow since the alternate could be being hit from behind if we brake too rapidly. However, if we are some distance from the intersection when the light turns from green to yellow we will most likely slow down and then *Stop.* In other words, we have exerted fuzzy control over our automobile based on vague information (''very close'' or ''some distance'' from the intersection). Chapters by Barnett and by Ohtsuki, et al, should be consulted for more detail on fuzzy logic, models, and control systems.

## MATHEMATICAL MODELS

Mathematical models may also be classified in many different ways among these being the purpose for which the model is to be used (planning, design or operation), the prediction to be made (performance, cost, reliability, etc.), the required accuracy of the predictions (qualitative or quantitative and all shades in between), the time behavior of the system (dynamic, steady state, continuous, discrete), and the mathematical or computer techniques to be used (algebra, differential equations, difference equations, statistics, logic, fuzzy math, etc.). Only a few classifications can be considered herein. Proponents of a specific type of model sometimes take the approach that their type of model is the ''only'' type of model to use so the word ''or'' is used herein to contrast different categories of models. However, the author is of the opinion that all of the different categories are useful under the appropriate circumstances and that practical models for application will ultimately make use of combinations of these different types. The categories of mathematical models to be discussed herein are given in Table 1.4.

### Dynamic or Steady State?

One of the most important distinctions for wastewater treatment processes is that between dynamic and steady state models. Many models

TABLE 1.4. Classification of Mathematical Models.

| |
|---|
| Dynamic or steady state? |
| Mechanistic or empirical? |
| Deterministic or stochastic? |

currently in use are based on the assumption of steady state which means that changes with respect to time cannot be predicted. The assumption of steady state can result in the reduction of differential equations to algebraic equations and thus simplify the model. Prior to the advent of computer simulation, such simplifications, although frequently unrealistic, were often the only technique for making the equations solvable. Consequently many models of wastewater treatment processes are still steady state models even though in practice the inputs are usually highly dynamic. Steady state models have been of considerable value in plant design by indicating average performance or performance at minimum or maximum loadings. However, by definition they cannot indicate how performance varies with time which is necessary for plant operation. Dynamic models are required for predicting time varying behavior and applying control systems and it is with such models that this paper is primarily concerned.

In order to simplify a dynamic model, some of the equations may be based on steady state or equilibrium relationships. For example, some chemical equilibria in solution, such as the ionization of an acid, take place very quickly (milliseconds) in comparison with biological rates (hours and days) so it is perfectly acceptable in models of biological processes to consider these chemical reactions to be in equilibrium. Also, some variables may change so slowly in comparison with the variables of interest that they may be considered as constants over the time period of interest. An example would be some sludge transport and disposal systems in which seasons are the time periods of interest so they could be considered as "steady state" with respect to the dynamics (hours and days) of treatment plants.

## Mechanistic or Empirical?

Mechanistic models, sometimes called theoretical models, are based on fundamental scientific and engineering knowledge about the physical, chemical, and biological phenomena which govern the system. Some of the more important phenomena for wastewater treatment processes are specific chemical and biological reactions, stoichiometry, equilibrium relationships, reaction kinetics, mass transfer, and conservation equations (mass, momentum, and energy balances). Mechanistic models give more insight into system behavior and may be more reliably extrapolated when data is not available. In addition, the use of fundamental principles enables one to search for systems with similar characteristics and thus draw on existing knowledge from other branches of engineering and science. In the next chapter, Stenstrom will give several examples of the development of simple mechanistic models.

When systems are complex the mechanisms involved may be either unknown or poorly understood and it may be necessary to use empirical relationships for all or parts of a model. An example is the biological oxygen demand (BOD) equation for expressing the exertion of oxygen demand with respect to time. Models based on such relationships have been used throughout the history of engineering and can be perfectly satisfactory as long as their limitations are recognized. Principal limitations are that they cannot be extended with any degree of confidence to similar systems or beyond the range over which data have been collected. The increasing use of computer monitoring in plant operations should result in more complete data bases thus extending the range of validity of empirical models.

A more recent example of an empirical model is the artificial neural network model which is inspired by biological models of the brain. Such models are able to modify their outputs in response to their inputs or "learn." The network is "taught" to give the appropriate responses by presenting it with a large number of examples along with the appropriate response to each example. Computer data bases developed during plant operation can provide such examples. More detailed information on neural networks will be provided later by Barnett.

Models used or developed by engineering researchers tend to be mechanistic because their primary purpose is to extend engineering knowledge whereas models used in practice are usually more empirical since practitioners often cannot wait for the explanation of why something occurs (the mechanisms). However, because engineering is practical in nature, most engineering models, whether for research or practice, are usually neither purely mechanistic or empirical but instead "blends" of mechanistic and empirical components. The long term trend in the history of engineering has been for models to become more fundamental and less empirical as more knowledge is gained about a system. The obtaining of fundamental knowledge is often slow, laborious, and unpredictable. The increasing availability of computer data bases for plant operations should be of considerable help in speeding up this process. However, from a practical viewpoint it is expected that a significant portion of the relationships used in models for wastewater treatment plants will continue to be empirical for some time.

### Deterministic or Stochastic?

Deterministic models are those in which the inputs, outputs, and system parameters can be assigned a definite fixed number, or series of fixed numbers, for any given set of conditions. In contrast, the principle of uncertainty is introduced in stochastic models, sometimes called probabilistic models, and statistical techniques are used to express the model in a mathematical form. An example would be a model to express the probability of low flows occurring in a stream as a function of time between low flow periods. For those not familiar with statistical procedures, Chapman [4] has discussed basic concepts and their application to the analysis of data from plant operations.

Just as models may have both mechanistic and empirical components, they may also have both deterministic and stochastic components. For example, the wastewater flow into a treatment plant is often approximated as a daily repeatable cycle (deterministic) with superimposed on this being a random or stochastic component. Time series analysis is a commonly used technique for separating deterministic (trends, amplitude and frequency) and stochastic components. These are developed by collecting large amounts of data and applying statistical techniques to fit empirical functions to the data. Hiraoka and coworkers [5,6] have made extensive use of time series analysis and computer data bases of plant operations in dynamic modeling and computer control of wastewater treatment plants and have reported on the application of these techniques at the Kawamata plant in Osaka, Japan.

## DEVELOPMENT OF MATHEMATICAL MODELS

Mathematical modeling is widely used by many different disciplines as well as for many different purposes. It is quite understandable that there would be disagreements between the different users and perhaps nowhere are these differences more obvious than those between engineering researchers and engineering practitioners. Some of the basic concepts involved in model development, as shown in Table 1.5, will be discussed with emphasis on these differences. It is hoped that this discussion will

TABLE 1.5. Some Basic Concepts Involved in Model Development.

| Concepts |
|---|
| Use of models |
| Information for development |
| Model testing |
| Required accuracy |

lead to improved understanding of the different viewpoints and more joint efforts by engineering researchers and practitioners in the application of mathematical modeling to both the design and operation of wastewater treatment plants.

### Use of Models

The primary objective of most engineering researchers is to extend engineering knowledge and they thus see modeling as a tool for guiding research with possibly application to practice at some future date. The researcher's initial models may be considered as mathematical versions of the verbal hypotheses used in the scientific method in which it is implicitly recognized that there is a possibility of being wrong. The researcher should thus expect and be prepared for the fact that the model may give inaccurate or even completely false predictions. This recognition of initial models as hypotheses makes it clear that the development of models is an iterative process (Figure 1.1) in which models are formulated, predictions compared with observed results, and the model revised as needed. This procedure is repeated until adequate agreement is obtained between predictions and observations and can be a time consuming process with years sometimes being required before a research model is suitable for practical application.

The practitioner must usually operate on a much shorter time scale and is more concerned with immediate application of models to assist in design, construction, or operation of real systems with advancement of engineering knowledge being secondary. Practitioners usually recognize the need for collecting data to obtain numerical values for some of the parameters and expect that there will be uncertainty in model predictions; however, they normally expect the basic structure of the model to be such that it will give satisfactory predictions without being modified. The practitioner cannot afford highly inaccurate predictions since the consequences (loss of life, reputation, money) can be severe. Also, practitioners do not have the time to go through several iterations of the modeling process. These differences in the use of and expectations from models by engineering researchers and practitioners can lead to disagreements as to the value of models unless their intended use is clearly defined.

### Information for Development

Other points of possible misunderstanding between engineering researchers and practitioners, as well as between scientific and engineering researchers, are the sources of information for developing mathematical models. Figure 1.1, which might also be called a block diagram of the scientific method, illustrates the use of four sources of information, these being; (1) existing knowledge, (2) observation of plant behavior, (3) model predictions, and (4) the results of planned experiments. The emphasis placed on each of these different sources of information is dependent upon the basic discipline and background of the developer or user of the model. For example, design engineers tend to stress sources of design information whereas operating engineers emphasize their experience, and that of others, in observing plant behavior. Also, some researchers emphasize the conduct of physical experiments while other researchers emphasize modeling and simulation.

The author is of the opinion that many different sources of information have value for the development of models. This is especially so for wastewater treatment plant models which require knowledge from a wide variety of scientific and engineering disciplines as well as practical knowledge from design engineers, operating engineers and equipment manufacturers. It is the synthesis of existing knowledge from other disciplines into models that may be one of the distinguishing characteristics between engineering and scientific research. Engineering researchers do consider new applications of existing knowledge from the sciences and other branches of engineering as original research whereas some scientists consider only the discovery of new knowledge as original research.

### Model Testing

The amount of testing needed will be a function of the purpose for which the model is to be used and the blend of existing knowledge and new hypotheses incorporated in the model. If the model is primarily based on well established principles from the sciences or other engineering disciplines, the structure of the model should be adequate and all that may be required is determination of numerical values for the parameters in the model. Moreover, first estimates of some of these may either be available or can be calculated from data available in the literature. Such models could well have been established in previous years since much of the basic knowledge needed to set up the equations comprising the model may have been available for some time. However, they may not exist for a reason previously given, this being that most engineers are practical people and inclined to say "Why set up equations for which it is not currently possible to obtain solutions?" In other words, advances in solution techniques may result in a renewed interest in old problems.

In the above case it might be more appropriate to say that the structure of the model requires no testing and the purpose of testing is to determine numerical values for the parameters. Such models are no longer in the realm of engineering research but are instead ready for engineering application.

At the other extreme are research models containing primarily new hypotheses for which only qualitative observations are available to support the hypotheses. Such models need extensive testing to prove or disprove the hypotheses and, if these are proven correct, to structure the model so that it can be applied in practice. Both modeling and simulation can play a significant role in this testing. Simulations can be used to design physical experiments with first experiments usually being at bench scale to prove or disprove hypotheses and obtain rough estimates of parameters. The model is modified to incorporate these increases in knowledge and then used in simulations to design experiments for pilot and/or full-scale testing.

Figure 1.1 can also be used to illustrate the difference in attitude toward the testing of models by researchers working on natural systems (streams, lakes, estuaries, etc.) and those working on man-made systems such as treatment plants. Experimentation is frequently not possible for natural systems and the usual model testing approach for these systems is to collect data over a long period of time to build a computer data base and then analyze the data to examine the fit between model predictions and the data. However, what if the dynamic phenomena of interest (a spill of a toxic chemical, for example) does not occur during this period of time? Or if the frequency of the measurements is inadequate to quantify the dynamic response?

With man-made systems such as treatment plants, experimentation is often possible and an integral part of model testing should be the design and conduct of carefully planned experiments including appropriate input disturbances and frequency of measurements so that data on the phenomena of interest can be more rapidly and accurately obtained. This type of information may be difficult to obtain from normal operating records since the control exerted in normal operation can remove much of the dynamic information of interest. During times of crisis, such as process upsets or pending process failure, significant dynamic phenomena are occurring. However, during such times operating engineers are more concerned with restoring good performance than they are with collecting data for the testing of models! This points out the need for the conduct of carefully planned experiments to evaluate such phenomena.

The testing of models also brings up the issue of measurements and illustrates another difference between the use of models by engineering researchers and practitioners. Models for research tend to be more complex and include more terms than those used for design, some of which may be difficult to measure. This can be an acceptable situation in research since the investigators usually have a good command of the basic theory involved and the ability to use the more sophisticated instruments needed for measurements. However, models for practical applications should be as simple as possible and focus on the inclusion of parameters or variables which are relatively easy to either measure or specify. A model which is too complex to understand or requires complicated measurements for implementation is subject to either "misuse" or "disuse."

### Required Accuracy

How accurate must be the model predictions? Is a "Yes" or "No" sufficient or must the answer be more quantitative? If it must be more quantitative, how closely must the predicted and observed behavior agree? The answers to these questions depend upon the purpose for which the model is to be used.

In the early stages of model development, a "Yes" or "No" answer may be quite adequate. For example, in the author's early work on the dynamics of anaerobic digestion [7] it became apparent that the model must be able to predict process failure by organic overloading since this phenomena was known to occur in the field and existing models could only predict failure by hydraulic overloading. Once the occurrence of the phenomena could be predicted on a "Yes" or "No" basis, the next step was to expand the model [8] so that it could predict the time-dependent behavior of the key variables normally monitored for evaluating process condition. The term "semi-quantitative", which is deliberately vague, was coined to indicate that predictions were in the appropriate direction (up or down) and of the right order of magnitude. Confirmation of these semi-quantitative predictions could be obtained from the literature and discussions with operating engineers. However, more quantitative model testing necessitates physical experimentation.

The differences in agreement between model predictions and observed results is due to uncertainty in measurements, the inputs to the model, the model itself, and any other information available for model testing. In design, uncertainty is often taken into account by the use of safety factors which can result in increases in size. For example, there is usually considerable uncertainty involved in predicting the population of a city at some time in the future (the design period) so the design engineer must use his best judgment to make this prediction. Increases in size to accommodate the uncertainty in these predictions should not be called overdesign as they are sometimes labeled, but instead attempts to protect against uncertainty. Since there is always uncertainty in attempts to predict the future, the possibility exists that the load on the plant at the end of the design period can be either larger or smaller than that predicted.

Process control based on model predictions is also subject to uncertainty. However, unlike design, it is possible in many cases to correct for this uncertainty by using automatic feedback control in which the amount of control exerted depends upon the difference between the desired and observed values of performance (the error signal). This means that a reasonable amount of uncertainty can be tolerated in a dynamic model for process control and models with considerable error in the predictions (semi-quantitative) can also be useful. It should be noted that feedback is also involved in process design in that when the design engineer becomes aware of deficiencies in his design, he uses these to correct future designs. However, the feedback is not automatic and is on a much longer time scale (years and sometimes decades) than feedback in process control. It may therefore not be very useful to the operating engineer who has to operate in a much shorter time scale with existing facilities.

## COMPUTER SIMULATION

After a dynamic model has been developed for a process, the equations which comprise the model must be solved in order to predict process behavior with respect to time. This is known as simulation and can be defined as the use of a model to explore the effects of changing conditions on the real system. Prior to the advent of computers, a computational bottleneck existed and efforts at mathematical modeling were frequently of little practical value since the equations comprising the model could not be solved. As previously mentioned, engineers being practical people would usually say, ''Why set up the equations when it is impossible to solve them?'' The ready availability of computers and simulation languages has largely eliminated this bottleneck with the current bottleneck being primarily the development of realistic models.

### HISTORY OF COMPUTER SIMULATION

The early use of digital computers for simulation was primarily restricted to specialists since a detailed knowledge of numerical techniques was required and a considerable amount of time was needed to become proficient in a low level programming language such as FORTRAN. These problems were overcome by the development of user friendly simulation languages which permit the engineer or scientist to concentrate on model development and interpretation of simulation results instead of programming details. However, users of simulation languages should always keep in mind that numerical solution of differential equations is an approximation and sometimes may give incorrect answers.

The early simulation languages were batch oriented since most computers of that era (1960–1975) were operated as batch systems in a central computer center. This impeded a more widespread use of computer simulation since the user was not able to interact directly with the computer. These obstacles were removed by low-cost personal computers (PCs) which became available in the early 1980's. A wide variety of simulation languages are now available for use on PCs. The Society for Computer Simulation publishes an annual directory of simulation software [9]. The 1995 edition of this directory lists 192 general and special purpose simulation software packages with 119 of these being suitable for use on PCs.

Simulation languages may be either general purpose languages, which can be used for any system that can be described in mathematical terms, or for specific applications such as for the activated sludge process. General purpose languages such as ACSL, Matlab/Simulink, Simnon, Simsci, etc. are usually preferred by researchers because of their flexibility and adaptability. Examples of the use of Matlab/Simulink are given in the chapter by Stenstrom. Languages specific to wastewater applications, such as GPS-X, BIOWIN, AQUASIM, EFOR, etc., may be preferred by practicing engineers since they already incorporate much of the process knowledge available. Examples of the use of GPS-X for wastewater treatment plants will be presented later by Takács, et al. It should be noted that GPS-X combines some of the features of both general purpose and specific wastewater application languages since it is based on ACSL, a general purpose language.

A paper which should be consulted by those considering the use of simulation is that of Vanrolleghem and Jeppsson [10] which is a review of a European Union working group meeting on the comparison of simulation software for wastewater treatment processes. Discussed in this review are some of the characteristics of simulation software packages such as the models included, programming languages, software tools for identification, optimization, etc. A list of software available at the time of the review (1994) is included together with recommendations as to the type of user for which each package is intended. Some general selection criteria for determining the usefulness of a package are given. The future of wastewater treatment simulation packages is discussed along with areas in which additional research is needed.

## ADVANTAGES AND DISADVANTAGES OF COMPUTER SIMULATION

Computer simulation has many of the same advantages, and disadvantages, as physical simulation. Considerable knowledge can be gained about a process by the development of a mathematical model and computer simulations using the model. Sensitivity analysis, or the response of the model to changes in specific variables, can be used for model improvement by indicating those variables which are most significant. Sensitivity analysis can also be used for model simplification by indicating those variables which have little effect on system performance. Simulation permits examination of large systems, such as river basins, where physical experimentation on the full scale system may not be possible. Time can be compressed on the computer with simulations being conducted in seconds or minutes. This is especially important for biological processes where rates are slow and physical experimentation may require weeks or months.

There are, of course, disadvantages to computer simulation. Poor models or poor data can give inaccurate or even completely misleading results. Also, some of the advantages of computer simulation can be disadvantages. Computer simulation can be easier, cheaper, and faster than physical experimentation. These factors can lead to one becoming overly enamored with the tools (modeling and computer simulation) and consequently neglecting physical experimentation or the collection of data.

## CONTROL SYSTEMS

Control systems are primarily involved with the handling of information. This may be done manually, automatically, or as a mixture of the two. Environmental engineers are familiar with the theory and technology involved in the handling of materials and energy but are not as accustomed to thinking of information in the same terms. We are in the midst of the ‘‘information revolution’’ which has been brought about by the ready availability of inexpensive computers, information storage devices and transmission systems. How can this technology be applied to wastewater treatment for the reliable production of high quality water at lower cost?

The following discussion is intended primarily as an introduction to control systems. For more detail on the basics of control systems and their application in wastewater treatment, the chapter by Olsson should be consulted. Respirometry, a valuable tool for use in both modeling and control of the activated sludge process is described in Chapter 4.

### INFORMATION HANDLING

The technology involved in information handling is of more recent vintage than that for materials and energy but many of the same concepts are applicable. The handling of materials, energy and information all involve collection, transporting, storing, processing and distribution (Table 1.6).

Flow diagrams are used in handling materials, energy and information, and an example of an information flow diagram is given in Figure 1.2 where the temperature of a process is to be automatically controlled. The temperature is changed from its desired value (setpoint) by some input disturbance such as a change in environmental temperature. This change

TABLE 1.6. Information Handling.

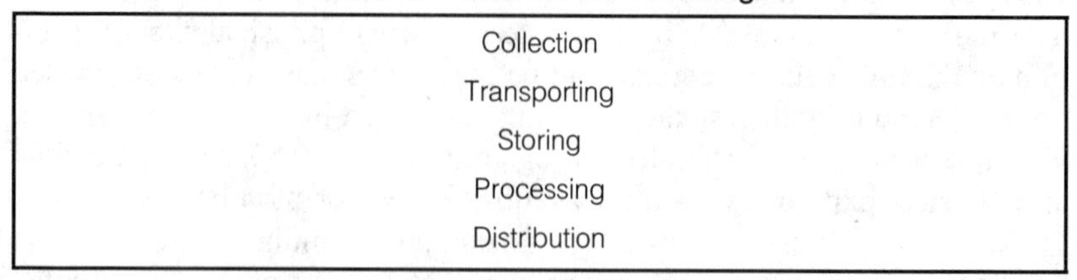

| Information Handling |
| --- |
| Collection |
| Transporting |
| Storing |
| Processing |
| Distribution |

(the output signal) is measured by a sensor such as a thermocouple. The sensor then transmits a signal to the controller.

The controller first compares the signal with the set point to determine if an error exists. If an error does exist, the controller then subtracts the signal from the set point with the resultant being called the error signal. The controller then performs various mathematical operations, using controller models, to calculate the value of the signal to be transmitted to the final control element, in this case a valve on a steam line. The controller has ‘‘closed the loop’’ by ‘‘feedback’’ of information from the process output to the process input.

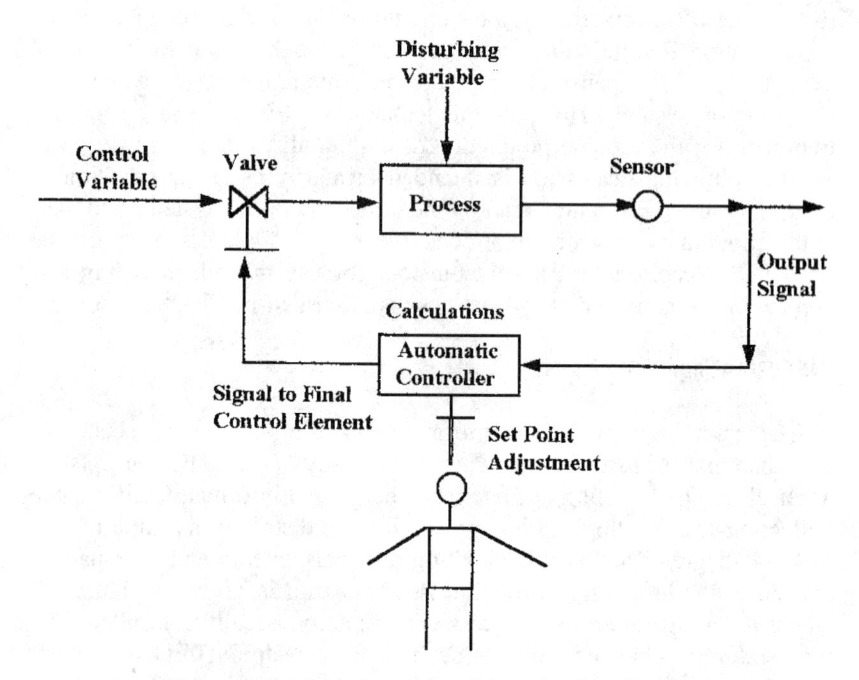


**Figure 1.2** Feedback control as an information flow diagram.

## BASIC QUESTIONS

Regardless of whether control is manual or automatic, some of the same basic questions, as shown in Table 1.7, must be answered in development of a control system.

It is in information processing that the computer has had the largest impact on control systems and attention will now be devoted to this topic.

## FEEDBACK CONTROL

Several controller models may be used in feedback control (Figure 1.2) with the simplest being On-Off control which means that the final control element cycles between completely open or completely closed. An example is the thermostat used for control of home heating systems. A modification of On-Off control commonly used in wastewater pumping stations is two-position On-Off control in which a pump is turned on at an upper limit of the water level in the wet well and is turned off when the water level reaches a lower level.

An improved (over On-Off control) and widely used controller model

TABLE 1.7. **Questions to Be Answered in Control System Development.**

| |
|---|
| **What are the objectives?** The general objectives of the plant control system listed in order of priority are; (1) keep the plant running, (2) meet the effluent requirements, and (3) reduce operating costs. Objectives of controllers for individual processes must be much more specific. |
| **What measurements should be made?** The dynamic behavior of the process and its control system must be considered in establishing information needs. Associated items are the required accuracy and frequency and the availability, reliability, and maintainability of on-line sensors. |
| **How should the information be transmitted?** Provision should be made for easy, reliable, and timely transmission of information from the measurements. Avoid breaking the control loop by filing information for the record instead of using it for control. |
| **What control actions should be taken?** Plant design philosophy in past years has been to provide a minimum number of variables for manipulation and attempt to take care of process dynamics by provision of additional capacity. The number of variables that can be manipulated is thus often limited. |
| **How should the information be processed?** Prior to the advent of the digital computer, the information processing capabilities of automatic controllers were very limited. These limits no longer exist, and a wide variety of controller models are now available. |

is the PID (proportional-integral-derivative) controller which calculates the amount of control as a function of four terms (Eq.1).

$$\underset{\text{Bias}}{C_A = K_B} + \underset{\substack{\text{Proportional}\\ \text{Present}}}{K_P{*}e} + \underset{\substack{\text{Integral}\\ \text{Past}}}{K_I{*}\text{INTGRL}(e{*}dt)} + \underset{\substack{\text{Derivative}\\ \text{Future}}}{K_D{*}\text{DERIV}(e)} \quad (1)$$

where:

$C_A$ = Amount of control to be exerted
$e$ = Error signal = measured signal – set point value
$K_B$ = Bias control coefficient
$K_P$ = Proportional control coefficient
$K_I$ = Integral control coefficient
$K_D$ = Derivative control coefficient

The reader is cautioned that different mathematical arrangements and different names for the terms in the equation are used in practice by control engineers. An excellent discussion of the PID controller, giving the several different forms in which it is used, is given in the chapter by Olsson. A simulated example of the application of on-off and proportional control to the anaerobic digester has been presented by Graef and Andrews [11].

The bias term ($K_B$) reflects the fact that some fixed amount of control is usually needed even when the error signal is zero. For example, some sludge recycle is needed in the activated sludge process even when the mixed liquor suspended solids (MLSS) concentration in the aeration basin is equal to that desired. The proportional term ($K_P$) reflects present conditions in that the amount of control is proportional to the current value of the error. The integral term ($K_I$) reflects the past in that the amount of control depends upon how long the error has persisted. The derivative term ($K_D$) attempts to predict the future by basing the amount of control on how rapidly the error is changing. The PID controller roughly mimics the way a human operator judges how much control to exert by using the position of a pointer on a meter to indicate present conditions and the trace of the signal on a strip chart recorder to indicate how long the error has persisted and how rapidly it is changing.

Numerical values for the four coefficients ($K_B$, $K_P$, $K_I$ and $K_D$) are needed in order to use a PID controller. Determination of these values is known as controller tuning and must take into account both the dynamics of the process as well as those of the different components of the control loop. In past years, controller tuning has normally been a manual operation requiring considerable expertise. However, in some cases this can now be done automatically using microprocessors embedded in "smart" or "self tuning" controllers.

## FEEDFORWARD CONTROL

This type of control is often called open-loop control since there is no feedback of information to the controller from the process outputs. Instead, one or more of the process inputs is measured and the amount of control to exert is based on these measurements (Figure 1.3). An example would be the ratio control system often used in the activated sludge process in which the recycled sludge flow rate is maintained as a set fraction of the influent wastewater flow rate to the aeration basin. This type of control is usually initiated in an attempt to maintain a more constant MLSS concentration in the aeration basin.

In the more general case of feedforward control, the amount and type of control to exert would be calculated from a high speed simulation using dynamic models of the process and the controller. Another name for this type of control is model reference control. Feedforward control is theoretically capable of perfect control since no error need exist, as for feedback control, before control action is initiated. This can be of special importance for processes with substantial time lag between the initiation of control and its measured effect on process outputs as is often the case in wastewater treatment plants.

In view of our previous discussion of the many types of dynamic models available, the question arises as to what type of model should be used. Use of almost all of the various types, and combinations thereof, is possible. For example, the ratio control previously discussed uses a very simple empirical model, this being a fixed number or ratio of the recycle flow to the influent flow. More accurate control of the recycle flow rate could

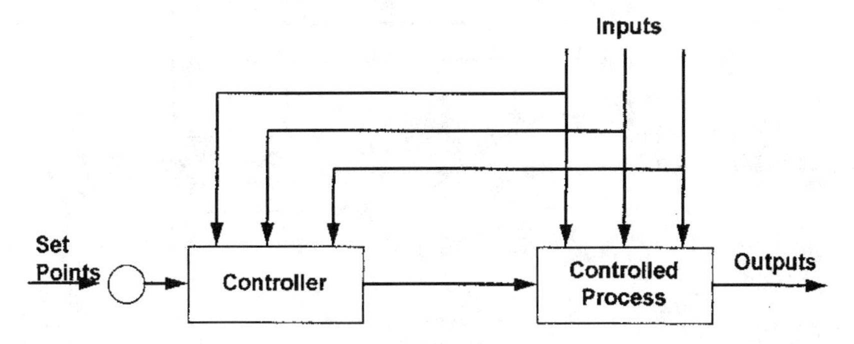


**Figure 1.3** Feedforward control.

be obtained by using a deterministic model of the solids distribution between the aeration basins and the secondary settler as demonstrated by Garrett et al. [12].

## FEEDFORWARD–FEEDBACK

Little process knowledge is required for feedback control; however, considerable time may elapse before proper operating conditions are restored. Although this disadvantage can be overcome by using feedforward control, a dynamic model of the process is needed. Since most models are only approximations of the real system, feedforward control systems usually incorporate some feedback control (Figure 1.4) to compensate for model inadequacies. This also points out that even semi-quantitative dynamic models can be of value for process control since feedforward control based on these models can be "trimmed" by feedback. Moreover, the models can often be continuously improved using on-line estimation techniques for updating numerical values of the model parameters.

An example of a feedforward controller with feedback trim would be the previously mentioned control of recycle sludge flow rate by ratio control but with the ratio of recycle flow rate to influent flow rate not being a constant but instead being "trimmed" or adjusted based on measurement of the MLSS concentration in the aeration basin.

## USE ALL AVAILABLE INFORMATION

So many different types of models and controllers have been mentioned

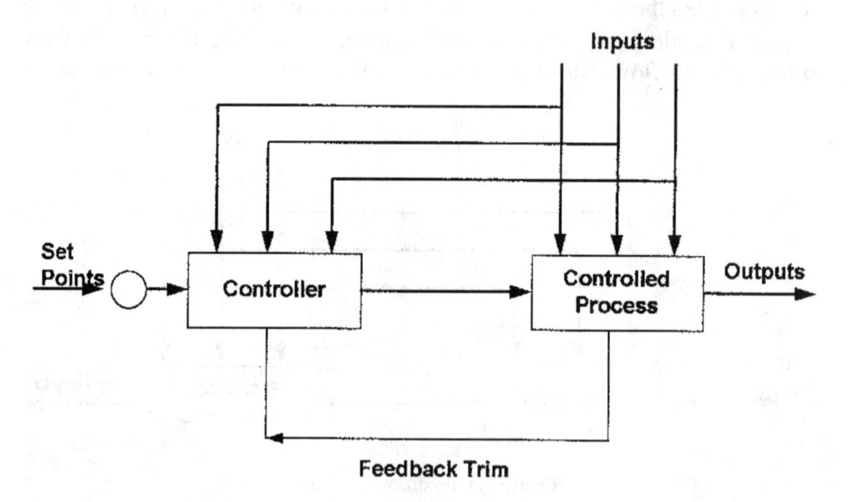


**Figure 1.4** Feedforward control with feedback trim.

TABLE 1.8. Use All Available Information.

| |
|---|
| Measurements from on-line sensors |
| Analyses made by operators |
| Computer data bases for plant operations |
| Laboratory analyses |
| Mechanistic models |
| Time series models |
| Neural network models |
| Expert systems |
| — Global knowledge |
| — Local knowledge |
| Fuzzy relationships |
| Visual information |
| Auditory information |

that the reader may be confused as to which type to use. This problem is compounded by the availability of different types of measuring instruments. The frequency and accuracy of the measurements as well as the dynamic characteristics of the final control elements must also be taken into account. The services of a control engineer are needed to address such problems. However, all of the types of models and controllers discussed can be of value under the appropriate circumstances and a combination of several different types should be considered.

In other words, all available information should be considered for use in process control. Some of the information which might be included is given in Table 1.8.

The expert system knowledge to be considered consists of at least two types. The first is global knowledge applicable to most treatment plants. Information of this type may be found in the professional literature, in mathematical models, reports of operating agencies, and in the minds of experienced operating engineers. The second type is local knowledge which applies primarily to the plant for which the expert system is being constructed. This is obtained from the plans and specifications of the plant and those who have experience with operating the plant, or a similar plant. When specific items of equipment are involved, the knowledge of engineers from the company which makes the equipment is also needed.

The reader may be surprised to see visual and auditory information listed in what has been a discussion of computer control of treatment plants. However, automation of the collection of this information and its transmission to a central location would of considerable value for operation of large plants as well as monitoring and control of a number of small plants from a central location. With tilt, pan, and zoom, video cameras can be used by an experienced operator for reading dials and inspecting equipment while microphones enable the operator to detect unusual sound patterns which could indicate potential equipment problems such as bad bearings on a pump, etc. A single skilled operator could therefore be in "several places at once." It is also possible to use neural networks, which are very good at pattern recognition, for automatic conversion of the reading of dials and gauges to numbers for the computer and computer analysis of audio signals for comparing the sound of pumps with bad bearings to those with good bearings. In the author's opinion we have only "scratched the surface" of what will be possible in the "Information Age."

## SUMMARY

There are many potential benefits from the use of dynamic models and control systems in wastewater treatment plants. One of these, the matching of the dynamics of plant effluents with those of receiving streams, should be given more consideration by both regulatory agencies and organizations responsible for plant operations.

Several different types of models can be useful in describing dynamic behavior and controlling treatment plants. Among these are (1) visual, (2) linguistic, (3) mental, (4) physical, (5) mathematical, and (6) fuzzy models. Each of these is briefly described and its possible use in wastewater treatment plants discussed. Each type can be further broken down into different categories and definitions and contrasts of the following categories of mathematical models are presented.

- dynamic or steady state?
- mechanistic or empirical?
- deterministic or stochastic?

A discussion of some of the basic concepts of mathematical modeling with emphasis on the relationship of these to engineering research and practice is presented. These are:

- differences in the use of, and expectations from, models by researchers and practitioners
- sources of information useful for model development
- the amount and type of testing needed for model validation and calibration
- the required accuracy

Dynamic mathematical models are solved using computer simulation to predict plant behavior with respect to time. Early simulation languages were batch oriented for use on relatively expensive computers which were controlled by computer centers. Rapid advances in both hardware and software for personal computers has resulted in inexpensive, user-friendly systems suitable for use in both large and small organizations as well as by individual engineers.

A discussion of the basic questions which must be answered in the development of control systems is presented. These questions are:

- What are the objectives of the control system?
- What measurements should be made for initiation of control?
- How should the information be transmitted?
- What control actions should be taken?
- How should the information be processed?

The author is of the opinion that all available information should be considered for use in control and a listing of several types of information which would be valuable has been presented. Visual and auditory information obtained from on-line video cameras and microphones are included in this list.

## LITERATURE SOURCES

The literature on the topics which have been discussed is much too voluminous to review herein. The field is highly interdisciplinary and pertinent articles may be found scattered throughout many different national and international publications. However, a good overview of the field may be obtained by consulting the proceedings [13–19] of a series of seven workshops on the instrumentation, automation, and control (ICA) of water and wastewater treatment and transport systems which have been conducted by the International Association on Water Quality. Participants in these workshops have come from operating agencies, consulting engineering firms, instrument and control system manufacturers, regulatory agencies, and government, university and industrial research organizations in many different countries.

U.S. organizations with publications in this field are the American Public Works Association (APWA), the Instrument Society of America (ISA) the Water Environment Federation (WEF), and the American Water Works Association (AWWA). These publications include manuals of practice [20,21] as well as proceedings of conferences.

## REFERENCES

1 Andrews, J. F. (1974) "Review Paper: Dynamic Models and Control Strategies for Wastewater Treatment Processes," *Water Research, 8,* 261–289.

2 Rossman, L. A. (1989) ''Wastewater Treatment and Receiving Water Body Interactions'', *Dynamic Modeling and Expert Systems in Wastewater Engineering* (Patry, G. G. and D. Chapman, Eds.), Lewis Publishers, Chelsea, MI.

3 Olsson, G. (1993) ''Advancing ICA Technology by Eliminating the Constraints,'' *Water Science and Technology, 28,* 1–7.

4 Chapman, D. T. (1992) ''Statistics for treatment plant operations,'' Chapt. 4, *Dynamics and Control of the Activated Sludge Process* (Andrews, J. F., Ed.), Technomics Publishing Co., Lancaster, PA.

5 Hiraoka, M., and T. Fujiwara (1992) ''The use of time series analysis in hierarchical control systems,'' Chapt. 5, *Dynamics and Control of the Activated Sludge Process* (Andrews, J. F., Ed.), Technomics Publishing Co., Lancaster, PA.

6 Hiraoka, M, and K. Tsumura (1992) ''Computer assisted operation of the Kawamata treatment plant, Osaka prefecture, Japan,'' Chapt. 6, *Dynamics and Control of the Activated Sludge Process* (Andrews, J. F., Ed.), Technomics Publishing Co., Lancaster, PA.

7 Andrews, J. F. (1969) ''Dynamic model of the anaerobic digestion process,'' *Journal Sanitary Engineering Division, American Society of Civil Engineers, 95,* 95–116.

8 Andrews, J. F. and S. P. Graef (1971) ''Dynamic modeling and simulation of the anaerobic digestion process,'' *Anaerobic Biological Treatment Processes, Advances in Chemistry Series No. 105,* 126–162, American Chemical Society, Washington, D.C.

9 Society for Computer Simulation (1995) *Directory of Simulation Software, Vol. 6,* Society for Computer Simulation, San Diego, California.

10 Vanrolleghem, P. A. and U. Jeppsson (1996), ''Report working group meeting COST-682 on simulators for modelling of WWTP,'' *IAWQ Specialist Group on Instrumentation, Control and Automation of Water and Wastewater Treatment and Transportation Systems, Newsletter No. 10,* International Association on Water Quality, London.

11 Graef, S. P. and J. F. Andrews (1974) ''Stability and control of anaerobic digestion,'' *Journal Water Pollution Control Federation, 46,* 666–683.

12 Garrett, M. T., J. Ma, W. Yang, G. Hyare, T. Norman and Z. Ahmad (1984) ''Improving the performance of Houston's Southwest wastewater treatment plant, U.S.A.'' *Water Science and Technology, 16,* 317–329.

13 Instrumentation, Control and Automation for Wastewater Treatment Systems (Andrews, J. F., R. Briggs and S. H. Jenkins, Eds.), *Proceedings 1st IAWPRC Workshop* held in London and Paris, May, 1973, Pergamon Press, Oxford, 1974; 570 p.

14 Instrumentation and Control for Water and Wastewater Treatment and Transport Systems, *Proceedings 2nd IAWPRC Workshop* held in London and Stockholm, May, 1977, Pergamon Press, Oxford, 1978, 646 p.

15 Practical Experiences of Control and Automation in Wastewater Treatment and Water Resources Management, *Proceedings 3rd IAWPRC Workshop* held in Munich and Rome, June, 1981, Pergamon Press, Oxford, 1981, 645 p.

16 Instrumentation and Control of Water and Wastewater Treatment and Transport Systems (Drake, R. A. R., Ed.), *Proceedings 4th IAWPRC Workshop* held in Houston and Denver, April–May, 1985, Pergamon Press, Oxford, 1985, 748 p.

17 Instrumentation, Control and Automation of Water and Wastewater Treatment and Transport Systems (Briggs, R. Ed.), *Proceedings 5th IAWPRC Workshop* held in Kyoto and Yokohama, July–August, 1990, Pergamon Press, Oxford, 1990, 781 p.

18 Instrumentation, Control and Automation of Water and Wastewater Treatment and Transport Systems (Jank, B. and others, Eds.) *Proceedings 6th IAWQ Workshop* held

in Banff and Hamilton, June, 1993, Wastewater Technology Centre, Burlington, Canada, 1993, 667 p.

**19** Instrumentation, Control and Automation of Water and Wastewater Treatment and Transport Systems, (Briggs, R., Ed.) *Proceedings of the 7th IAWQ Workshop* held in Brighton, UK, July, 1997, International Association on Water Quality, London, UK, 1997, 593 p.

**20** Water Environment Federation (1993), *Instrumentation in Wastewater Treatment Facilities—Manual of Practice No. 21,* Water Environment Federation, Washington, D.C. 332 p.

**21** Water Environment Federation (1997), *Automated Process Control Strategies, (Special Publication),* Water Environment Federation, Washington, D.C., 236 p.

CHAPTER 2

# Mathematical Modeling and Computer Simulation

**MODELING** and simulation are not new. Both have a long history of use in engineering and science. Scaled-down physical models have long been used for simulation in such diverse areas as astronomy (planetariums), hydraulic engineering (river models), architecture (building models), and chemical engineering (pilot plants). Biologists, chemists, and sociologists have studied model organisms (the famous *E. coli* for example), model compounds, and model communities, respectively. Even the hypothesis that is formulated in applying the scientific method is a verbal model, and simulation may be used to test the validity of a hypothesis when it is not feasible to test the real system.

Mathematical models are commonly used for more quantitative descriptions and consist of one or more equations relating the important inputs, outputs, and characteristics of a system. To those not familiar with systems engineering terminology, the term ''mathematical model'' is sometimes frightening since it may bring to mind large sets of complex equations and the use of sophisticated mathematical techniques. However, this is not always the case since most of the common engineering design formulae are mathematical models. For example, as simple an expression as $y = mx + b$ is a mathematical model where, in systems engineering terminology, the system output ($y$) is related to the system input ($x$) by the system parameters ($m,b$).

The literature on mathematical modeling and computer simulation is much too voluminous to review herein. Emphasis will therefore be placed

Michael K. Stenstrom, Hsueh-Hwa Lee, Jinn-Shiu Ma, Civil and Environmental Engineering Department, University of California, Los Angeles, Los Angeles, CA; and Michael W. Barnett, Gensym Corporation, Cambridge, MA.

on discussing some of the basic concepts of modeling and simulation with emphasis on how these relate to engineering research and practice. The approach that has been used by the author in the development of models for biological wastewater treatment processes will then be described. This will be followed by specific examples of some simple process models programmed in Matlab (Matlab User's Guide, 1992). Matlab has the advantage of a very compact representation for mathematical models and a rich set of readily-available tools for working with models. The user needs to know very little about methods for solution of differential equations or specific programming languages, however, experience with programming will facilitate the use of Matlab.

## AN APPROACH TO PROCESS MODELING

Modeling is still more of an art than a science and several different approaches are possible. However, as an example the author would like to illustrate the approach that he has used over the years in the development of several models for biological processes used in wastewater treatment. These are engineering research models with a primary objective of improving process understanding and an ultimate practical objective of being useful for process control as well as process design. The approach is primarily applicable to the development of dynamic mechanistic models expressed in a deterministic manner. However, both empirical and stochastic components can be added to reflect a lack of knowledge of mechanism or a statistical discrepancy between predicted and observed results.

### IDENTIFY THE SIGNIFICANT MASS TRANSFER/REACTION TERMS

Identification of these terms requires a fundamental understanding of the process being modeled with the key word being "significant." The significant mass transfer/reaction terms are those that have the greatest impact on model predictions. Inclusion of relatively insignificant terms can make the model difficult to comprehend and increase the number of measurements required for testing the model. On the other hand, not including a significant term can result in faulty predictions. An example is the common BOD test in which both carbonaceous and nitrogenous oxygen demand may occur. The use of BOD values in stream models, as if they reflected only carbonaceous demand whereas in actuality nitrogenous demand was also exerted, can lead to faulty predictions of dissolved oxygen concentration.

A simple reaction is given in Equation (2.1) and a simple mass transfer step in Equation (2.2). More specific examples are the reaction in which organic matter and oxygen are converted to cell mass, $CO_2$, and $H_2O$, and the mass transfer step in which the oxygen in air is transported to the dissolved oxygen in water.

*Reaction*

$$A \rightarrow B \tag{2.1}$$

where

$A$ = concentration of A
$B$ = concentration of B

*Mass Transfer*

Phase 1 | Phase 2

$$A1 \longrightarrow A2 \tag{2.2}$$

Phase Boundary

where

$A1$ = concentration of A in phase 1
$A2$ = concentration of A in phase 2

## ESTABLISH THE STOICHIOMETRY

Stoichiometric coefficients establish "how much" of each reactant is utilized and the amounts of the different products formed. Examples are the amount of oxygen utilized per unit of BOD removed or the mass of sludge produced per unit of BOD removed, which are important numbers for both plant design and operation. An example of how stoichiometric coefficients may be expressed mathematically is shown in Equation (2.3):

*Stoichiometric Equation*

$$A \rightarrow Y_{B/A} * B \tag{2.3}$$

$Y_{B/A} = B/A$ = mass $B$ produced/mass $A$ utilized.

For some processes it is also necessary to determine the thermochemistry which defines the energy utilized or produced in the reaction. An example is afforded by the autothermal aerobic digestion process in which knowledge of the heat energy released is needed to calculate the temperature at which the reactor will operate.

## DETERMINE THE KINETICS

Whereas the stoichiometry establishes ''how much'' of the reactants or products are utilized or produced, the kinetics describe ''how fast'' the reactions proceed. Both reaction and mass transfer kinetics may be of importance as exemplified by the Streeter-Phelps model for determining the dissolved oxygen concentration in a stream subject to a point discharge of wastewater. In this model, the reaction kinetics for disappearance of BOD are assumed to be first order and the transport of oxygen into the stream from the atmosphere is assumed to be in accordance with the two-film mass transfer equation.

If the disappearance of $A$ in Equation (2.1) is assumed to be first order with respect to $A$, the rate expression for the disappearance of $A$ is:

*Rate Equation*

$$dA/dt = -K * A \tag{2.4}$$

where

$t$ = time
$K$ = rate coefficient

The disappearance of $A$ is linked to the appearance of $B$ by the yield coefficient as shown below.

$$dB/dt = Y_{B/A} * K * A \tag{2.5}$$

where

$B$ = concentration of B.

Although simple, Equations (2.4) and (2.5) comprise a dynamic mathematical model for a batch reactor (see the following) since the disappearance of $A$ and appearance of $B$ are described by differential equations involving time.

## SPECIFY OR CHARACTERIZE THE REACTOR

The reactions involved may take place in many different types of vessels (reactors). These may be characterized with respect to inflow and outflow patterns, degree of mixing and location of the reaction within the vessel. With respect to inflow-outflow patterns, two common types of reactors are the batch reactor (no flow) and the continuous flow reactor as illustrated in Figure 2.1. In the batch reactor [Figure 2.1(a)], concentrations of reactants (*S*,*X*) are continually changing with respect to time. In a continuous-flow, stirred-tank reactor (CFSTR) [Figures 2.1(b), 2.1(c)] the reactants are being continuously added to and/or removed from the reactor. Flow may be steady [Figure 2.1(b)], where the flow rate is constant with respect to time, or unsteady [Figure 2.1(c)] in which case the flow rate varies with respect to time. Steady flow may create a condition called steady state in which the concentrations of reactants in the effluent from the

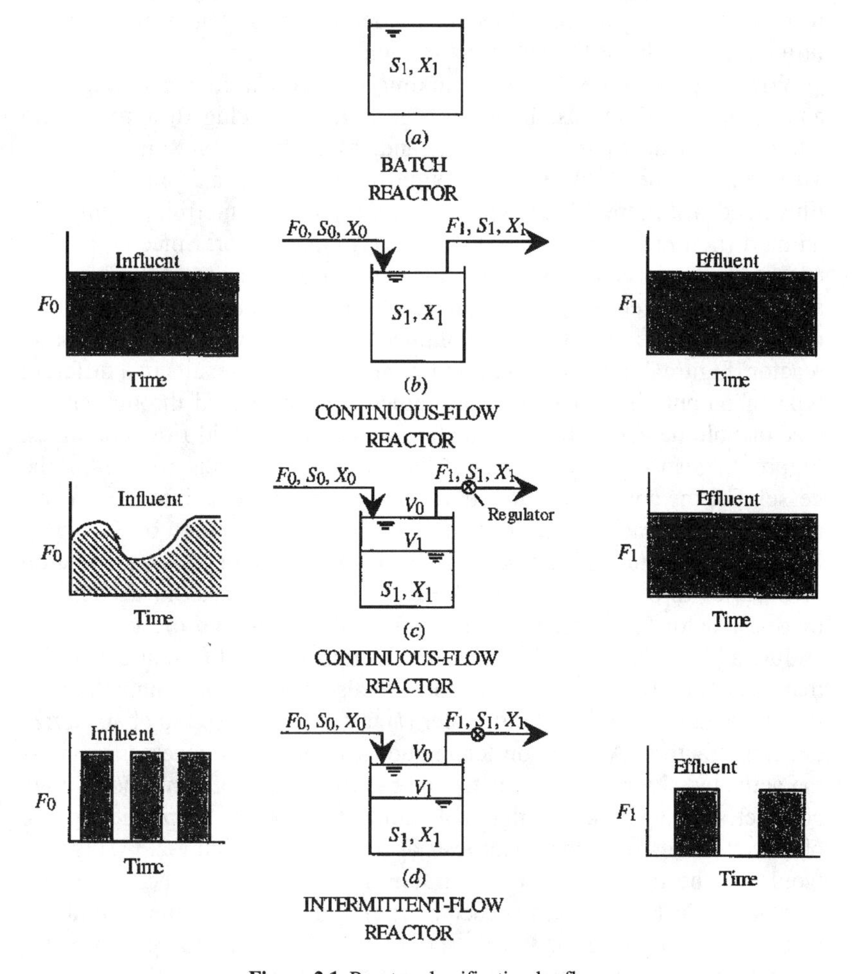


**Figure 2.1** Reactor classification by flow.

reactor do not change with respect to time. Figure 2.1(c) also illustrates that the influent and effluent flow rates need not be the same and in this instance the unsteady influent flow is changed to a steady effluent flow by using a flow regulator and a variable volume reactor. The influent flow rate to most wastewater treatment processes is unsteady; however, hydraulic design is such that the reactor volume is essentially constant and the effluent flow rate is therefore also unsteady.

Obviously, there are reactors that are neither unambiguously batch or continuous flow and an example of this is encountered in many anaerobic digesters where both the influent and effluent flow rates are intermittent as shown in Figure 2.1(d). Still another example is the sequencing batch reactor (SBR) version of the activated sludge process where two or more reactors are used to accommodate a continuous influent flow with the influent and effluent to each reactor being intermittent.

With respect to the degree of mixing, the two extremes are plug flow and complete mixing as shown in Figure 2.2. In a plug flow reactor no attempt is made to induce mixing and, as a first approximation, it is assumed that the fluid moves through the reactor as a "plug." This is illustrated in Figure 2.2(a), which shows a pulse or intermittent input of an inert (non-reacting substance) tracer, *I*, to the reactor. Since no mixing occurs, the tracer is transported as a "plug" through the reactor. The pulse appears unchanged in shape in the reactor effluent in a time equal to the reactor residence time (reactor volume/flow rate). In a complete mixing reactor [Figure 2.2(b)], a pulse input of an inert tracer results in a different type of output since the tracer is immediately dispersed throughout the reactor volume and is then gradually diluted out as fluid flow continues. Figure 2.2 also shows the effect of the location of the reaction within the vessel. For the continuous flow of a reacting substance, *S*, into the reactor, there is a difference at steady state in the spatial distribution of *S* for plug flow and complete mixing reactors. In a plug flow reactor, the concentration of *S* varies throughout the length of the reactor whereas in a complete mixing reactor the concentration of *S* is uniform through the vessel.

Just as models are frequently combinations of the different categories that have been listed, "real" reactors are also frequently combinations of the above classifications. Inert tracer (*I*) tests may be used to characterize existing reactors. A common technique for biological processes, such as the activated sludge process, is to find the "number" of continuous-flow, complete-mixing reactors that will come the closest to predicting the observed results from the tracer tests as illustrated in Figure 2.3. This then would be the model for the "hydraulic regime" of the reactor and has both mechanistic (the basic reactor type) and empirical (the "number" of reactors) components. It will also have a stochastic component since the data points from the test will not all fall on the predicted line.

COMPUTER SIMULATION

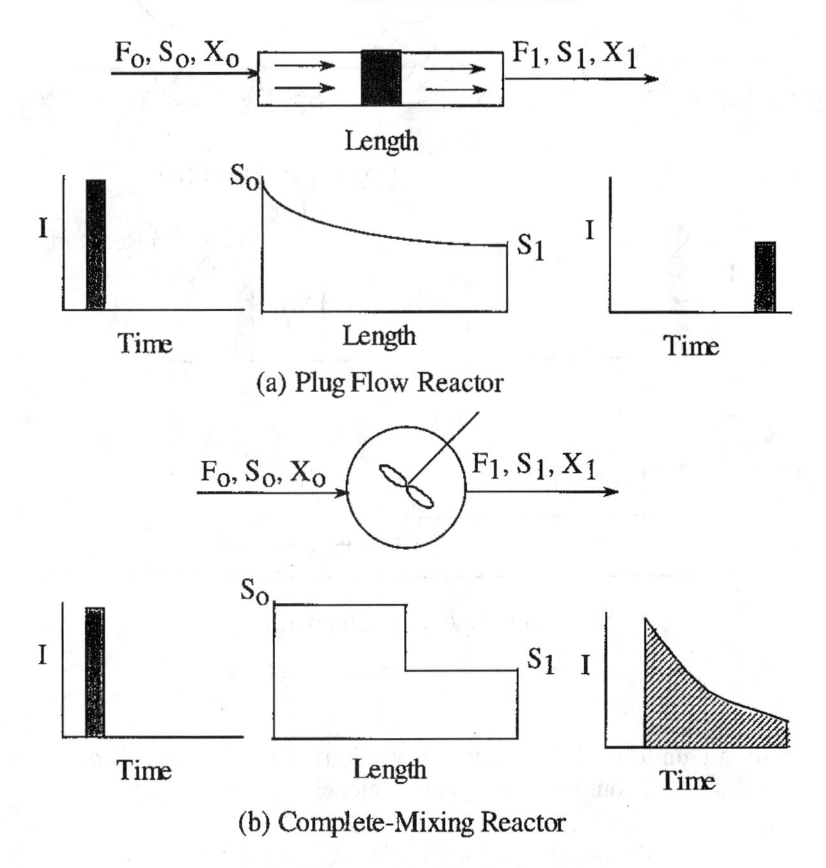


**Figure 2.2** Reactor classification by mixing.

## DEVELOP MODEL EQUATIONS

The previous steps provide the information needed to prepare the model equations. These are prepared by performing material balances on the substances of interest using the type of reactor selected. The general form for a material balance is:

$$\text{Rate of accumulation} = \text{Rate on inflow} - \text{Rate of outflow} + \text{Rate of Generation or Utilization} \quad (2.6a)$$

For the reaction shown in Equation (2.1), the material balances on *A* and *B* for a continuous-flow, complete-mixing reactor (CFSTR) can be made over the whole reactor since the concentrations of materials in the

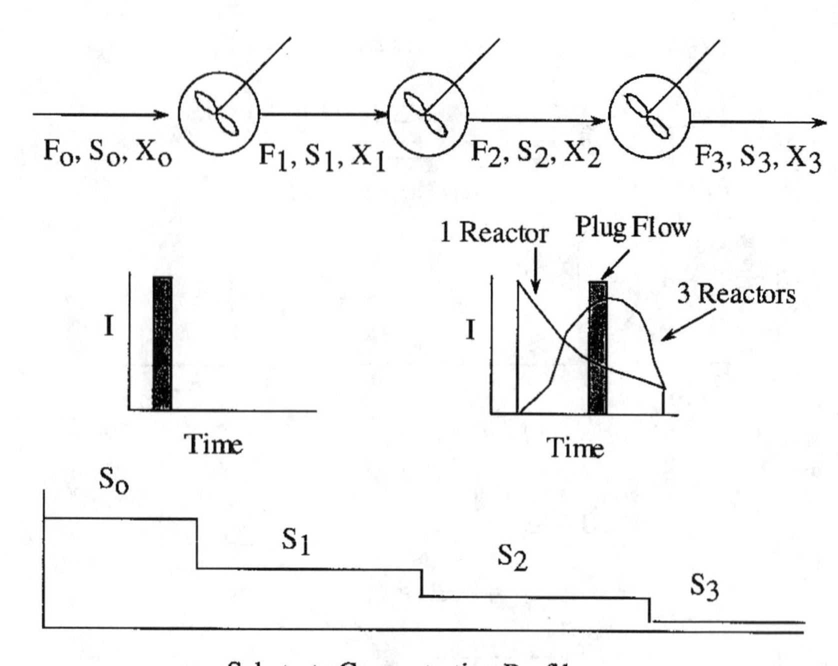


**Figure 2.3** Complete mix reactors in series.

reactor are uniform (independent of position). These balances [Equations (2.7) and (2.8)] comprise the dynamic model.

$$(V)(dA_1/dt) = (F)(A_0) - (F)(A_1) - (K)(A_1)(V) \tag{2.7}$$

$$(V)(dB_1/dt) = (F)(B_0) - (F)(B_1) + (Y_{B/A})(K)(A_1)(V) \tag{2.8}$$

where

$V$ = reactor volume

$F$ = flow rate

0, 1 = subscripts, denoting influent and effluent, respectively

When the concentrations are not uniform, as for a plug flow reactor, the balances must be made over a differential element of reactor volume. This makes it possible to consider spatial distribution in a reactor. For a multiphase reactor, it may be necessary to make material balances for substances in each phase as well as to consider transport of the components between phases [for example, Equation (2.2)].

One or more of the terms given in the general material balance [Equation

(2.6)] may be zero, thus simplifying the resultant model. For example, in a batch reactor the second and third terms are zero since there is no material flow into or out of the reactor. The resulting model therefore consists of Equations (2.4) and (2.5). For a CFSTR with an inert substance as tracer, term four is zero since there is no reaction term. If the tracer is added as an impulse (a ''slug'' of tracer dumped into the reactor), term two also becomes zero immediately after addition of the tracer. Term one, the derivative term which expresses the accumulation or depletion of a substance in the reactor, becomes zero for a CFSTR at steady state. The elimination of term one, or the assumption of steady state, has been widely used since it reduces the mathematical complexity of the resultant model. For a CFSTR, this assumption usually reduces the differential equations in the model to algebraic equations and for a plug-flow reactor the partial differential equations are reduced to differential equations.

Each of the terms presented in the general material balance equation may have more than one component. For example, there may be several streams containing the material of interest which enter or leave the reactor. In the activated sludge process, a portion of the carbon dioxide produced will leave the reactor in the gas with another portion leaving the reactor as dissolved carbon dioxide in the liquid stream.

Although elementary, the reader is cautioned that each term in a material balance must have the same units. Frequent checking of units is desirable even for those with experience in making material balances and is especially so for the beginning student.

If the reactor temperature is to be predicted, an energy balance, using knowledge of thermochemistry, will also be required.

After a dynamic mathematical model has been developed for a process, the equations that comprise the model must be solved in order to predict the behavior of the process with respect to time. This procedure is known as simulation and can be defined as the use of a model to explore the effects of changing conditions on the real system. Obviously, the model must be a reasonable representation of the real system in order for the results to be meaningful since the simulation results can be no better than the mathematical model and data on which they are based.

## REACTOR MODELING AND SIMULATION USING MATLAB

The development of models and their solution through computer simulation is best learned by practice with examples. Two examples are presented in detail in this section. The examples are presented in Matlab, a well known, commercially available, mathematical modeling and analysis software package (produced by Mathworks, Inc., 24 Prime Park Way, Natick,

MA 01760). The previous edition of this text used Simnon [1], which is also a very useful tool. Matlab was chosen over Simnon for the new edition of the text because of its greater popularity and availability. A full-featured version is available as well as an inexpensive student version. The Matlab User's Guide (Matlab User's Guide, 1992) has detailed explanation of the available commands and auxiliary tool kits.

In Matlab, equations are entered in a fashion similar to other programming languages. Parameters are first declared with the "global" statement. Values are assigned by equating the parameter to a fixed value. The syntax for these commands is intuitive and simple to master. In the examples below, several Matlab scripts are presented to demonstrate the representation of a mathematical model in a computer simulation language. You can run these examples by typing the text directly, in sequence, at the Matlab command line or you can create a simple text file with a ".m" extension containing this text. You can then read the file into Matlab, typically with a menu selection. Refer to the Matlab User's Guide for more information on setting up and running Matlab.

## WATER TANK MODELING AND SIMULATION

Two simple models are presented below. The system to be modeled is comprised of a single completely-mixed tank, possibly containing one or more reactants, with an inflow of water and reactants and an outflow of contents from the tank.

### Problem Description

A cylindrical tank discharges water at its bottom (*F*out) through a circular orifice with an area (*Ao*) of 0.8 sq. ft. and a coefficient of discharge (*C*) of 0.7. A pump with a flow rate of 10 cfs commences discharge of water into the tank (*F*in) when the height of water in the tank (*Ht*) is 10 ft. Prepare both tabular and graphical output showing the height of water in the tank versus time after the pump commences operation. Repeat the simulation for pump flow rates of 20 and 25 cfs.

#### *Symbols and Units*

*At* = cross-sectional area of the tank, sq. ft.
*Ao* = cross-sectional area of the orifice, sq. ft.
*C* = orifice coefficient
*F*in = pumped flow of water into the tank, cfs.
*F*out = flow of water out of the tank through the orifice, cfs.
$F\text{out} = C * Ao * (2 * g * Ht)^{0.5}$
*Ht* = depth of water in the tank at any time, ft

$t$ = time, seconds
$g$ = acceleration due to gravity, ft/sec$^2$
$V$ = volume of water in the tank, cu. ft.

### *Mathematical Model*

The only step needed to develop the model is to write a material balance for a single material—water. There are no reactants or reaction kinetics and stoichiometry to consider. The general form of a material balance [Equation (2.6] is repeated below for your convenience. The material balance for water is given by Equation (2.9).

| Rate of Change in the Reactor | = Flow in– Flow out +or– | Appearance or Disappearance by Reaction | (2.6b) |
|---|---|---|---|

$$dV/dt = Fin - Fout \tag{2.9}$$

$$At * (dH/dt) = Fin - Fout \tag{2.10}$$

$$DHt/dt = (Fin - Fout)/At \tag{2.11}$$

### *Matlab Code*

The following Matlab code solves the differential equations shown above. Note that the "%" sign denotes a comment statement.

```
function dy = wtank (t,y);

% This is a MATLAB function. Place this code in a single file
% named 'wtank.m' and copy the file to a directory that is in
% MATLAB's search path.
% The function wtank (wtank.m) returns the derivative of state
%  variable Ht defined as;
% dHt = (Fin - Fout)/At
%        where Fout = C * A0 * (2 * g * Ht)^ 0.5

% Global parameters
global Fin;
global At;
global A0;
```

```
global C,
global g,

dy = (Fin – C * A0 * (2 * g * y)^ 0.5)/At;
```

---

```
% This is a script file to calculate the state variable Ht and
% plot the results.
% Place this code in a single file named 'calht1.m' and copy the
% file to a directory that is in MATLAB's search path. To run this
% script, type 'calht1' at the MATLAB prompt. NOTE: this script
% uses the 'wtank' function above (see comments in the code for
% wtank).

% Global parameters
global At,
global A0,
global C,
global g,
global Fin,

% Numerical values for the parameters
At = 1000;
A0 = 0.8;
C = 0.7;
g = 32;
Fin = 10;

% Using ode45 method to solve the differential equation.
% Ode45 integrates a system of ordinary differential equations
% using 4th and 5th order Runge-Kutta formulas.

[t,y] = ode45 ('wtank', 0, 4000, 10); %Initial condition: Ht = 10
% wtank is a function contained in the M-file: wtank.m.
% This file must be in the MATLAB search path.

Ht = y;

% Plot the results.
plot (t, Ht, 'r-')
xlabel ('Time (sec)', 'FontSize', 12)
ylabel ('Ht (ft)', 'FontSize', 12)
title ('Ht vs. Time', 'FontSize', 18)
```

```
axis ([0, 4000, 0, 20])
grid on
```

---

When you run the Matlab script file (an M-file) calht1.m you will obtain the plot in Figure 2.4.

Often it is desirable to run a simulation for a range of parameter values. The following code can be used to solve the proceeding model equations with different values of Fin.

---

```
% calht2.m is a script file to calculate state variable Ht for
% different Fin values.
% Place this code in a single file named 'calht2.m' and copy the
% file to a directory that is in MATLAB's search path. To run this
% script, type 'calht2' at the MATLAB prompt. NOTE: this script
% uses the 'wtank' function above (see comments in the code
% for wtank).

% Global parameters
```

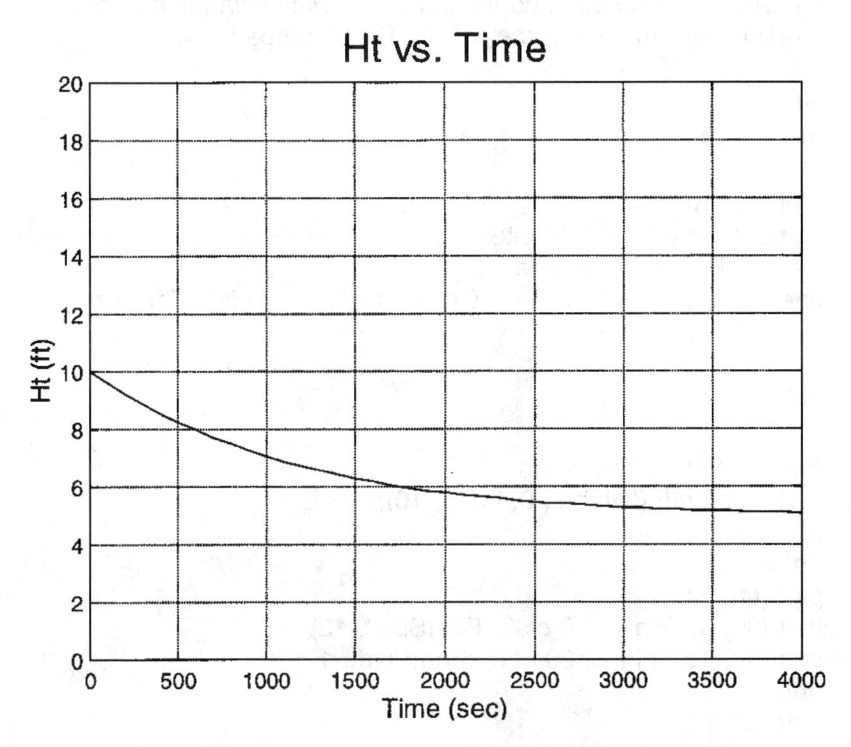


**Figure 2.4** Height of water in tank with constant inflow of 10 cfs.

```
global At,
global A0,
global C,
global g,
global Fin,

% Numerical values for the parameters
At = 1000;
A0 = 0.8;
C = 0.7;
g = 32;
Fin = 10;

% Using ode45 method to solve the differential equation.
% Ode45 integrates a system of ordinary differential equations
% using 4th and 5th order Runge-Kutta formulas.

[t,y] = ode45 ('wtank', 0, 4000, 10); %Initial condition: Ht = 10
% wtank is a function contained in the M-file: wtank.m.
% This file must be in the MATLAB search path.

Ht = y;
tm = t/60;

plot (tm, Ht, 'r-')
xlabel ('Time (min)', 'FontSize', 12)
ylabel ('Ht (ft)', 'FontSize', 12)
title ('Ht vs. Time for Fin = 10 (-) and 20 (*—) cfs', 'FontSize', 18)
axis ([0, inf, 0, 20])

hold on

Fin = 20; % change Fin
[t,y] = ode45 ('wtank', 0, 4000, 10);
Ht = y;
tm = t/60;
plot (tm, Ht, 'r*')
text (30, 4, 'Fin = 10 cfs', 'FontSize', 12)
text (30, 15, 'Fin = 20 cfs', 'FontSize', 12)
grid on
hold off
```

When you run the Matlab script file calht2.m you will obtain the plot in Figure 2.5. Finally, we can easily simulate flows of 20 and 25 cfs using the following code.

```
% calht3.m is a script file to calculate state variable Ht at
% different Fin values.
% Place this code in a single file named 'calht3.m' and copy the
% file to a directory that is in MATLAB's search path. To run this
% script, type 'calht3' at the MATLAB prompt. NOTE: this script
% uses the 'wtank' function above (see comments in the code for
% wtank).

%  Global parameters
global At,
global A0,
global C,
global g,
```

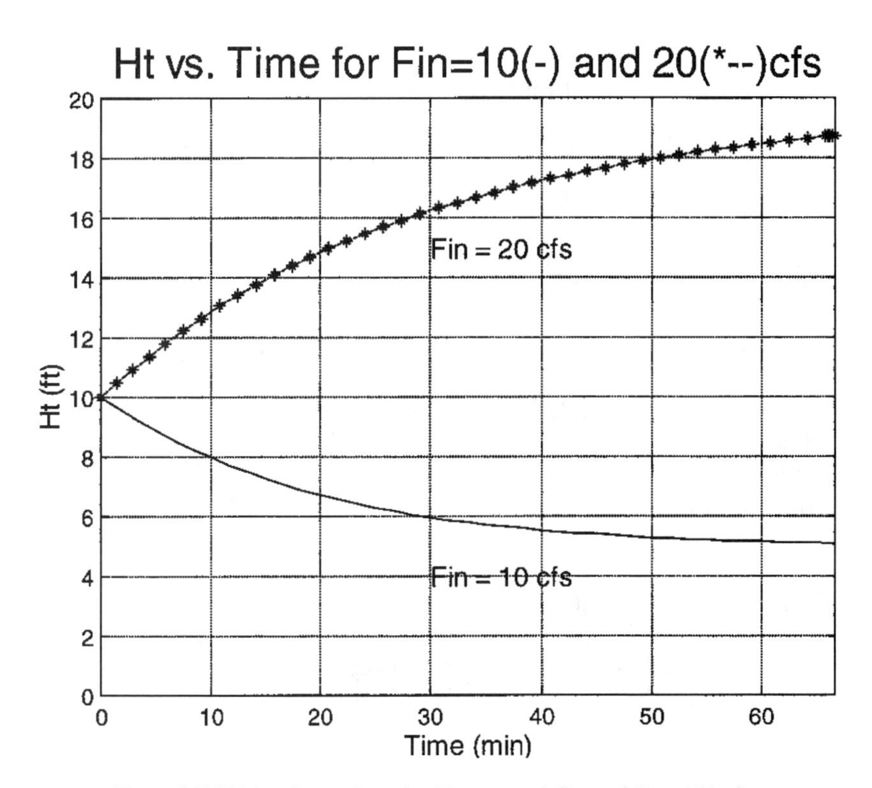


**Figure 2.5** Height of water in tank with constant inflows of 10 and 20 cfs.

```
global Fin,

% Numerical values for the parameters
At = 1000;
A0 = 0.8;
C = 0.7;
g = 32;
Fin = 20;

% Using ode45 method to solve the differential equation.
% Ode45 integrates a system of ordinary differential equations
% using 4th and 5th order Runge-Kutta formulas.

[t,y] = ode45 ('wtank', 0, 4000, 10); %Initial condition: Ht = 10
% wtank is a function contained in the M-file: wtank.m.
% This file must be in the MATLAB search path.

Ht = y;

plot (t, Ht, 'r-')
xlabel ('Time (sec)', 'FontSize', 12)
ylabel ('Ht (ft)', 'FontSize', 12)
title ('Ht vs. Time for Fin = 20(-) and 25(*—) cfs', 'FontSize', 18)

hold on

Fin = 25; % change Fin

% Solve the equation for these conditions.

[t,y] = ode45 ('wtank', 0, 4000, 10);
Ht = y;
plot(t, Ht, 'r*')
grid on, box on
axis ([0 4000 10 30])

% Add text to the graph

text(1500, 20, 'Fin = 25 cfs', 'FontSize', 12)
text(1500, 15, 'Fin = 20 cfs', 'FontSize', 12)

hold off
```

When you run the Matlab script file calht3.m you will obtain the plot in Figure 2.6.

Problem Description

The second example addresses a more advanced problem and further illustrates the utility of a simulation program. In this example, compound A reacts with compound B to produce more of compound B plus other products. The reaction is first order with respect to the concentrations of both A and B. It takes place in a continuous-flow, stirred-tank reactor (CFSTR) with two input feed streams, one of which contains A and the other B. The flow rate of the stream containing A varies sinusoidally while that containing B is constant. A flow diagram is given in Figure 2.7.

The reactor (Figure 2.7) is a CFSTR with a constant volume. The influent stream containing A has a constant concentration of A with the flow rate F varying sinusoidally while the stream containing b has both a constant concentration of B and a constant flow rate FR.

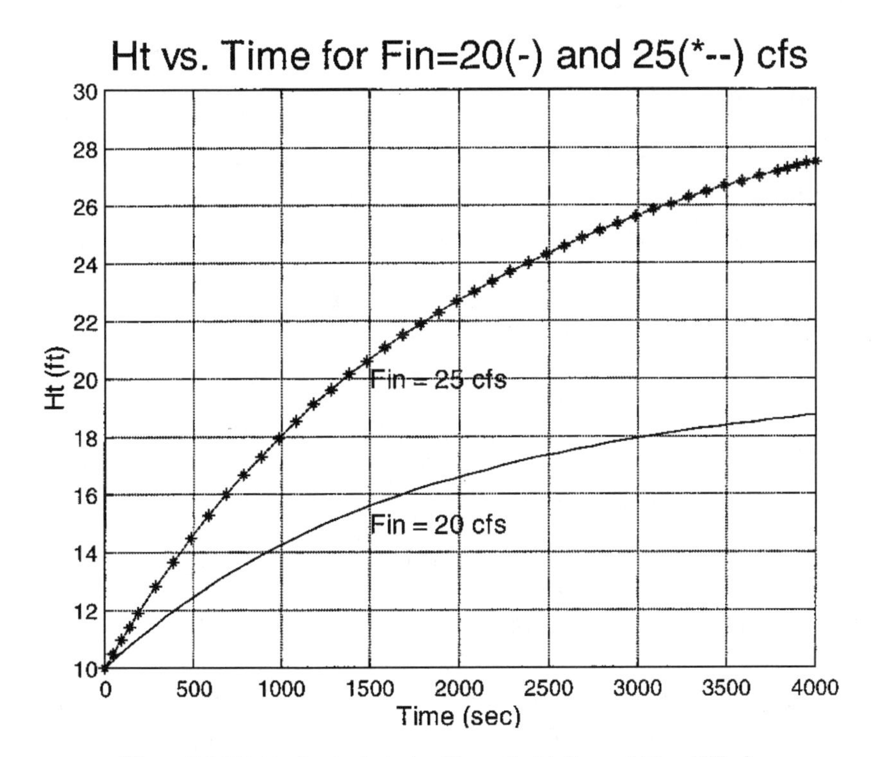


**Figure 2.6** Height of water in tank with constant inflows of 20 and 25 cfs.

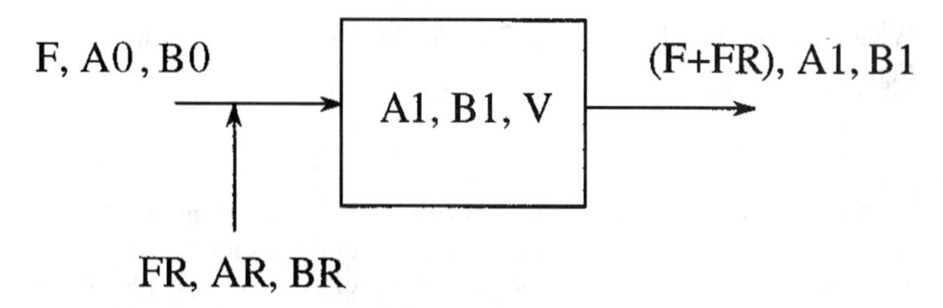


**Figure 2.7** CFSTR with sinusoidal input.

*Symbols and Units*

$A$ = concentration of compound A, g/cu m
$B$ = concentration of compound B, g/cu m
$K$ = reaction rate coefficient, 1/(h-g B/cu m)
$t$ = time, h
$V$ = reactor volume, cu m
$Y_{B/A}$ = stoichiometric coefficient, g B produced/g A utilized
0, 1, $R$ = subscripts denoting influent, effluent, and R streams

*Mathematical Model*

Going through the five steps presented earlier in this chapter develops the model. These are:

(1) Identify the significant reactions.

$$A + B \rightarrow B + \text{Other Products} \tag{2.12}$$

(2) Establish the stoichiometry.

$$A + B \rightarrow Y_{B/A} * B + \text{Other Products} \tag{2.13}$$

(3) Determine the kinetics.

$$dA/dt = -K * A * B \tag{2.14}$$

$$dB/dt = Y_{B/A} * K * A * B \tag{2.15}$$

(4) Specify the reactor. The reactor type is CFSTR as given in the problem statement. Note that any number of CFSTRs in series can simulate a plug-flow reactor.

(5) Material balances on A and B: The material balances for A and B comprise the mathematical model for the reactor. They are simplified by assuming that the influent stream $F$ containing A contains no B ($B_0 = 0$) and the stream $FR$ containing B contains no A ($AR = 0$).

$$V * \frac{dA_1}{dt} = F * A0 - (F + FR) * A1 - K * A1 * B1 * V \quad (2.16)$$

$$V * \frac{dB_1}{dt} = FR * BR - (F + FR) * B1 + Y_{B/A} * K * A1 * B1 * V \quad (2.17)$$

The following Matlab code simulates the reactor described above.

```
function dy = react (t,y)
% This is a MATLAB function. Place this code in a single file
% named 'react.m' and copy the file to a directory that is in
% MATLAB's search path.
% The function react return derivatives of state variables At and
% Bt for a CFSTR system in which A reacts with B:
% A + B → B + Other Products

% Global parameters
global FA;
global FC;
global W;
global P;

% Calculate sinusoidal flowrate FW
FW = FA + FC * sin(W * t + P);

% Global parameters
global A0;
global BR;
global K;
global YBA;
global FR;
global V;

% The following code defines the derivatives
dy = zeros (2,1);
dy (1) = (FW * A0 - (FW + FR) * y(1))/V - K * y(1) * y(2);
dy (2) = (FR * BR - (FW + FR) * y(2))/V + YBA * K * y(1) * y(2);
```

```
% calconc.m is a script file to calculate both concentrations
% of A and B in a CFSTR.
% Place this code in a single file named 'calconc.m' and copy the
% file to a directory that is in MATLAB's search path. To run this
% script, type 'calconc' at the MATLAB prompt. NOTE: this script
% uses the 'wtank' function above (see comments in the code for
% wtank).

% Sinusoidal flow rate parameters
global FA;
global FC;
global W;
global P;

% Influent and recycle stream concentrations
global A0;
global BR;
% Kinetic and stoichiometric parameters
global K;
global YBA;
% Reactor parameters
global FR;
global V;

% Numerical values of parameters
A0 = 200;
BR = 8000;
K = 0.01;
YBA = 0.667;
FR = 25;
V = 100;
FA = 100;
FC = 50;
W = 0.2618;
P = -1.5708;

% Using ode45 method to solve the differential equation.
% Ode45 integrates a system of ordinary differential equations
% using 4th and 5th order Runge-Kutta formulas.
% Set relative and absolute error tolerance to 1e - 4.
tol = le - 4;

[t,y] = ode45 ('react', 0, 72, [10, 1600], tol);
% 'react' is a function contained in the M-file: react.m.
```

```
% This file must be in the MATLAB search path.

At = y (:, 1);
Bt = y (:, 2);

figure (1);
F = FA + FC * sin (W * t + P); % Input flow rate F varies
sinusoidally

% Plot the results.
plot(t, F, 'r-')
xlabel ('Hours', 'FontSize', 12)
ylabel ('cubic meters/hour', 'FontSize', 12)
title ('F (Input flow rate) vs. Time', 'FontSize', 18)
axis ([0, inf, 0, 180])
grid on
figure (2);
plot (t, At, 'r-')
xlabel ('Hours', 'FontSize', 12)
ylabel ('gms/cubic meter', 'FontSize', 12)
title ('A1 (state variable) vs. Time', 'FontSize', 18)
axis ([0, inf, 0, 25])
grid on
figure (3);
plot (t, Bt, 'r-')
xlabel ('Hours', 'FontSize', 12)
ylabel ('gms/cubic meter', 'FontSize', 12)
title ('B1 (state variable) vs. Time', 'FontSize', 18)
axis ([0, inf, 0, 3000])
grid on
```

---

When you run the Matlab script file cakonc.m, you will obtain the plots shown in Figures 2.8 to 2.10.

The simulation can be run for different integration increments and error criteria. The value used for 'tol' (tolerance) determines the required accuracy of the simulation results. If this value is too low (e.g., try 'tol = 0.01'), the results become noisy. It is important to understand the simulation conditions that cause model inaccuracies as this has an impact on the interpretation of simulation results.

## SUMMARY

This chapter presented a short tutorial on mathematical modeling and the development of dynamic models using Matlab.

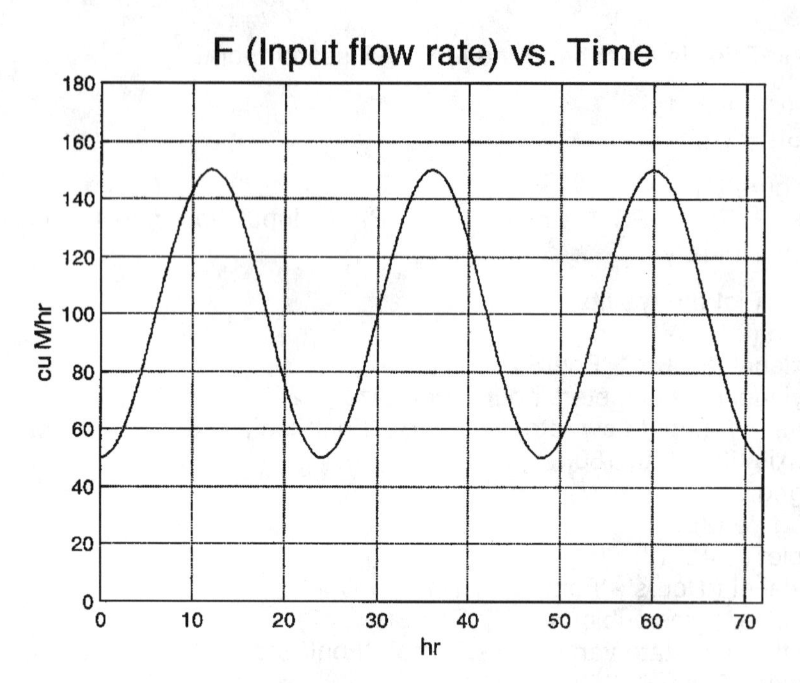


**Figure 2.8** Sinusoidally varying flow rate.

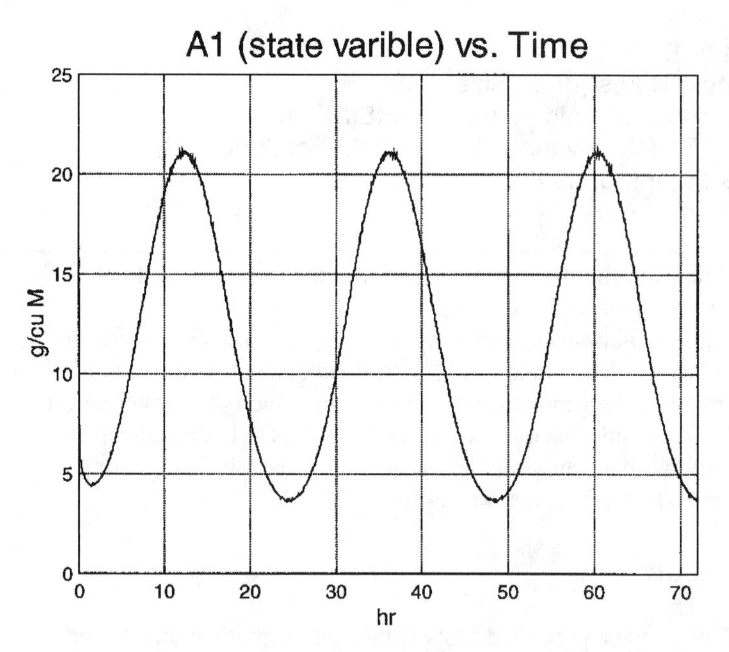


**Figure 2.9** Concentration of reactant *A* versus time.

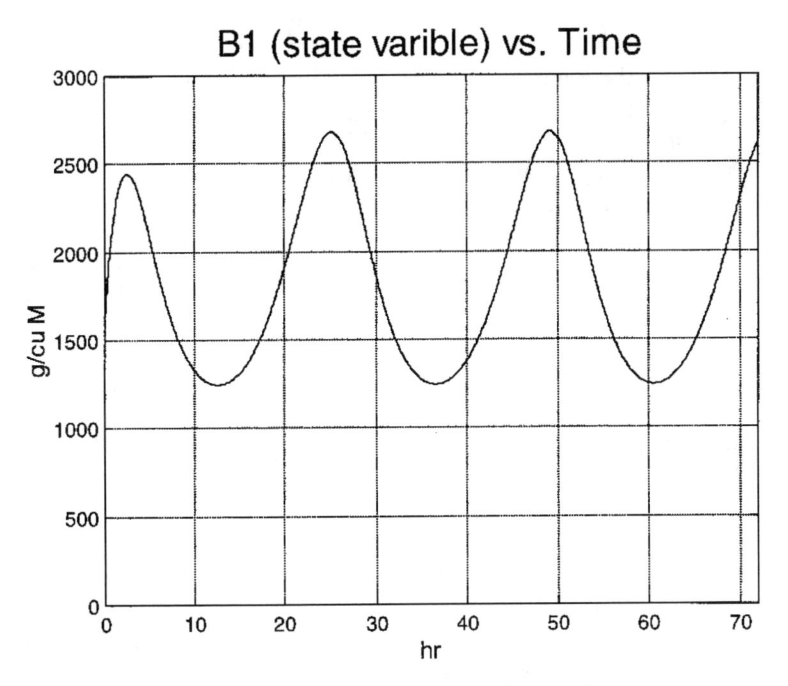


**Figure 2.10** Concentration of reactant *B* versus time.

The approach used for the development of dynamic mechanistic models for biological processes is presented and discussed. This approach consists of applying the following steps:

(1) Identify the significant mass transfer/reaction terms.
(2) Establish the stoichiometry.
(3) Determine the kinetics.
(4) Specify or characterize the reactor.
(5) Develop model equations from material balances.

Early simulation languages were batch-oriented for use on relatively expensive computers. The availability of low-cost personal computers has removed these obstacles and a wide variety of interactive simulation languages are now available for use on PCs, including Matlab, the program used for the examples in this chapter.

The development of mathematical models and their solution through computer simulation is best learned by practice with examples. Two detailed examples are presented these being:

- dynamic modeling and simulation of the height of water in a tank

with a constant pumped inflow and variable outflow through an orifice in the bottom of the tank.
- dynamic modeling and simulation of the reaction of two substances in a continuous-flow, stirred tank reactor. One substance is contained in a stream with a sinusoidally varying flow rate while the other is in a stream with a constant flow rate.

## REFERENCES


1 Elmqvist, H., K. J. Astrom, T. Schonthal and B. Wittenmark. 1990. *Simnon Vse6 Guide for MS DOS Computers, V. 3.0,* SSPA Systems, Box 24001, S-400 22 Goteborg, Sweden.

2 Matlab User's Guide (1992) The Math Works Inc., 24 Prime Park Way, Natick, Mass. 01760-1500.

CHAPTER 3

# Process Control in Biological Wastewater Treatment

## OVERVIEW

SOME basic ideas of process control applied to the activated sludge process are presented. A computer control system has several tasks, such as disturbance detection, measurement noise filtering, calculation of non-measurable process variables, condensation of measurement information and calculation of control strategies.

The activated sludge process is subject to disturbances all the time, and this gives the main motivation for control. The consequences of disturbances have to be sufficiently damped. A proper understanding of the dynamics is important, if a relevant control is to be realized. There is no single model that adequately describes a process for all needs. Instead special models have to be derived for control purposes.

Even if the obvious goal for a plant is to produce a good effluent, it is far from trivial to translate that goal into operational and control goals. Given the goal one has to examine adequate measurement variables and possible sensors together with suitable control variables to properly manipulate the plant.

Most activated sludge plants are controlled by a large number of simple regulators. Simple controllers are described, and the most common controller, the proportional-integral-derivative (PID) controller, is discussed in detail. After that, various control schemes for quality control of an activated sludge plant are shown.

Several practical aspects have to be considered for implementation of

Gustaf Olsson, Department of Industrial Electrical Engineering and Automation (IEA), Lund Institute of Technology (LTH), Box 118, S-221 00 Lund, Sweden.

a controller, such as the sampling rate in a computerized system, and problems related to limited control signals. There are processes, however, where simple controllers are inadequate, and some of these systems are discussed.

Computer implementations of controllers are discussed and programmable logical controllers and some structural aspects of industrial controllers are described.

## GOALS AND INCENTIVES FOR CONTROL

The goal of the operation of a wastewater treatment plant is to satisfy the effluent requirements at all times at minimum cost. The effluent quality and a good performance have to be guaranteed *consistently,* which ought to have a great impact on instrumentation, control and automation. Effluent requirements are typically formulated in terms of daily, weekly or annual averages of suspended solids, BOD, total nitrogen and total phosphorus. In order to comply with the limits of effluent concentrations one wishes to have certain safety margins. The goal of automatic control is to keep the effluent quality consistently good, despite disturbances. This allows the margins to be smaller, thus saving costs.

The state of the plant is described by a large number of process variables, such as concentrations, flow rates, levels, pressures, temperatures, and pH. All these variables would relate to the effluent quality in some way. Many variables, however, can be controlled separately since the process can be decoupled into many small loops with small couplings between them. The ultimate objective is to produce an acceptable effluent for a minimum cost. This objective has to be translated into adequate criteria for each control loop.

### CRITERIA

The capital costs are usually the dominant part of the total cost. Traditionally many plants have been overdesigned with large tank volumes to meet the varying loads to the system. One purpose of control and operation is to make a maximum use of available tank storage. Thus, the total cost can be decreased, if the balance between design and operational costs is systematically considered. The latter are made up of many components, such as consumption of electric energy (for pumping, aeration, etc.), chemicals for phosphorus removal or polymers for sludge conditioning. Obviously labor costs are important. *Consistency* of the water quality, however, is probably the main reason for automation.

Some attributes can be added which, while not unique to our processes, do significantly affect our approach to control:

- The daily volume of wastewater treated can be huge.
- The disturbances in the influent are enormous compared to most industries.
- The influent must be accepted and treated, no returning it to the supplier.

These attributes are the essence of our problem—major disturbances. Furthermore:

- The process has significant nonlinearities limiting the usefulness of simple controllers.
- There is a very wide range of response times, both a problem and a blessing.
- There are many interactions within the process with major recycles.
- There are many external couplings from the sewer to the receiving waters.
- Designers often leave too few manipulated variables and too little control authority.
- The concentrations of nutrients (pollutants) are very small, even challenging sensors.
- Sensors don't exist for many states and are expensive for others.
- The value of the product in the marketplace is remarkably low.

These attributes are why control is far from straight forward to design or to implement:

- The microorganisms change their behavior and their population distributions.
- The separation of the effluent from the biomass is challenging and easily disturbed.
- Future effluent standards will be tight and based on spot checks.

The difficulties in implementing control, and these attributes are why disturbance rejection must be the major goal of operation. Large disturbances must be attenuated, and we must do it as early in the process as we can. This will not only meet effluent requirements, but ease the task of controllers later in the process.

The development of instrumentation, control and automation (ICA) is *not* to be driven by technology but by the needs. Some of the driving forces can be categorized as (Olsson, 1993):

- effluent quality standards
- economic

- increasing plant complexity
- available tools (sensors, computers, models, etc.)

In some countries there is a water shortage which in itself is a driving force for improved wastewater treatment, e.g., for recycling. Moreover, the desire for municipalities to be leaders is often a driving force for better wastewater treatment standards. Many process industries are becoming environment-conscious and wish to improve their environmental record.

## PLANT DYNAMICS

A biological wastewater treatment plant is a complex dynamical system, containing both biological, chemical and physical phenomena. Any operator is aware of its dynamical character by observing time varying flow rates, concentrations and compositions. Those variations may appear abruptly and be quite significant. The control actions have to be related to the dynamics, since the result of such an action cannot be observed momentarily. Therefore it is often not trivial to find the right cause-effect relationships. Furthermore, some couplings between the different process units are quite strong, such as the settler-aerator coupling. This means, that a unit process can not be considered in isolation. Rather, its interaction with other unit processes has to be part of the control design.

### TIME SCALE OF THE PROCESS

The time scale is probably the most important single factor in the characterization of a dynamic process. Many industrial plants—and in particular the activated sludge process—include a wide spectrum of response times. It is important to consider which ones are relevant for the actual purpose. The wide span of response times makes it possible to decouple many unit processes, i.e. many variables can be controlled by separate, local controllers [see Olsson (1985), Olsson-Piani (1992) and Olsson-Newell (1997)].

The inherent dynamical properties of the plant determine how quickly and significantly the system reacts on external disturbances or control actions. The wide span of dynamics of a typical treatment plant can be illustrated by the following examples:

- Pump operations can create disturbances in different time scales. The start of a primary pump causes a disturbance that propagates through the plant within 20 to 40 minutes. It causes the clarifier effluent suspended solids concentration to change (Olsson-Chapman, 1985). In a slower time scale, several hours, an

increasing flow rate will give a dilution effect. This response time depends on the tank volume rather than the tank surface area.
- A change of the air flow rate will not be noticed momentarily in the dissolved oxygen concentration since the transfer of gaseous to dissolved oxygen takes place within 15 to 30 minutes.
- Cell growth will be noticed in days and cell decay in weeks. Anaerobic growth needs even longer time. This means that cell population cannot be controlled in an hour-to-hour time frame. In order to calculate the sludge retention time it is typically necessary to make daily averages of the sludge mass balances.
- Seasonal variations produce temperature influence on growth, particularly for nitrification. In order to find out the relation between temperature and cell growth it is necessary to measure during several months.

## DISTURBANCES

A wastewater treatment system is driven by disturbances, so for all practical reasons it is never in steady state. The disturbances are related to

- flow rate changes
- concentration variations
- composition variations

The disturbances are not only related to the influent flow characteristics, but may be generated internally in the plant due to operational procedures. All of these disturbances will more or less influence the effluent water quality. Again, the ultimate operational goal is to minimize the influence of disturbances to satisfy the effluent quality. Therefore, wastewater treatment operation is a typical *disturbance rejection problem.*

### Hydraulic Disturbances

Hydraulic disturbances are the most common and the most difficult ones to control. The most apparent reason is, that the flow rates are so large, that equalization becomes very costly. Usually the influent hydraulic variations follow a diurnal pattern, but a sudden rainstorm may influence the plant within an hour, if the sewer system is small.

Not only influent flow rate variations but also poor pumping operations contribute to operational disturbances (Olsson et al., 1989). Clarifier upsets are often caused by inadequate pump control. A sudden flow increase caused by a pump start will influence the flow propagation along the plant within a fraction of an hour. A proper settler operation requires, that the influent pumping has to be smooth. If the primary pumps are operated on/off in too large steps, then the result is exactly like having sudden rainstorms coming to the plant all the time.

Unfortunately internally generated disturbances are quite common, often due to unsuitable pumping of influent flow or sludge recycle pumps. The use of fixed speed pumps create great risks for unsuitable changes. Variable speed pumping is a proven technology today and is a crucial component to create a smooth hydraulic operation of the plant.

The design of the sewer system has a profound influence on the plant hydraulics. A widely distributed sewer system contributes to attenuate many flow peaks. A proper operation of the pumping in the sewer system can significantly improve the operation of the treatment plant (Gustafsson et al., 1993; Nyberg et al., 1996).

### Concentration Disturbances

Variations in concentration and composition may be both rapid (e.g., caused by pulses or recirculation of sludge) and significant in amplitude. A sudden change in biodegradable substrate concentration should naturally be met by control actions, such as dissolved oxygen control or various recycle flows. Influent concentrations of inhibitory or toxic material ought to be treated differently. Preferably, they should never enter the plant untreated, but be detected as far upstream as possible.

Concentration disturbances are also generated within the plant itself. For example, sludge thickener overflow is usually returned to the plant inlet. This overflow is usually operated with on/off pumping without considering the influent load. Since the recirculated flow may have a large concentration it will make a significant contribution to the overall load, both in terms of hydraulics and substrate loading.

It is usually impossible to remove the source of the disturbance, even if this is an important option, especially for industrial effluents, where production control in the industry may be improved. Sometimes the magnitude of the disturbance can be reduced *before* it reaches the plant. An integrated sewer-treatment plant control can attenuate hydraulic disturbances, and some waters can be pre-treated in order to avoid harmful effects on the plant, such as neutralization of wastes with high or low pH. The control can also compensate for effects *within* the process. A common example is dissolved oxygen control.

## MODELS FOR CONTROL

A model is of fundamental importance for control. Any feedback scheme is based on some understanding of how the physical process will react to an input signal. Therefore, the ability to model dynamic systems and to analyze them are two basic prerequisites for successful feedback control. A thorough description of dynamic models of different complexity in wastewater treatment systems as well as the problem of model verification is presented in Olsson-Newell (1997), Chapters 2–9.

Clearly, there are different model approaches depending on how to use the model. Different controllers need different process models. There is a common misunderstanding that only one model can ultimately describe the process. Instead, the model complexity and structure has to relate to the actual purpose of it.

## WHO NEEDS THE MODEL?

A mathematical model can be considered as a package of knowledge of the process dynamics. Given appropriate models, they may serve as powerful tools for process designers or operators. They can give advice for possible control actions and be used also for predictions. Probably the model accuracy will never be such that reliable quantitative predictions can be made over a long time (weeks), but still the couplings between different process units could be reflected and the values or just the trends would be in the right order of magnitude.

In wastewater treatment the need for dynamical models is apparent for several groups of people:

- the designer, who wishes to explore not only the average properties of a plant, but also its robustness to dynamical disturbances: Such analysis has to be performed before the plant is built;
- the process engineer, who wishes to explore different process configurations or operating principles of an existing plant
- the operator, who needs a decision support system, where he can explore different *what-if* situations
- the educator, who will use the model for teaching plant dynamics for different categories of people, ranging from operators to researchers
- the researcher, who will use the model as a condensed version of his knowledge: This will give possibilities to further investigate different properties of the model.

It is quite apparent that the different users want to find different answers from the models. Some of them have to calibrate the model to an existing plant, while others have to use best available guesses of the plant parameters. The different users are interested in different time scales and have different demand for simulator flexibility. While the researcher wants to be able to change almost everything in the model, the operator at the plant emphasizes the user-friendliness and the reliability of the model.

## DYNAMICAL MODELS

There are many different ways to describe systems. The choice of one way or another is a matter of information at disposal, possibility to collect further information and—most important of all—the purpose for which modeling is done. Differently to science, where the purpose of modeling is to gain insight in a system, a model in control engineering is adequate if the control process based on it leads to the desired purpose (small variations around a given value, reproducibility of an input signal, etc.).

In control applications we are interested in dynamical systems and operational models as opposed to design models. There are many different ways to model them, and the most important types of models are:

- continuous time description of the system in terms of linear or nonlinear differential equations, giving a quantitative description of mass, energy, force or momentum balances of a physical process: In many cases, nonlinear equations can be linearized under reasonable assumptions.
- sampled time description in terms of linear or nonlinear difference equations: This means that information is available only at specified discrete time instants. Sampling is necessary, when using process computers working sequentially in time. Choice of the sampling period is part of the modeling.
- discrete event, or sequential systems: Typically the amplitudes of the inputs and outputs of the system are discrete and are often of *on-off* type.
- systems with uncertainty: The system itself or the measurements are corrupted by undesired noise. In some cases this noise can be given a statistical interpretation. The statistical description may model either real random disturbances or modeling imperfections. In other cases the uncertainty can be described by a more linguistic approach rather than numerical or mathematical terms. One way is to use a special algebra, called fuzzy algebra, to describe the uncertainties. Another way is to apply non-numerical descriptions of the type "if-then-else" rules.

Clearly, there are different model approaches depending on how to use the model. Different controllers need different process models. Since a computer works in time discrete mode it is important to formulate the physical process descriptions accordingly.

Most often modeling practice is a combination of physical modeling and identification. With more insight into the fundamental properties of the process there is a greater potential to obtain an accurate dynamical description. Note, however, than even the most elaborate model based on physical insight has to be verified by experimentation. The topic of modeling is elaborated more in Olsson-Piani (1992) or Åström-Wittenmark (1990).

## MODEL LIMITATIONS

The biological part of the activated sludge process has been subject to a lot of research. The influent wastewater is characterized by its composition, not only in carbonaceous and nitrogen fractions, but also in soluble, particulate, biodegradable and unbiodegradable fractions etc. The organisms are represented by at least one type of heterotrophic and one or two species of nitrification organisms. The main reactions are added for each type of species and each type of available substrate.

The best known models are probably the IAWQ model no. 1, describing the dynamics of organic and nitrogen removal (Henze et al., 1987) and the IAWQ model no. 2, that also models biological phosphorus removal (Gujer et al., 1994). The main emphasis of the models are the biological reactor, while the settler dynamics are treated comparatively superficially. The IAWQ models describe the growth and decay of organisms in aerobic, anoxic and anaerobic environments as well as alkalinity balances and alkalinity effects on the kinetics. Each state includes several reaction rate and stochiometric parameters. It is obvious, that the verification of such a model is an awkward task.

In most models available the clarifier has been treated as a pure concentrator, sometimes with a time lag or time delay (Allsop et al., 1990). More structured models that incorporate both the clarification and the thickening phenomena have been presented (Vitasovic, 1986; Takacs et al., 1991; Patry-Takacs, 1992; Jeppsson, 1996). Still the dependence of the settling parameters on the biological conditions of the sludge is not straightforward. Usually it is assumed that there is no bioactivity outside the aerator. There are, however, indications that some biodegradation takes place in the settler (Sorour et al., 1992).

Most of the growth models are generally described by Monod type of equations,

$$\mu = \hat{\mu} \frac{s}{K_s + s}$$

where $\mu$ and $\hat{\mu}$ are the specific growth rate and its maximum value respectively, $s$ the substrate concentration, and $K_s$ the half saturation constant. Already the parameter identification of $\hat{\mu}$ and $K_s$ in a system with only carbonaceous removal is difficult, as shown for example by Holmberg (1982), Ossenbruggen et al. (1991) and Jeppsson-Olsson (1993). Even under almost ideal measurement conditions the two parameters can not be found simultaneously with any reasonable accuracy.

To find the influent wastewater composition presents another difficulty. Complex models containing many parameters can almost always (but not easily!) be adjusted to fit any given data set. The problem is, that this model adjustment can be made in many different ways, either by changing the wastewater characteristics or by changing some model parameters. It becomes apparent that the models have to be simplified in order to obtain unique model verifications.

If a model has too many parameters, it may be adjusted to a certain data set satisfactorily. However, if the model is used later for another data set from the same plant, some parameters may be changed in order to fit the new data. However, sometimes such a model fit can also be made by changing some influent wastewater characteristics. This shows, that some further constraints are needed in order to uniquely determine the model.

Why are model verifications necessary? Obviously the models can be used in simulation studies of the plant systems. Off-line simulations are important tools for studying the plant responses to possible disturbances. On-line simulators are becoming more and more interesting for operator guidance. There the emphasis for model verification becomes critical, since the model will be used for operator decisions. Furthermore there has to be methods to update the model parameters in a unique fashion as the plant operating conditions change with time.

## PARTICULAR DIFFICULTIES

It is well known that accurate modeling (and thus also model based control) is hampered by the following major problems, which call for adequate engineering solutions:

- The process kinetics are most often poorly understood nonlinear functions, while the corresponding parameters are time varying.
- Up until now there is a lack of reliable sensors suited to real time monitoring of process variables which are needed in advanced control algorithms.

Therefore, the earliest attempts at control of a biotechnological process used no model at all. Successful state trajectories from previous runs which had been stored in the process computer were tracked using open loop control. Many industrial fermentations are still operated using this method.

There are two trends for the control of biotechnological processes that have emerged, the optimal control approach and the adaptive control approach.

In the optimal control approach, the difficulties in obtaining an accurate mathematical process model are ignored. In numerous papers classical methods, such as Kalman filtering, optimal control theory etc., are applied under the assumption that the model is perfectly known. Many of these papers are devoted to the theoretical investigation of optimal control problems in fed-batch fermentation. Control opportunities in fed-batch operated fermentations have been reviewed in detail in a number of articles [e.g. Parulekar and Lim (1985)]. Most of the optimal control papers considered so far have been limited to low order processes.

Optimal control based methods all suffer from problems with respect to practical implementation. Since optimal control is a very model sensitive technique, and accurate mathematical models for bioprocesses are extremely hard to obtain, it is very unlikely that a real life implementation of such controllers would result in the predicted simulation results. Furthermore the implementations are already hampered by sensor problems.

In an adaptive control approach the aim is to design specific monitoring and control algorithms without the need for a complete knowledge of the process model, using concepts from adaptive control. This is further discussed later.

## GOALS OF OPERATION

The formulation of the goal of a wastewater treatment operation is much more demanding than to define the effluent standards. Having a goal or a performance criterion defined the necessary control actions can be analyzed.

There are several goals of wastewater treatment, and one has to recognize the different levels and interest groups. The specific goals of a wastewater treatment plant are set so that it contributes to meeting the more general corporate and community goals. It includes at least

- to meet effluent discharge requirements
- to achieve good disturbance rejection
- to optimize operation to minimize the operating cost

The first goal of course includes keeping the plant running. This is the major task for many operations, at least in terms of man-power. It is not a trivial task to maintain advanced instrumentation and equipment, to make sure it runs smoothly and to detect any faults in its operation. Much of this is ''traditional'' automation and process control, and is not specific to running wastewater treatment systems. The environment may be very specific and harsh for much equipment, but that is also the case in other industries as well, so from a strict control point of view the problems are traditional.

However, keeping the plant running and simply recording effluent quality is too often taken as the only goal. This has been and is still largely due to a lack of process knowledge on the part of plant operators. Too many operators of plants rely on the consultant for expertise in both design and operation. To satisfy the effluent quality requirements, it is crucial to ask: is this task solved primarily by design or does instrumentation, control and automation play any significant role in this part? The traditional answer among designers used to be *no.* When only carbonaceous removal was considered the plant was considered ''self-regulating'' and could take care of load disturbances and still produce a satisfactory effluent. The traditional method was large volumes, excess aeration and chemical dosage. Consequently a lot of plants have been oversized. No incentive was given for improved operation, since operational costs were not compared with investment costs. Often they were met from different financial sources.

Due to the significant load variations and disturbances, typical for wastewater treatment systems, the plant is hardly ever in steady state. Despite the wide swings of load, the plant still has to *consistently* produce a satisfactory effluent. If the volumes of the tanks are not oversized, this means that some control system has to minimize the influence of disturbances and ensure that the operation is satisfactory. Pressures to fully utilize volumes and the complexities of nutrient removal are making process control essential. This of course includes disturbance attenuation.

Minimizing the operating cost and maximizing the use of available capacity is becoming necessary. Here control and optimization are the key tools. Costs to be minimized include air [improved dissolved oxygen (DO) control and levels], chemicals (more use of the biomass), pumping and carbon addition. Of course, if control and optimization is part of the answer it is also part of the cost. Some control actions present little cost:

- waste sludge flowrate changes
- return sludge flowrate changes
- step feed changes
- recycle schemes

Other control actions may be costly such as chemical addition for P removal or sludge conditioning.

Operational objectives are much more specific instructions for operation. They are developed so that a specific plant will meet the process or plant goals. Sometimes they are general in that many plants may use the same objective, but often they are specific to a plant and its influent and its receiving waters. Examples of more general operational objectives are:

- grow the right biomass population
- maintain good mixing where appropriate
- adequate loading and DO concentration
- adequate air flow
- good settling properties
- avoid clarifier overload
- avoid denitrification in clarifier

## MEASUREMENT AND CONTROL VARIABLES

Any controller needs information about the process, so sensors and instrumentation are essential for any control. To realize any control algorithm there has to be possibilities to control or manipulate the process. Here we briefly review the potential for measurements and for manipulation of an activated sludge process.

### MEASUREMENT VARIABLES AND SENSOR TECHNOLOGY

Sensor technology is crucial for the success of control and operation. It is true that great progress has been made over the last few years. Still, however, there is a lack of robust, reliable sensors for essential parameters. Sensor technology was given a special emphasis on a IAWQ Specialized Conference in Copenhagen (IAWQ, 1995).

For the control of an activated sludge system it is necessary to understand which measurements to make and how they relate to the process behavior. Moreover, it is important to define the purpose of the measurement. Is it going to be used for control purposes, for quality check or for reporting to the regulatory agencies? The sampling rates may be quite different depending on the purpose. Important types of measurements are:

- flow rates in different plant units
- air flow rates and air pressure
- sludge flow rates
- temperature in water basins and in digesters
- gas (e.g. methane) flow rates and temperatures
- alkalinity
- pH
- dissolved oxygen in different locations
- sludge levels
- suspended solids in different places
- BOD, COD, TOC
- phosphorus fractions
- nitrogen fractions

- respiration rate

There are many instruments which satisfactorily measure flows, levels, pressures, dissolved oxygen in liquids and suspended solids and dry solids concentrations in sludge. Even if an instrument is deemed to be reliable, proper maintenance is still required. Maintenance requirements will vary greatly between different applications.

There is a group of instruments for measuring quality variables to judge the performance of a wastewater treatment plant with respect to effluent permits. Included in this group are instruments for measuring organic material, phosphorus and different nitrogen fractions. These instruments and measurement systems are commercially available today although a good performance requires considerable maintenance and operator competence to obtain reliable measurement signals. Therefore, even if an instrument is commercially available, it is not necessarily suitable for use at a treatment plant.

The development of respirometry has been significant. Within IAWQ there is a special Task Group preparing a Scientific and Technical Report during 1997. Some of the progress was described in Singapore at the Biennial IAWQ Conference in 1996 (Spanjers et al., 1996).

New technology may lead to the development of new instruments such as ion selective solid state components, optical sensors, bio-sensors and acoustical sensors. However, much development work remains before such instruments will be commonplace at treatment plants.

## MANIPULATED VARIABLES

There are quite a few variables to manipulate an activated sludge process. Still the possibilities to control the plant in a flexible way are quite limited. In many plants, however, there may be a potential to make a better use of the manipulated variables that are available. We categorize them in the following groups:

- hydraulic, including sludge inventory variables and recirculations
- additions of chemicals or carbon sources
- air or oxygen supply
- pre-treatment of influent wastewater

There are several other manipulated variables in a plant, that are related to the equipment and to basic control loops in the process, such as local flow controllers, level controllers etc. They are not included in this discussion. Further details are found in Olsson-Jeppsson (1994).

A majority of the manipulated variables are meant to change the hydraulic pattern in the plant. The different flow rates will influence the retention times in the different units. Moreover, the clarification and thickening processes are particularly sensitive to the *rate* of change. The hydraulic flows also determine the interaction between different unit processes. We classify the hydraulic manipulated variables into four major groups:

- variables controlling the influent flow rate
- variables controlling the sludge inventory and its distribution
- internal recirculations within the activated sludge system
- external recycle streams, influencing the interactions between sludge and liquid processes

Sequential batch reactors (SBR) offer other control possibilities due to the flexibility associated with working in time rather than in space. Some of the features of SBRs are described in some detail in Irvine-Ketchum (1989). The Danish Bio-Denitro and Bio-Denipho processes are not the traditional fill-and-draw process but are alternating processes with continuous discharge [see Bundgaard (1988) and Einfeldt (1992)]. Pilot scale control experiences have been reported in Isaacs et al. (1992), and Zhao et al. (1994). Favorable full scale experiences and control system descriptions have also been reported by Thornberg et al. (1993). They will not be further discussed in this chapter.

Some of the control variables discussed above are sometimes called *selection mechanisms* for the viable bacteria in the system (see Gujer-Henze, 1990). They are classified in the groups:

- electron acceptors (e.g., oxygen or nitrate)
- substrate
- thickening or clarification properties
- temperature
- growth rate
- free swimming organisms

Not all of these can be manipulated. Only the electron acceptors (mostly oxygen) and (to some extent) the thickening and clarification properties can be controlled (by chemical precipitation). Moreover, substrate can be added (e.g. carbon sources). The other selection mechanisms are considered as disturbances.

## SIMPLE CONTROLLERS

As noted in previous sections there are different objectives of different control loops. Many are used to keep the plant running. Others are used more directly to control the water quality. We will first describe some typical local variables for the operation of the physical plant parts, and in Section 8 we will look at some common quality control loops.

## BASIC FEEDBACK CONTROL STRUCTURE

In the simplest case the input to the controller is the output error, or the difference between the reference value (set-point) $u_c(t)$ and the measurement $y(t)$

$$e(t) = u_c(t) - y(t) \tag{3.1}$$

For example, for a DO controller the error is the difference between the desired DO concentration and the measured one. The controller output $u(t)$ is a simple function of the control error. In the DO control case the $u(t)$ is the desired air flow rate. This form is the most common form of simple feedback systems and is represented by the block diagram in Figure 3.1.

It is reasonable to assume that with more controller parameters there would be more degrees of freedom to change the system behavior. Later, I will illustrate when more complex controllers are needed in order to achieve the desired result.

## EXAMPLES OF LOCAL FEEDBACK CONTROL

There are many so called local feedback loops in a wastewater treatment processes. The term *local* refers to the fact, that the controller keeps just

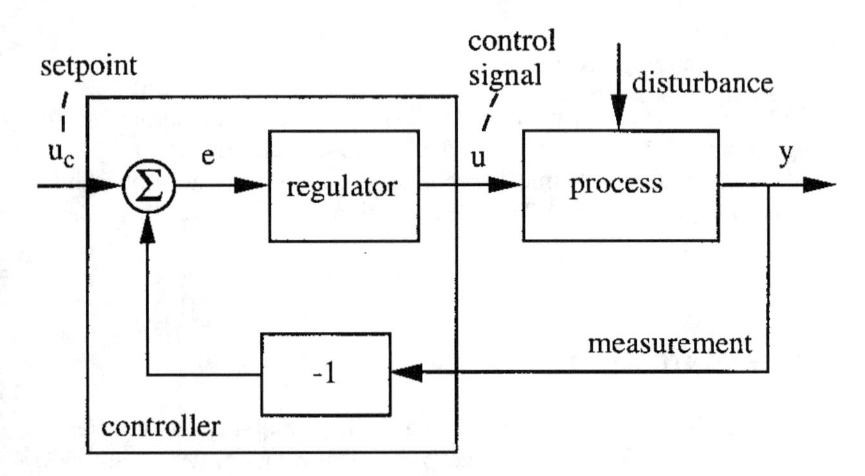


**Figure 3.1** The simplest controller structure.

one variable at some set-point value. This value is usually chosen constant, but can be changed manually or by some higher level calculation. Some examples:

- *Flow control* is applied to both liquid and gas flows. Pump control is traditionally based on flow rate measurements or some level measurement. In a wastewater treatment plant the flow rate control may be based on a wet well level.
- *Liquid flow splitting* between parallel trains is important in order to ensure good operation. A common practical problem is that weirs and splitter boxes are fixed, while the flow splitting is a dynamical process and has to be based on true flow measurements in order to work satisfactorily.
- Local control is used to produce a well-defined *air flow rate* to the aerators.
- *Dissolved oxygen control* in its simplest form is a conventional cascade control and is discussed below.

## OPEN LOOP SEQUENCING

Open loop sequencing is common. In such a scheme there is no feedback from the measurement, but the control is based on a timer or a predetermined action (Figure 3.2).

Some examples are:

- *Air compressors* are often switched on and off according to timers. Sometimes a DO sensor is used to feed back information for on/off control of the compressors. Feedback DO control is further discussed below.
- *Sludge removal,* particularly in primary sedimentation, is often based on timers instead of some sludge density or sludge blanket measurements. However, measurements of the sludge density can ensure that a consistent solids concentration is obtained.
- *Waste activated sludge* pumping is often based on a timer. A better action is to base the wastage rate on conventional sludge age calculations. Then it becomes a feedback control.
- *Bar screen cleaning* is sometimes based on a timer operation. The loop, however, can be closed if the cleaning process is started when a high differential level over the screen occurs.

## ON-OFF CONTROL

On-off controllers are simple, inexpensive feedback controllers that are commonly used in simple applications such as thermostats in heating

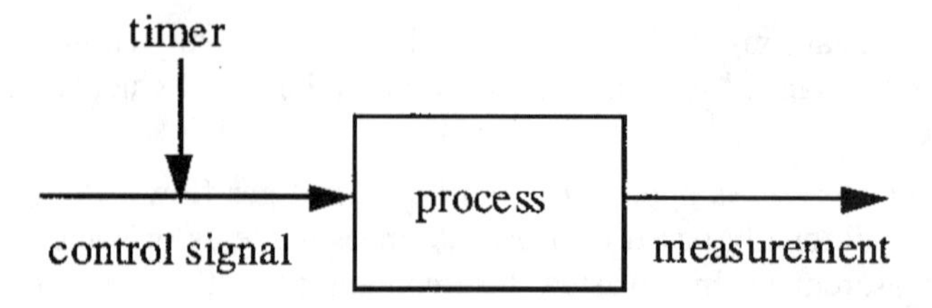


**Figure 3.2** Open loop control.

systems and domestic refrigerators. They are also used in industrial processes such as level control systems or dosage controllers. For example, if the oven gets too hot, turn off the power; if it gets too cold, turn on the power.

On-off controllers are usually modified to include a deadband for the error signal to reduce the sensitivity to measurement noise. The on-off control is sometimes referred to as *two-position* or *bang-bang* control. An *on-off* controller causes an oscillation about a constant set-point, since the control variable jumps between the two possible values. Therefore it produces an excessive wear on the final control element. If a valve is used as an actuator this is a significant disadvantage, while it is not a serious drawback if the element is a solenoid switch. The speed of the cycling depends on the process, how fast it responds to an on or off control action, as shown in Figure 3.3.

Another aspect of on-off control is that it can create a disturbance to other parts of the process. We discussed this problem earlier, where the control of sewer or receiving basin level by on-off switching of influent pumps can create a major disturbance to the secondary clarifiers.

It is not uncommon to see on-off control in use for dissolved oxygen

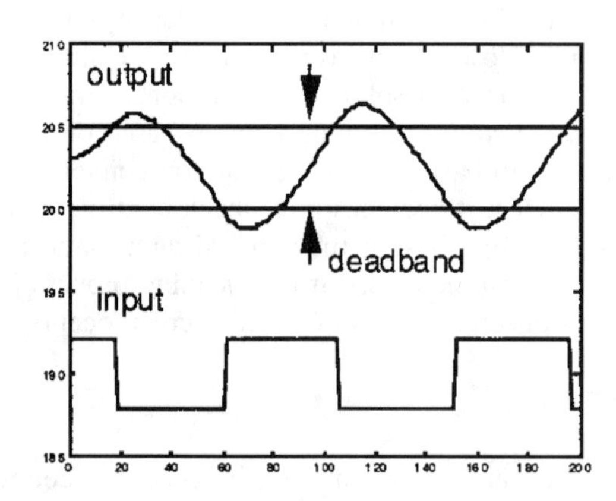


**Figure 3.3** On-off control.

control. Either valves in the air supply line or even the blowers are switched on and off. There is one interesting side benefit of this. When the air is turned off, the rate at which the DO level falls can be related to the oxygen uptake rate (OUR) of the biomass. When the air turns on, the rate at which the DO level increases is related to the OUR and the $K_La$ parameter (oxygen mass transfer rate coefficient). This can be a useful way of obtaining estimates of these two parameters on a continuing basis. OUR changes particularly are useful in gauging the activity of the biomass, which can be related to changing nutrient load or to the health of the microorganisms. Unexpected falls can indicate toxic shocks.

## THE PID CONTROLLER

In the PID controller—the most common controller structure—the controller output is the sum of three parts, that are proportional to the error, to its time integral, and to the error derivative. The idealized PID controller looks like

$$\begin{aligned} u(t) &= u_0 + K \cdot \left[ e(t) + \frac{1}{T_i} \int_0^t e(\tau) d\tau + T_d \cdot \frac{de}{dt} \right] \\ &= u_0 + u_P(t) + u_I(t) + u_D(t) \end{aligned} \tag{3.2}$$

where $u_P(t)$ is proportional to the error, the second part $u_I(t)$ proportional to the time integral of the error and the third part $u_D(t)$ is proportional to the error derivative. The parameter $K$ is the controller gain, $T_i$ the integral time and $T_d$ the derivative time. The value $u_0$ is a bias value that gives the controller its proper average signal amplitude. A idealized controller does not show any physical limits on its output. A real controller saturates when its output reaches a physical limit, either $u_{max}$ or $u_{min}$.

In process control one may find several special cases of the PID controller, such as P (only proportional), PI (proportional plus integral) or PD (proportional plus derivative). The integral part $u_I(t)$ of the controller is used to eliminate steady state errors. Its function can be explained in an intuitive way. Assume that the system is in steady state so that all the signals are constant, particularly $e(t)$ and $u(t)$. Steady state can only remain if the integral part $u_I(t)$ is constant, otherwise $u(t)$ would change. This is only possible if $e(t)$ is zero. Note that the integral time coefficient appears in the denominator in (3.2). This makes the dimensions of the controller terms proper and has a practical interpretation. To see this, consider a step change of the error $e(t)$ and its response in a PI controller (Figure 3.4). Immediately after the step the controller output is $K \cdot e$. After the time $T_i$ the controller output has doubled. A PI controller is often symbolized by its stcp rcsponsc.

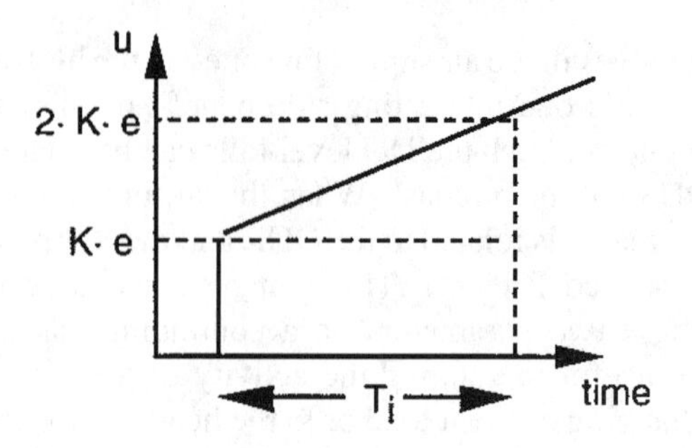


**Figure 3.4** Step response of a continuous PI controller.

There is a cost for the steady state elimination of the integral part. This is illustrated in Figure 3.5, where a system is responding to a change in the desired output or setpoint from 200 to a value of 205.

The PI controller is eliminating the offset of about 0.5, but at the expense of a larger overshoot, showing the decrease in relative stability (the proportional gain was the same in both cases). The loop gain must be decreased to maintain the same relative stability, by decreasing the proportional gain.

Another potential problem arises if the manipulated variable hits a constraint for some time, such as the valve becoming fully shut or wide open. Because of the constraint, it is impossible to achieve a zero error and the error integral just grows and grows. This problem is called *integral* or *reset windup.* Commercial controllers generally solve the problem either by putting an upper limit on the integral term, or by turning off the integral action when the controller output saturates. A problem with the latter approach is that the actuator or the manipulated variable itself may saturate before the controller output.

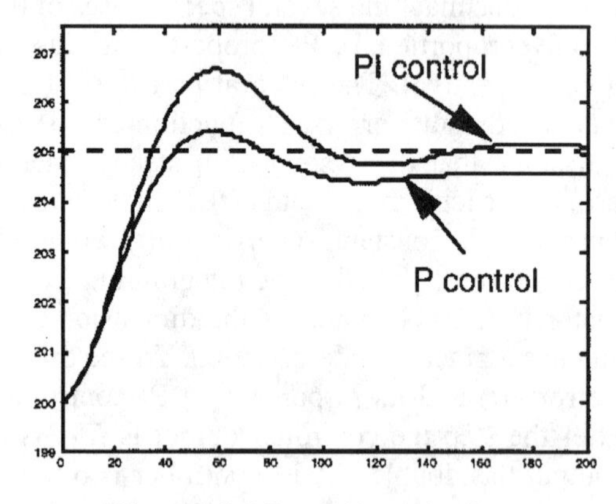


**Figure 3.5** Offset and integral action.

The purpose of the derivative part is to make the controlled system more stable. This is intuitively realized, since the derivative is a simple predictor of the control error. There is, however, one difficulty related to the derivative term. To calculate the derivative of a noisy measurement signal is tricky. It picks up the instantaneous slope of the measurement. Therefore any fast noise components have to be smoothed before the derivative is calculated. The result of the derivative (*D*) action can be seen in Figure 3.6, where *D* action was added to the same P and I action that we saw in Figure 5.

## PLANT QUALITY CONTROL

Any control that extends beyond local control actions discussed previously has to consider quality related or model related behavior. This means that the controller has to get information from several sensors or from estimated variables. The choice of the setpoint is often non-trivial and is closely related to the choice of operating conditions.

In order to improve operation and establish a relation between control and product quality there has to be a connection between quality measures and the control actions. The effluent water limiting concentrations are the obvious signs of the quality of the plant operation. However, there are few control systems that relate these quality numbers to controller performance criteria. Some control loops have setpoints defined that in some way are related to the quality, such as DO or sludge retention time (SRT) setpoints. There are also other measures that are interesting to incorporate into control actions, such as respiration rate (or specific oxygen utilization rate),sludge volume index (SVI), initial settling velocity, oxidation/reduction potential (ORP), alkalinity change (in connection with nitrification), pH change etc. Here we will discuss DO control and sludge inventory control.

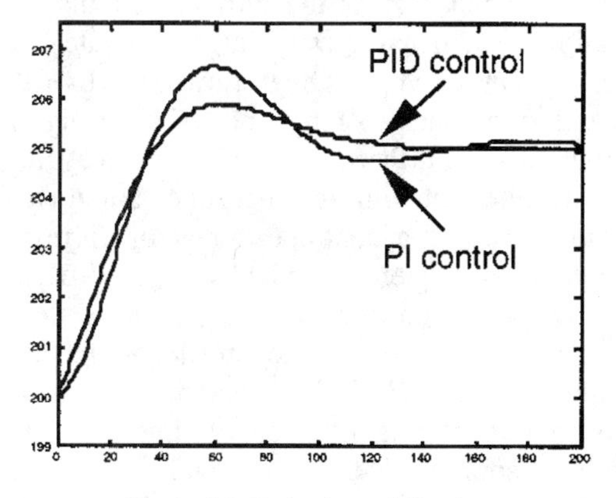


**Figure 3.6** Derivative stabilization.

## CONTROL GOALS FOR THE ACTIVATED SLUDGE ANOXIC ZONE

This will be used as an example of how to specify the control goals for one of the unit processes in an activated sludge system. The primary purpose of this unit is denitrification (DN), the reduction of nitrate to free nitrogen by the oxidation of organic carbon. The dominating organisms in the process are the heterotrophs, a proportion of which reduce nitrate in an (ideally) oxygen free environment. Therefore, the dissolved oxygen level has to be kept at a minimum. If not, the organisms will use oxygen instead of nitrate as the electron acceptor, and the zone will work at least partly like an aerobic reactor. The DN not only reduces the nitrate to nitrogen, it is also important for other reasons:

- carbon is consumed without using oxygen, which means an energy saving;
- nitrification causes a pH decrease and part of this is recovered by the DN.

The DN can be structured either as post- or pre-denitrification systems (Figure 3.7). In the figure we have indicated the major liquid streams that will influence the operation of the DN zone.

From a control point of view, it is favorable to consider different time frames. This will make it possible to decouple the problem and make the control task manageable. We consider three time frames for DN:

- fraction of hours—mixing and reaction rate control: The DN reaction rate can be affected in minutes by at least two manipulated variables, the dissolved oxygen concentration in the influent liquid streams, and the available carbon source for the DN.
- fractions of an hour to hours—hydraulic control: The retention time of the anoxic zone can be rapidly changed by several liquid streams. In particular the nitrate recirculation in a pre-denitrification system has a very large flowrate, typically four times the influent flowrate. Other flowrates, like the influent flowrate and filter backwashing, may change significantly on an hour-to-hour basis. Furthermore the sludge recycle streams will influence the retention time of the zone, but to a lesser extent. The retention time of a continuous reactor corresponds to the phase length of a sequencing batch reactor (SBR).
- hours to days—reaction control: The result of the DN reaction can be determined by the nitrate profile or the nitrate concentration at the outlet of the anoxic zone. In an SBR the nitrate concentration as a function of time can be monitored. Ideally it will approach zero at the end of the phase or at the outlet of the reactor.

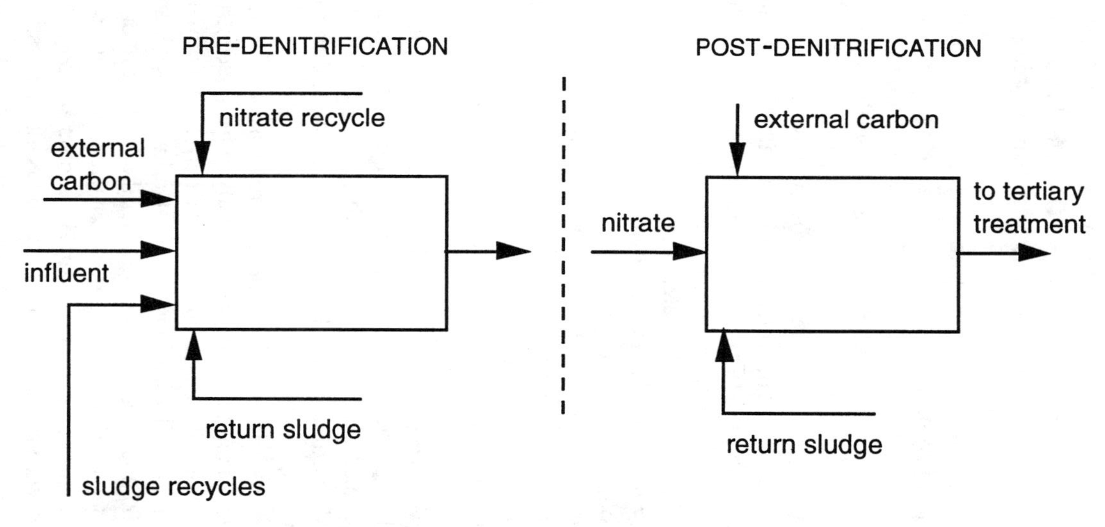


**Figure 3.7** The major influent streams influencing denitrification.

Due to limited space, let us just consider the fast time scale, where mixing and reaction rate control are dominating. The DN requires an oxygen free environment, but there are several disturbances that may violate this condition. The various streams may have a high dissolved oxygen concentration, that will rapidly influence the dissolved oxygen concentration of the anoxic zone, thus the DN rate. In particular one has to consider the nitrate recirculation and filter backwashing streams, that may be quite oxygen rich.

In a pre-denitrification system (Figure 3.7 left) oxygen and nitrate rich water is recirculated from the outlet part of the aerator. It is apparent that from a DN point of view, the dissolved oxygen concentration at the outlet of the aerator has to be kept as low as possible, so that no more oxygen than necessary is recirculated.

Recirculation of nitrate in a pre-denitrification system is a fast process, since the flowrate is so high. Consequently the nitrate concentration can be changed within minutes. This in turn will rapidly change the DN rate.

Another source of disturbance may be backwashing streams from filters downstream in the plant. Such water may disturb the process from two points of view. One is the hydraulic load, causing the retention time to decrease. The other is the oxygen addition, since the backwash water usually is oxygen rich. The influent wastewater may be oxygen rich as well (during rain storms or during snow melting periods) and will influence the denitrification in the same way.

Using a DN rate measurement or estimate, some control actions can be taken:

- The dissolved oxygen content of the nitrate recirculation has to be minimized. It has to be set as a compromise between the need for nitrification in the aeration zone and the desired limitation for the DN zone. The control signal is a setpoint command to the nitrification DO controller.
- The denitrification organisms need carbon as an energy source, so it is important to have a sufficiently high concentration of organic carbon in the system. This is delivered with the influent flow in a pre-denitrification system. During a low load period extra carbon may have to be added, like methanol or ethanol. In a post-denitrification system external carbon has to be added. To obtain a proper carbon dosage is a crucial operational task. Insufficient carbon causes not only limits denitrification, but influences the biomass floc formation in a negative way. Too much carbon in a pre-DN system will be oxidized in the aerated zone (and demand extra air). The cost for external carbon is high and any overdose should be avoided. In a post-DN system there is often an extra aerator at the end to remove

surplus carbon. Naturally such removal is costly and should be minimized.
- Filter backwashing can be disastrous for the operation, if it is done improperly. A smooth recycle from the filters can minimize any negative influence on the DN process.

The possible control structure in the fast time scale can be summarized in Figure 3.8. The DO set-point command has to be calculated as a compromise between the needs of the aeration tank and the anoxic tank. The carbon addition strategy is quite simple once the DN rate can be established, while the filter backwashing strategy may be a supervisory control.

## DISSOLVED OXYGEN CONTROL

The control of the DO concentration as a physical variable does not require any in-depth knowledge of the microbial dynamics. There is an extensive experience of DO, in most cases with traditional PI control. Despite the straightforward task of DO control, several difficulties are involved. The DO dynamics contain both non-linear and time varying

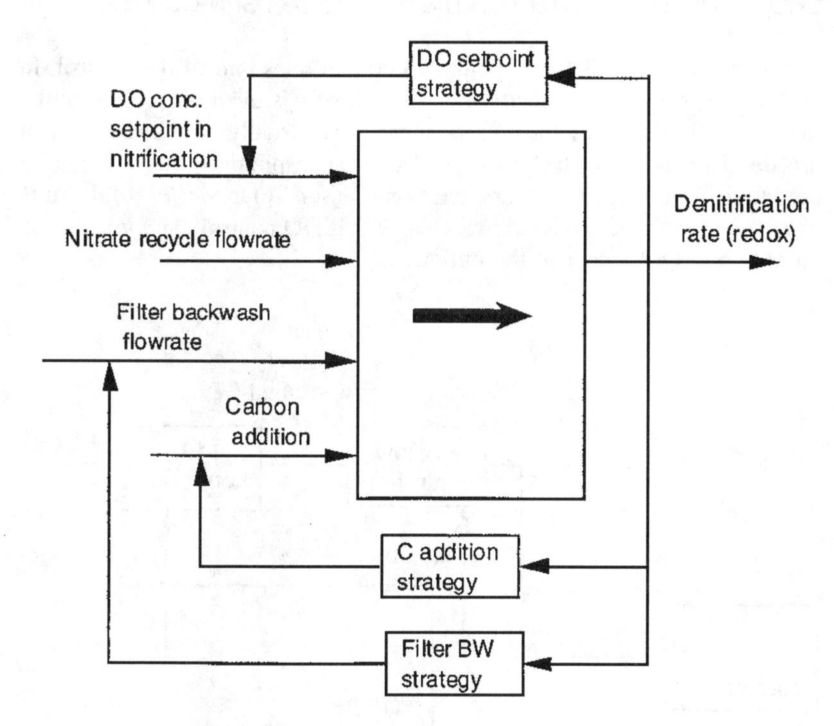


**Figure 3.8** Possible control strategies for the DN rate.

parts. The long time constants and random influent disturbances makes the tuning of a conventional controller tedious. Figure 3.9 shows a typical control structure of DO control in a large wastewater treatment plant (Olsson et al., 1985).

The DO controller output calculates a desired air flow rate that is compared with the actual air flow rate. The air flow error drives a PI controller in cascade that adjusts the air valve position. As the throttle valve is opened or closed, the air pressure will change. The air pressure is in turn controlled by a PI controller (not shown in the figure); the desired pressure could be chosen to be constant. The pressure error drives a controller with the guide vane position or the desired compressor speed as its output.

In order to save energy, the pressure may be purposefully varied. The set-point of the pressure is set at a minimum. It is decreased until the throttle valve is almost fully opened. In doing so, the losses over the air flow valve are kept at a minimum and energy is saved. With a cascaded control system, the DO concentration could be kept well within ±0.1 mg/l from the set-point (see Olsson et al., 1985; Rundqwist, 1988).

## CHOICE OF SET-POINTS FOR DISSOLVED OXYGEN CONTROL

The choice of the DO set-point is a crucial question for the control. In a nitrifying system the sensitivity to DO levels is even more noticeable. In a plug flow system the DO concentration usually displays a typical profile. The shape of the profile reflects the respiration rate and can be used to determine the necessary aeration (Olsson-Andrews, 1978). Results from profile analysis have shown clearly that DO control can*not* be based on only one DO sensor at the outlet.

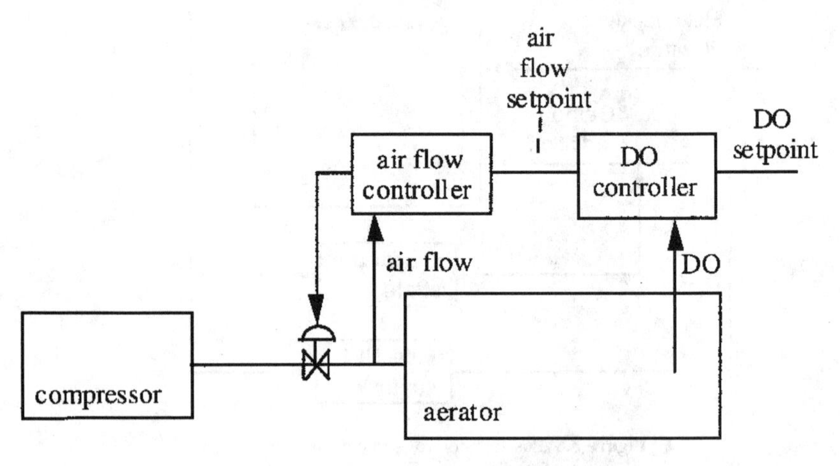


**Figure 3.9** Cascade control of the DO.

The traditional assumption of a fixed DO set-point may not be optimal. Instead, there are indications that some periodic set-point change may influence the process favorably so that the sludge obtains a combination of good clarification and thickening properties (Olsson et al., 1985). The results indicate that a time-varying set-point may be favorable for a high quality effluent. Similar results have been reported by Speirs et al. (1985). A time-varying set-point can be understood by a mixed population model having two competing organism types, floc-forming and filamentous organisms. A constant DO concentration would favor one or the other population type. However, a time varying DO concentration can result in the co-existence of both population types (Hermanowicz, 1987).

When carbonaceous removal is combined with nitrification and denitrification, the problem of a proper DO set-point is even more emphasized. A plant with pre-denitrification as in Figure 3.10 may illustrate the point.

In the aerated zones there is a combination of carbonaceous removal and nitrification. The nitrified water is recirculated into the head end of the aerator where the conditions have to be anoxic in order to favor denitrification. It is obvious that the oxygen uptake rate variations due to carbonaceous removal and nitrification, respectively, will be reflected in the air flow demand in the different cells. Due to the higher growth rate of heterotrophic organisms carbonaceous removal is completed prior to nitrification. Since the carbon removal is completed in the second or third cell, any DO control has to be based on at least two sensors. One of them is located where the carbon removal is almost complete, while the other one is located at the tail end of the aerator where the nitrification is expected to be complete.

It is obvious that the DO level is crucial for both nitrification and denitrification. The recirculation from the aerator outlet to its inlet makes the DO level in the tail end important and will influence both nitrification and denitrification. The DO level has to be kept sufficiently high for nitrification and still as low as possible in order not to recirculate oxygen-rich water to the anoxic zone (Aspegren et al., 1990).

Another possibility of using a time-varying setpoint of the DO has been presented by Lindberg-Carlsson (1996a). The DO concentration in a nitrifying system can be determined by the ammonium removal. The basic idea is to control the DO set-point by measuring the ammonium concentration in the last zone of the nitrifying reactor. The principle can be illustrated by Figure 3.11 The first controller varies the DO-setpoint and tries to maintain a specified ammonium level.

## SLUDGE INVENTORY CONTROL

There are basically three control variables for the sludge inventory in an activated sludge system. The waste activated sludge (WAS) flow rate

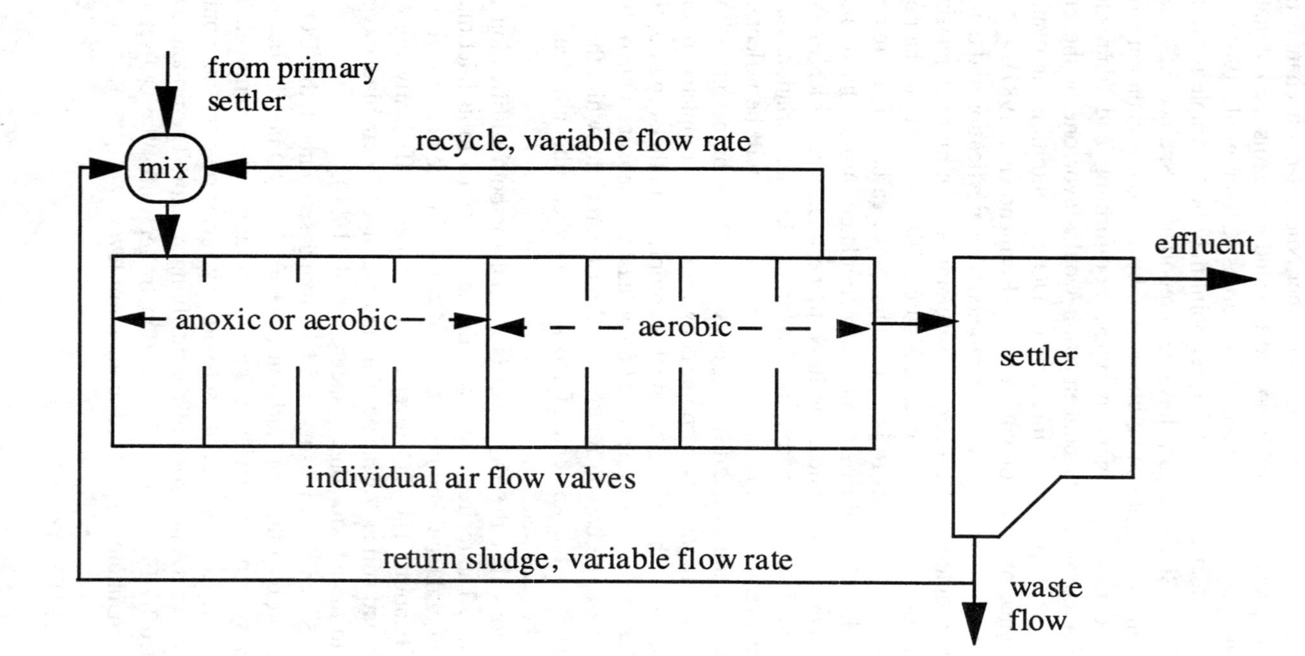


**Figure 3.10** Outline of a nitrogen removal plant with pre-denitrification.

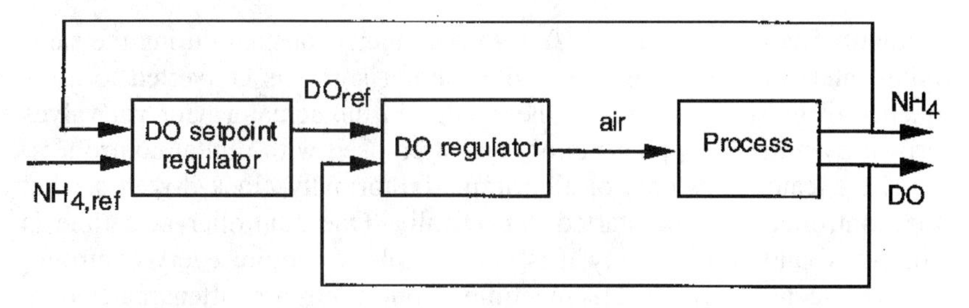


**Figure 3.11** Block diagram of the variable DO setpoint control.

controls the total sludge mass in the system and the sludge retention time (SRT) can be kept at a desired level. The traditional sludge age formula is a steady state calculation and does not take short term fluctuations into consideration. Therefore it has to emphasized, that the SRT calculation has to be based on sludge concentration and flow rates averaged over several days.

The sludge distribution *within* the system is controlled by the step feed flow distribution or the return activated sludge (RAS) flow rate. The former can dynamically redistribute the sludge within the aerator while the latter can shuffle sludge between the settler and the aerator. More details are presented in Olsson-Newell (1997).

## CONTROLLER IMPLEMENTATION

The activated sludge process is characterized by several inputs and outputs. In most cases, however, the internal couplings are not significant and the process can be controlled by many local controllers, one input/ output pair at a time. This is the normal structure in Direct Digital Control (DDC) systems. The PID or a similar controller is the most common type, and its computer implementation is discussed here in some detail.

A computer makes it reasonable to realize other control structures, such as non-linear or self-tuning controllers. Once the controller parameters are known, the implementation of the control algorithms is usually quite straightforward. However, every implementation has to be supplied with a "safety umbrella" of routines to test the performance of the control.

### SAMPLED CONTROL

When a feedback control strategy is implemented digitally, the continuous signal from the sensor is sampled, a control action is calculated, and the control signal is sent via the computer output port to the final control element. The control signal $u(t)$ usually remains constant during the sampling interval. Sometimes the digital output signal is converted to a sequence of pulses representing the change in the actuator. Control valves driven by pulsed stepping motors are often used with digital controllers.

The execution of control algorithms is normally clock driven so that the controller must be started periodically. One controller at a time is computed and consequently it is not suitable to require every controller to execute at exactly the same time. Controllers are often realized in dedicated computers close to the physical process.

We will demonstrate how a PID controller can be realized in a computer.

## DISCRETIZATION OF THE PID CONTROLLER

The controller in a digital system must be discretized at some stage in order to be implemented on a computer. Given a sufficiently short sampling interval, the time derivative can be approximated by a finite difference and the integral by a summation. Let us consider one term at a time (see Figure 3.1).

The error (1) is calculated at each sampling interval

$$e(kh) = u_c(kh) - y(kh) \tag{3.3}$$

The sampling period $h$ is assumed to be constant and the signal variations during the sampling interval are neglected. The parameter $k$ is an integer (usually equal to 1) that adjusts the sampling interval. The time discrete form of the PID controller is

$$u(kh) = u_0(kh) + u_P(kh) + u_I(kh) + u_D(kh) \tag{3.4}$$

The proportional part of the controller (2) is

$$u_P(kh) = K \cdot e(kh) \tag{3.5}$$

The integral is approximated by finite differences,

$$\begin{aligned} u_I(kh) &= u_I[(k-1)h] + K \frac{h}{T_i} e(kh) \\ &= u_I[(k-1)h] + K \cdot \alpha \cdot e(kh) \end{aligned} \tag{3.6}$$

where $\alpha = h/T_i$.

The integral part forms a recursive expression and is updated at every sampling interval. The derivative part is also approximated by finite differences,

$$u_D(kh) = K \frac{T_D}{h} [e(kh) - e[(k - 1)h]] \tag{3.7}$$

There are more elaborate approximation possible, and we refer to Olsson-Piani (1992) for more details.

## PRACTICAL CONSIDERATIONS FOR CONTROLLERS

In the implementation of a PID controller algorithm one has to consider several practical issues. The sampling time is crucial for the operation. Since the controller always produces a limited signal one must know how to deal with problems due to the signal limitation. Other problems, like integral windup and bumpless transfer are further discussed in Olsson-Piani (1992) and in Olsson-Newell (1997).

Usually PID controllers are already preprogrammed in process computers and the user has to give its parameters suitable values in so called block oriented languages.

## SAMPLING RATE IN CONTROL SYSTEMS

The sampling time of a measurement is a crucial problem. In order to detect a short pulse disturbance the sampling rate must be sufficiently large. Since many measurements are costly there is a trade-off between the cost of measurements and the cost of not detecting the actual disturbance. Basically the sampling has to be so frequent that the continuous signal can be reconstructed from the sampled signal without loss of essential information. Note, that the sampling rate here is defined by the stored information in the computer. Each stored value may very well be an average of several values, acquired more frequently. For example, 6-minute values are often stored in the computer, based on 1 minute measurements.

It is not trivial to choose a suitable sampling rate for control; in fact, finding the right sampling frequency still remains more like an art than a science. Too long a sampling period can reduce the effectiveness of feedback control, especially its ability to cope with disturbances. In an extreme case, if the sampling period is longer than the process response time, then a disturbance can affect the process and will disappear before the controller can take corrective action. Thus, it is important to consider both the process dynamics and the disturbance characteristics in selecting the sampling period.

Commercial digital controllers which handle a small number of control loops (e.g., 8–16) typically employ a fixed sampling period of a fraction of a second. Thus the performance of these controllers closely approximates continuous (analog) controllers.

The signal-to-noise ratio also influences the selection of the sampling period. For low signal-to-noise ratios, fast sampling should be avoided because changes in the measured variable from one sampling interval to the next will be mainly due to high frequency noise rather than to slow process changes.

### Example: Sampling Time in DO Control

The DO concentration may vary in a minute-to-minute time scale due to load disturbances. However, the air flow rate can not influence the DO control more quickly than typically 10–20 minutes. The consequence is, that it is impossible to control the fast variations by air flow control. It does not pay to change the air flow signal more often than typically 10–12 minutes. In fact, a controller with a shorter sampling interval may very well behave poorer, since it is trying to cancel too fast disturbances (see Olsson et al., 1985).

### CONTROL SIGNAL LIMITATIONS

The controller output value must be limited, at least for two reasons. The desired amplitude cannot exceed the digital-to-analog (D/A) converter range, and can not become larger than the actuator range. A valve is limited to 100% full open, or a motor current has to be limited. Thus, the control algorithm needs to include some limit function.

In several control loops it is important to introduce a dead band. If an incremental controller is used, each increment may be so small, that it is not significantly larger than other disturbances. It is of interest not to wear out the actuators. Consequently the control variable increments are accumulated until the control signal reaches a certain value. Naturally the deadband has to be larger than the resolution of the D/A converter.

## WHEN ARE SIMPLE CONTROLLERS NOT SUFFICIENT?

Simple controllers, like the ones previously described are useful for a majority of the control loops in a plant. Many control tasks in the activated sludge process will demand more elaborate solutions than control structures based on simple controllers. The main reasons are

- inherent time delays
- strong couplings between unit processes
- non trivial choices of controller set-points
- non-linear parameters
- time-varying and unknown parameters

An overview of digital controllers is made in Olsson-Piani (1992) while a detailed treatment is found in Åström-Wittenmark (1990).

## TIME DELAYS

Time delays are common because of the presence of distance lags, recycle loops, or the dead time associated with composition analysis. The time delay makes the information from the true process variable change arrive at the controller later than desired. This limits the performance of the control system and may lead to system instability. A system with time delay that is controlled by a PID controller usually behaves quite sluggishly. The inherent dead times in the system can be compensated for with digital controllers. Old control values have to be stored; i.e., the controller is of the form

$$u(t) = -r_1 \cdot u(t-h) - \cdots - r_n \cdot u(t-nh) + t_0 \cdot u_c(t) + \cdots + t_n \cdot u_c(t-nh) - s_0 \cdot y(t) - \cdots - s_n \cdot y(t-nh) \quad (3.8)$$

where $u$ is the controller output, $y$ is the measurement and $u_c$ is the set-point value and $h$ the sampling interval. The time-discrete version of the PID controller can be seen as a special case of this general digital controller (Olsson-Piani, 1992). The DO controller described in Olsson et al. (1985) has the structure of (3.8).

## PARAMETER VARIATIONS

Parameter variations can be either predictable or assumed completely unknown. This leads to different controller structures.

### Predictable Parameter Variations—Gain-Scheduling Control

In many processes, the process parameters change with operating conditions. The aeration dynamics can serve as an example. The oxygen transfer rate $k_La$ is a non-linear function of the air flow rate and can be considered linear only for small air flow variations. At a high load the sensitivity of $k_La$ to air flow changes is smaller than at a low load. Consequently, the controller gain needs to be larger at high loads.

The DO concentration dynamics are also non-linear. To demonstrate this we consider the dissolved oxygen mass balance in a complete mix aerator,

$$\frac{dc}{dt} = d_i \cdot c_i - d \cdot c + k_La \cdot (c^s - c) - R \quad (3.9)$$

where $c$ is the DO concentration, $c_i$ the influent and $c^s$ the saturation DO concentrations respectively, $k_La$ the oxygen transfer rate and $R$ the oxygen uptake rate. The parameters $d$ and $d_i$ are dilution terms. The oxygen transfer rate $k_La$ is nonlinear function of the control variable (the air flow rate). Furthermore $k_La$ is multiplied with $c$.

Due to the nonlinear behavior the controller gain has to be different at different operating levels. Since both the air flow rate and the dissolved oxygen concentration can be measured, the process gain can be modeled and the controller gain can be stored in a table as a function of the operating condition. A nonlinear controller for DO has been derived and tested in pilot scale by Lindberg-Carlsson (1996a). The advance of such a controller is, that it is well tuned for both low and high plant loads.

### Unknown Parameter Variations—Self-Tuning Control

In the activated sludge system, the aeration equipment may be clogged or the composition of the sludge may vary so that the DO dynamics are gradually changed. Thus, the process parameters need to be gradually updated in order to maintain satisfactory control. If the parameters are continuously updated, the controller may be called *self-tuning*. Such controllers have two distinctive parts, an *estimation* part and a *controller* part (Figure 3.12).

The estimation part measures the process output and input signals to recursively update the process parameters. The design algorithm then updates the controller parameters. The controller part can be a discrete controller of the form (3.8). There are several variants of this general scheme. Instead of updating the process parameters, the controller parameters can be updated directly. Even if the basic algorithms look quite simple from a programming point of view, the self-tuning control system needs a "safety network" of rules to avoid operational problems. Certainly,

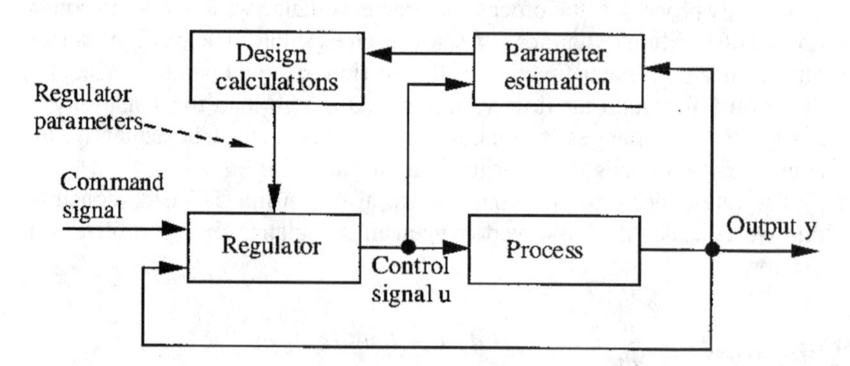


**Figure 3.12** The principal parts of an adaptive controller.

the adaptive controller does not automatically solve all difficult control problems. If it is used with caution and knowledge, however, it offers new possibilities to solve complex control tasks. A comprehensive treatment of adaptive controllers is found Åström-Wittenmark (1989).

Self-tuning control was implemented in a full-scale plant, the Käppala Sewage works close to Stockholm, Sweden (Olsson et al. 1985) to examine the potential of self-tuning control in activated sludge systems. The DO controller, in the structure of Figure 3.9, was replaced by the self-tuner. The Käppala controller has been in use for several years and performs satisfactorily (Rundqwist, 1988).

It is of great interest to continuously know the value of the oxygen uptake rate $R$ (3.9) while controlling the DO concentration. The problem of simultaneous estimation and control was solved and tested at the Malmö Sewage works in southern Sweden (Holmberg et al. 1988, Holmberg 1990). The $k_La$ was linearized around the operating point. Further improvements were made by Carlsson et al. (1994) and Lindberg-Carlsson (1996b) where they allowed the $k_La$ function to be nonlinear.

## PROGRAMMABLE CONTROLLERS

Hitherto we have considered the realization of controllers for DO control. In this section we will consider some implementation aspects of them:

- the language representation of the controllers
- the hardware for the controllers
- logical sequencing around the controllers
- communication between programmable controllers

More comprehensive description of PLCs and their function and applications are found in Warnock (1988) and in Olsson (1996).

### A BLOCK LANGUAGE IMPLEMENTATION

The control algorithms can of course be described in any sequential language. However, in practical implementations more process oriented high level languages are common. Often controller functions are represented in blocks, where only the input and output signals are marked while the algorithm itself is hidden. The parameters are available for tuning. Figure 3.13 shows a typical structure of a PID controller. The symbols are shown on the screen and the programmer has to label the inputs and outputs with proper variable names to connect the controllers to other elements.

The diagram shows two PID controllers connected to a switch. By a binary signal to the switch one of the two regulator outputs will be chosen

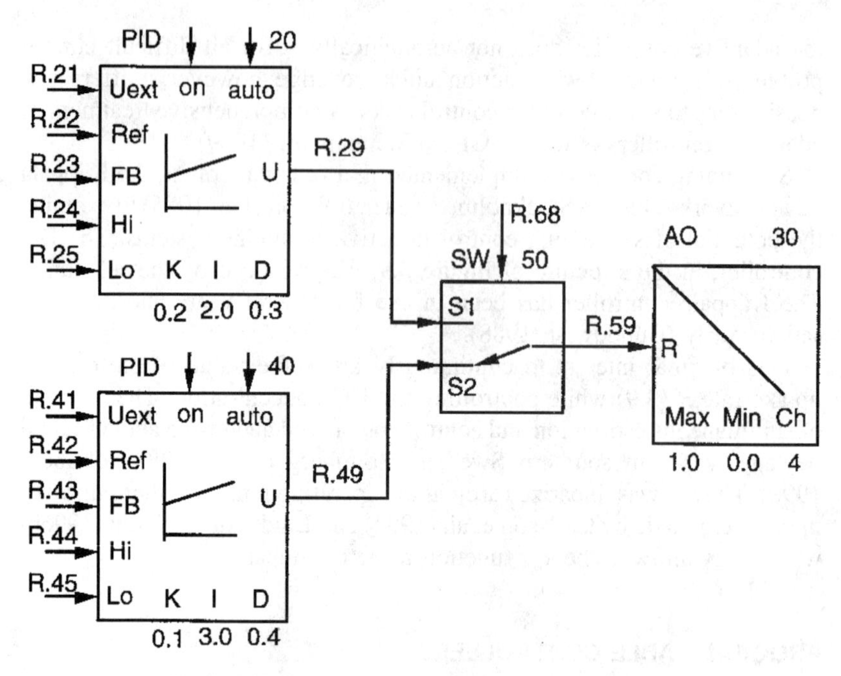


**Figure 3.13** Block program symbol of two PID controllers.

and sent to the analog output unit. The AUTO input is a binary variable for manual/automatic mode. The reference value is sent into REF while the measurement is connected to FB (feedback). The control signal limits are marked by the two parameters HI and LO. The tuning parameters K, I, and D correspond to the gain, integral time and derivative time and their values are displayed at the bottom line of the symbol. The analog output unit is defined by its range and channel number.

Note that the control system code is written in a format similar to logical gates. In most industrial process computer software the controller blocks are incorporated in the logical programming to easily add different logical conditions for the signals. In a modern package there are blocks not only for logical sequences and controllers. The user may define his own blocks from algebraic or difference equations. Thus it is possible to incorporate any of the more general controller structures, described in the following sections. The logical programming is further structured by using function charts for sequencing.

## SEQUENCING AND LOGICAL CIRCUITS

Sequencing has traditionally been realized with relay techniques. Until the beginning of the 1970s electromechanical relays and pneumatic couplings were dominating in the market. During the 1970s *programmable logical controllers* (PLCs) became more and more common, and today sequencing is normally implemented in software. Even if ladder diagrams are being phased out of many automation systems they still are used to describe and document sequencing control implemented in software.

Programmable logical controllers are microcomputers developed to handle Boolean operations. A PLC produces *on/off* voltage outputs and can actuate such elements as electric motors, solenoids (and thus pneumatic and hydraulic valves), fans, heaters and light switches. They are vital parts of industrial automation equipment found in all kinds of industries.

The early PLCs were designed only for logic-based sequencing jobs (on/off signals). Today there are hundreds of different PLC models on the market. They differ in their memory size (from a few hundred bytes to several kilobytes) and I/O capacity (from a few lines to thousands). The difference also lies in the features they offer. The smallest PLCs serve just as relay replacers with added timer and counter capabilities. Many modern PLCs also accept proportional signals and they can perform simple arithmetic calculations and handle analog input and output signals and PID controllers. This is the reason why the letter *L* was dropped from PLC, but the term PC may cause confusion with personal computers so we keep the *L* here.

The logical decisions and calculations may be simple in detail, but the decision chains in large plants are very complex. This naturally raises the demand for structuring the problem and its implementation. Sequencing networks operate asynchronously, i.e. the execution is not directly controlled by a clock. The chain of execution may branch for different conditions and concurrent operations are common. This makes it crucial to structure the programming and the programming languages. Grafcet, an international standard, is an important notation to describe binary sequences, including concurrent processes; it is used both as a documentation tool and a programming language. Applications of function charts in industrial control problems are becoming more and more common.

## ANALOG CONTROLLERS

PLCs nowadays can handle not only binary signals. Analog-to-digital (A/D) and digital-to-analog (D/A) converters can be added to the PLC rack. The resolution, i.e. the number of bits to represent the analog signal, varies between the systems. The converted analog signal is placed in a digital register in the same way as a binary signal and is available for the standard PLC arithmetic and logical instructions.

In the event that plant signals do not correspond to any of the standard analog ranges, most manufacturers provide signal conditioning modules.

Such a module provides buffering and scaling of plant signals to standard signals (typically 0–5 V or 0–10 V).

A PLC equipped with analog input channels may perform mathematical operations on the input values and pass the results directly to analog output modules to directly drive continuous actuators in the process. The sophistication of the control algorithms may vary, depending on the complexity of the PLC, but most systems today offer PID controller modules. The user has to provide the tuning of the regulator parameters. In order to obtain sufficient capacity many systems provide add-on PID function modules that contain both input and output analog channels together with dedicated processors to carry out the necessary control calculations. This processor operates in parallel with the main CPU. When the main CPU requires status data from the PID module, it reads the relevant locations in the I/O memory where the PID processor places this information each time it completes a control cycle.

Typically a PLC system may provide programming panels with a menu of questions and options relating to the set-up of the control modules, such as gain parameters, integral and derivative times, filter time constants, sampling rate and engineering unit conversions.

## COMMUNICATION

A distributed system is more than connecting different units in a network. Certainly, the units in such a system can communicate, but the price is too much unnecessary communication, while the capacity of the systems cannot be fully used. Therefore, the architecture of the communication is essential. Reasons for installing networks instead of point-to-point links are:

- All devices can access and share data and programs.
- Cabling for point-to-point becomes impractical and prohibitively expensive.
- A network provides a flexible base for contributing communication architectures.

To overcome the difficulties of having to deal with a large number of incompatible standards, the International Organization for Standardization (ISO) has defined the *open systems interconnection* (OSI) scheme. OSI itself is not a standard, but offers a framework to identify and separate the different conceptual parts of the communication process. In practice, OSI does not indicate what voltage levels, which transfer speeds or which protocols need to be used to achieve compatibility between systems. It says that there *has* to be compatibility for voltage levels, speed and protocols as well as for a large number of other factors. The practical goal of OSI is optimal network interconnection, in which data can be transferred between different locations without having to waste resources for conversion purposes with the related delays and errors.

PLC systems are an essential part of most industrial control systems. Below we will illustrate how they are connected at different levels of a plant network, illustrated by Figure 3.14.

### Fieldbus—Communication at the Sensor Level

It is a trend to replace conventional cables from sensors with a single digital connection. Thus, a single digital loop can replace a large number of 4–20 mA conductors. This has been implemented not only in manufacturing plants but also in aircraft and automobiles. It is obvious that each sensor needs an interface to the bus, and standardization is necessary. This structure is known as Fieldbus. There is no single Fieldbus yet, but different solutions have been presented by the industry and by research institutions.

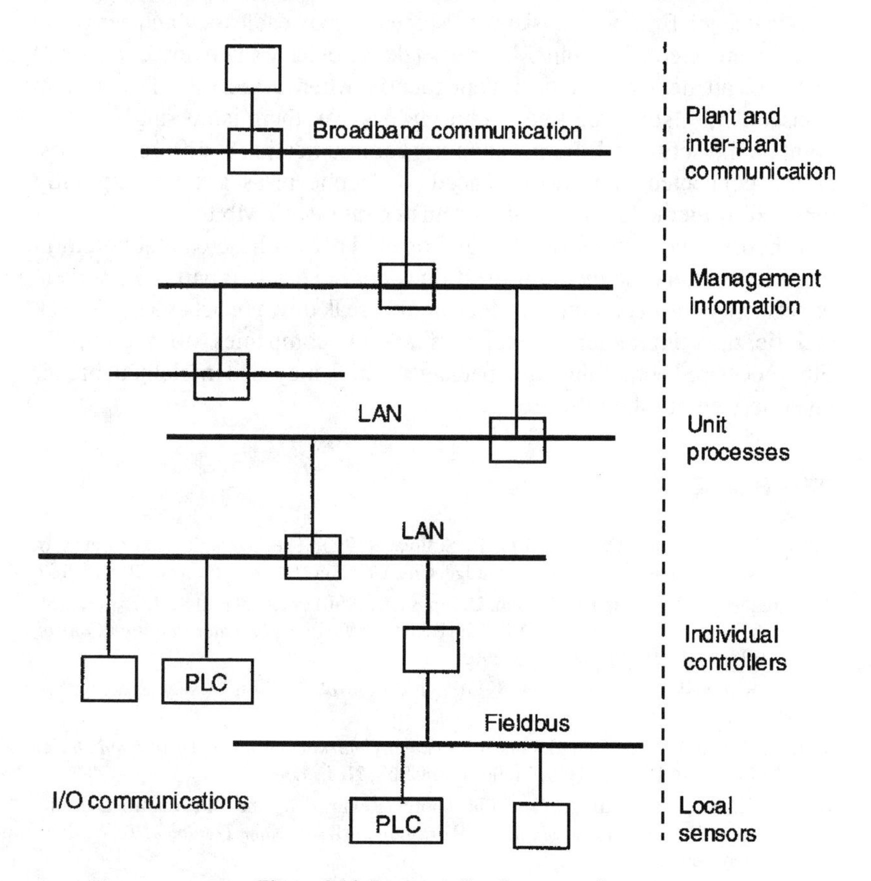


**Figure 3.14** Structure of a plant network.

In the course of time, what has been proposed and is operating in the field will crystallize around one or few technologies that will then become part of a more general Fieldbus.

The possibilities opened by fieldbuses are notable. A large share of the intelligence required for process control is moved out to the field. The maintenance of sensors becomes much easier because operations like test and calibration can be remotely controlled and require less direct intervention by maintenance personnel. The quality of the collected data influences directly the quality of process control.

### Local Area Networks

In order to communicate between different PLC systems and computers within a plant there is a clear trend to use Ethernet as the medium. Ethernet is a widely used local area network (LAN) for both industrial and office applications. Ethernet has a bus topology with branch connections. At physical level, Ethernet consists of a screened coax cable to which peripherals are connected with ''taps.'' Ethernet does not have a network controlling unit and all devices decide independently when to access the medium. Consequently, since the line is entirely passive, there is no single-failure point on the network. Ethernet supports communication at different speeds, as the connected units do not need to decode messages not explicitly directed to them. Maximum data transfer rate is 10 Mbit/s.

Ethernet's concept is flexible and open. There is little capital bound in the medium, and the medium itself does not have active parts like servers or network control computers which could break down or act as a bottleneck and tie up communication capacity. Some companies offer complete Ethernet-based communication packages which may also implement higher layer services in the OSI hierarchy.

## REFERENCES

Allsop, P. J., M. Moo-Young and G. R. Sullivan (1990) The Dynamics and Control of Substrate Inhibition in Activated Sludge. *CRC Crit. Rev. Environ. Control, 20,* 115–167.

Aspegren, H., B. Andersson, G. Olsson, U. Jeppsson (1990) Practical Full Scale Experiences of the Dynamics of Biological Nitrogen Removal, *Advances in Water Pollution Control,* IAWPRC, (R. Briggs, ed.) 283–290.

Åström, K. J. and B. Wittenmark (1989) *Adaptive Control,* Addison-Wesley, Reading, MA (USA).

Åström, K. J. and B. Wittenmark (1990) *Computer Controlled Systems, Theory and Design* (2nd edition), Prentice-Hall, Englewood Cliffs, N. J., USA.

Bundgaard, E. (1988) Nitrogen and Phosphorus Removal by the Bio-Denitro and Bio-Denipho Processes. *Proc. of the Int. Workshop on Wastewater Treatment Technology,* Copenhagen.

Carlsson, B. and C-F. Lindberg, S. Hasselblad and S. Xu (1994) On-line Estimation of the Respiration Rate and the Oxygen Transfer Rate at Kungsängen Wastewater Plant in Uppsala, *Wat. Sci. Tech., 30* (4), 255–263.

Einfeldt, J. (1992) The Implementation of Biological Phosphorus and Nitrogen Removal with the Bio-Denipho Process on a 265,000 PE Treatment Plant, *Wat. Sci. Tech., 25,* 161–168.

Gujer, W. and M. Henze (1990) Activated Sludge Modelling and Simulation, *Wat. Sci. Tech., 23,* 1011–1023.

Gujer, W., M. Henze, T. Mino, T. Matsuo, M. C. Wentzel and G. v. R. Marais (1994). Basic Concepts of the Activated Sludge Model No 2: Biological Phosphorus Removal Processes, *IAWQ Specialized Seminar, Modelling and Control of Activated Sludge Processes,* Copenhagen, 22–24 August.

Gustafsson, L. G., D. J. Lumley, C. Lindeborg and J. Haraldsson (1993) Integrating a Catchment Simulator into Wastewater Treatment Plant Operation, *Wat. Sci. Tech., 28* (10–11), 45–54.

Henze, M., Grady Jr., C. P. L., Gujer, W., Marais, G. v. R. and Matsuo, T. (1987) Activated Sludge Model No. 1, IAWPRC Scientific and Technical Reports No. 1, IAWPRC, London, UK.

Hermanowicz, S. W. (1987) Dynamic Changes in Populations of the Activated Sludge Community: Effects of Dissolved Oxygen Variations. *Journal of Water Science & Technology, 19,* 889–895.

Holmberg, A. (1982) On the Practical Identifiability of Microbial Growth Models Incorporating Michaelis-Menten Type Nonlinearities, *Math. Bioscience, 62,* 23–43.

Holmberg, U. (1990) On Identification of Dissolved Oxygen, *Advances in Water Pollution Control,* IAWPRC (R. Briggs, ed.) 113–120.

Holmberg, U., G. Olsson and B. Andersson (1988) Simultaneous DO Control and Respiration Estimation, *Wat. Sci. Tech., 21,* 1185–1195.

IAWQ (1995) (A. Lynggaard & P Harremoes, Eds.) Sensors in Waste Water Technology, *Wat. Sci. Tech., 33* (1).

Irvine, R. L. and L. H. Ketchum, Jr. (1989) Sequencing Batch Reactors for Biological Wastewater Treatment. *CRC Critical Reviews in Environmental Control, 18,* Issue 4, 255–294.

Isaacs, S. H., H. Zhao, H. Søeberg and M. Kümmel (1992) On the Monitoring and Control of a Biological Nutrient Removal Process. *Proc. of the 6th Forum of Applied Biotechnology,* Bruges, Belgium.

Jeppsson, Ulf (1996) Modelling Aspects of Wastewater Treatment Processes, Ph.D. Thesis, (ISBN 91-88934-00-4), Report TEIE-1010, Dept. of Industrial Electrical Engineering and Automation, Lund Institute of Technology, Lund, Sweden.

Jeppsson, U. and Olsson, G. (1993) Reduced Order Models for on-line Parameter Identification of the Activated Sludge Process. *Wat. Sci. Tech., 28* (11–12), 173–183.

Lindberg, C. F. and B. Carlsson (1996a) Nonlinear and Set-point Control of the Dissolved Oxygen Dynamics in an Activated Sludge Process, *Wat. Sci. Tech., 34* (3–4), 173–180.

Lindberg, C. F. and B. Carlsson (1996b) Estimation of the Respiration Rate and Oxygen Transfer Function Utilizing a Slow DO Sensor. *Wat. Sci. Tech., 33,* 1, 325–333.

Nielsen, M. K. and T. B. Önnerth (1995) Strategies for Handling of On-line Information for Optimizing Nutrient Removal, *Wat. Sci. Tech., 33,* 1, 211–222.

Nyberg, U., B. Andersson and H. Aspegren (1996) Real Time Control for Minimizing Effluent Concentrations during Stormwater Events, *Wat. Sci. Tech., 34* (3–4), 127–134.

Olsson, G. (1977) State of the Art in Sewage Treatment Control. *American Inst. of Chemical Engineers, Symp Series, 159,* no. 72, 52–76.

Olsson, G. (1985) Control Strategies for the Activated Sludge Process, *Comprehensive Biotechnology,* Pergamon Press, chapter 65, 1107–1119.

Olsson, G. (1993) Advancing ICA Technology by Eliminating the Constraints, *Wat. Sci. Tech., 28* (11–12), 1–7.

Olsson, G. (1996) Programmable Controllers. Chapter 18, *Control Handbook* (William Levine, Editor) CRC Press and IEEE Press.

Olsson, G. and J. F. Andrews (1978) The Dissolved Oxygen Profile—A Valuable Tool for the Control of the Activated Sludge Process, *Water Research, 12,* 985–1004.

Olsson, G. and D. Chapman (1985) Modelling the Dynamics of Clarifier Behaviour in Activated Sludge Systems, *Advances in Water Pollution Control,* R. A. R. Drake, ed., IAWPRC, Pergamon Press, 405–412.

Olsson, G., L. Rundqwist, L. Eriksson and L. Hall (1985) Self-tuning Control of the Dissolved Oxygen Concentration in Activated Sludge Systems, *Advances in Water Pollution Control,* R. A. R. Drake, ed., IAWPRC, Pergamon Press, 473–480.

Olsson, G., B. Andersson, B. G. Hellström, H. Holmström, L. G. Reinius, and P. Vopatek (1989) Measurements, Data Analysis and Control Methods in Wastewater Treatment Plants—State of the Art and Future Trends, *Wat. Sci. Tech., 21,* 1333–1345.

Olsson, G. and G. Piani (1992) *Computer Systems for Automation and Control,* Prentice Hall Int., N. J. Also in German: *Steuern, Regeln und Automatisieren,* Carl Hanser, München, 1993 (revised and extended revision of the English version).

Olsson, G. and U. Jeppsson (1994) Establishing Cause-effect Relationships in Activated Sludge Plants—What Can Be Controlled, Invited paper, *Forum for Applied Biotechnology* (FAB), Bruges, Belgium, 28–30 Sep. 1994.

Olsson, G. and B. Newell (1997) *Control of Biological Wastewater Treatment Plants.* Book manuscript.

Önnerth, T. B., M. K. Nielsen, and C. Stamer (1995) Advanced Computer Control Based on Real and Software Sensors, *Wat. Sci. Tech., 33* (1), 237–246.

Ossenbruggen, P. J., H. Spanjers, H. Aspegren and A. Klapwijk (1991) Designing Experiments for Model Identification of the Nitrification Process, IAWPRC Watermatex '91, New Hampshire, USA, *Wat. Sci. Tech., 24* (6), 9–16.

Parulekar, S. J. and H. C. Lim (1985) Modeling, Optimization and Control of Semi-Batch for various Fed-Batch Fermentation Processes, *Biotechnol. Bioeng., 28,* 1396–1407.

Patry, G. G. and I. Takacs (1992) Settling of Flocculent Suspensions in a Secondary Clarifier, *Water Research, 26.*

Rundqwist, L. (1988) Self-tuning Control of the Dissolved Oxygen Concentration in the Käppala Plant, TFRT-7383, Dept. of Automatic Control, Lund Inst. of Technology, Lund, Sweden.

Sorour, M. T., G. Olsson and L. Somlyody (1992) Potential use of Step Feed Control using the Biomass in the Settler, *6th IAWQ Workshop on Instrumentation, Control and Automation of Water and Wastewater Treatment and Transport Systems,* Banff-Hamilton, Canada.

Spanjers H., P. Vanrolleghem, G. Olsson and P. Dold (1996) Respirometry in Control of Activated Sludge Processes, *Wat. Sci. Tech., 34* (3–4), 117–126.

Speirs, G. W., R. D. Hill and P. J. Laughton (1985) Wastewater Treatment Plant Selection and Refit for Demonstration of Direct Computer Control—The Tillsonburg, Ontario

Experience. *Proc. of a Seminar on Instrumentation and Control Systems in Wastewater Treatment Plants,* Pollution Control Association of Ontario, Burlington, Ontario.

Takács, I., Patry, G. G. and Nolasco, D. (1991) A Dynamic Model of the Clarification/ Thickening Process. *Water Research, 25* (10), 1261–1273.

Thornberg, D. E., M. K. Nielsen and K. L. Andersen (1993) Nutrient Removal. On-line Measurements and Control Strategies, *Wat. Sci. Tech., 28* (10–11), 549–560.

Vitasovic, Z. Z. (1986) An integrated control strategy for the activated sludge process. Ph.D. Dissertation, Rice University, Houston, Texas, 288 pp.

Warnock, I. G. (1988) *Programmable Controllers Operation and Application,* Prentice-Hall, Englewood Cliffs, N. J. (USA).

Zhao, H., S. H. Isaacs, H. Søeberg and M. Kümmel (1994) A Novel Control Strategy for Improved Nitrogen Removal in an Alternating Activated Sludge Process. Part I: Process Analysis. Part II: Control Development. *Water Research, 28,* 521–542.

CHAPTER 4

# Respirometry in Modeling and Control of Wastewater Treatment

## INTRODUCTION

**RESPIROMETRY** is the measurement of the oxygen consumption rate of activated sludge under well defined experimental conditions. Because oxygen consumption is directly associated with both biomass growth and substrate removal, respirometry is a useful technique for modeling and operating the activated sludge process. In the early years application of the technique was focused mainly on measurement of Biochemical Oxygen Demand (BOD) of (waste) water (Montgomery, 1967), where respirometry was mainly seen as an instrument-based alternative for the BOD-test. Although, the standard five-day dilution BOD test itself can be considered as a respirometric procedure. Later, starting in the sixties, respirometry began to generate much interest as a process control variable (Spanjers et al., 1998). The past decade respirometry increasingly is being employed for obtaining kinetic data, and it is considered as one of the most important variables for activated sludge process modeling (Rozich and Gaudy, 1992; Henze and Gujer, 1995).

Already since the discovery of the activated sludge process in the beginning of the century it has been recognized that the rate at which activated sludge consumes oxygen, the *respiration rate,* is an important indicator of the process state, and efforts have been made to measure this variable. Respiration rate is usually measured by using *respirometers.* These range from very simple manually operated BOD-bottles to fully self-operating instruments that automatically perform sampling, calibration,

H. Spanjers, Department of Environmental Technology, Wageningen Agricultural University, NL-6700 EV Wageningen, The Netherlands.

and calculation of respiration rate. All respirometers are based on some technique for measuring the rate at which biomass takes up *dissolved oxygen* (DO) from the liquid. This can be done directly by measuring DO or indirectly by measuring *gaseous oxygen.* Although DO can be measured chemically (eg. the Winkler method), electrochemical DO measurements (eg. based on the Clark-cell) are almost uniquely applied in DO-based respirometers. Gaseous oxygen concentration can be measured by physical techniques such as the paramagnetic method. Other physical techniques such as manometric and volumetric methods measure rather the change in gaseous oxygen concentration.

While in the early years all respirometers were based on measuring gaseous methods, after the introduction of the electrochemical Clark-type DO measuring cell in 1959 (Clarke, 1959) this type of sensor became more and more common in respirometers. Currently about 50% of the commercial respirometers are based on a DO-sensor. While commercial respirometers are chiefly being used in wastewater treatment practice, a considerable number of home-made respirometers, of which the majority is of the DO-sensor type, is operational in research environments.

This chapter deals with the biochemical background of respiration, the measurement of respiration rate and the application in modeling and control of the activated sludge process.

## FUNDAMENTALS OF RESPIRATION

### BIOCHEMICAL BACKGROUND

In general terms *respiration* is the ATP generating metabolic process in which either organic or inorganic compounds serve as electron donor and inorganic compounds such as $O_2$, $NO_2^-$, $NO_3^-$, $SO_4^{2-}$ etc. serve as the ultimate electron acceptor. If oxygen is the ultimate electron acceptor the process is called *aerobic respiration.* The ATP is generated during the process of removing electrons from the substrate and then passing them from one metabolic carrier to the next in the *electron transport chain.* This process can be schematically depicted as follows:

$$AH_2 \rightarrow A + 2H^+ + 2e^- - \Delta G$$

The compound $AH_2$ is oxidized with associated release of free energy, and this is accompanied by a reduction:

$$B + 2H^+ + 2e^- \rightarrow BH_2 + \Delta G$$

The reactions proceed with a net yield of energy which is converted into biological utilisable energy by microbial metabolism and stored inside the organism in chemical form as ATP:

$$ADP + P + H^+ \rightarrow ATP + H_2O \qquad \Delta G = +30 \text{ kJ/mol}$$

The stored energy is released by hydrolysing the phosphoryl bond in the ATP:

$$ATP + H_2O \rightarrow ADP + P + H^+ \qquad \Delta G = -30 \text{ kJ/mol}$$

## OXYGEN DEMANDING PROCESSES IN ACTIVATED SLUDGE

The bacteria use the energy from ATP for growth, reproduction and maintenance. Only a portion of the substrate is oxidized to provide the energy. The remainder (the yield $Y$) is reorganized into new cell material. *Heterotrophic bacteria* which use substrate consisting of carbonaceous material include a bit more than half of the substrate (on a weight/weight basis) into new biomass. *Nitrifying bacteria* include only a minor part of the substrate ammonia into new biomass whereas most of the substrate is oxidized for energy production. These autotrophic bacteria use dissolved carbon dioxide as a carbon source for new biomass. In comparison with heterotrophic bacteria, nitrifiers need more oxygen for their growth.

The overall conversion equation of organic substrate (approximate chemical composition $C_{18}H_{19}O_9N$) by heterotrophic bacteria (organotrophs, approximate chemical composition $C_5H_7O_2N$) is:

$$C_{18}H_{19}O_9N + 0.74\ NH_4^+ + 8.8\ O_2 \rightarrow$$
$$1.74\ C_5H_7O_2N + 9.3\ CO_2 + 0.74\ H^+ + 4.52\ H_2O$$

From this equation it follows that per g heterotrophic biomass formed $8.8 \cdot 32/(1.74 \cdot 113) = 1.43$ g $O_2$ is consumed. The overall equation of reaction for nitrifiers (chemical composition $C_5H_7O_2N$) is:

$$NH_4^+ + 1.86\ O_2 + 0.1\ HCO_3^- \rightarrow$$
$$0.02\ C_5H_7O_2N + 0.98\ NO_3^- + 1.88\ H^+ + 1.04\ H_2O$$

From this equation it can be calculated that per g nitrifying biomass formed $1.86 \cdot 32/(0.02 \cdot 113) = 26.3$ g $O_2$ is consumed. It also follows that per g $NH_4^+$-N converted $1.86 \cdot 32/14 = 4.25$ g $O_2$ is consumed.

When there is no substrate, the bacteria oxidize their own cellular material to obtain the energy necessary for the *endogenous metabolism:*

$$C_5H_7O_2N + 5O_2 \rightarrow 5CO_2 + NH_3 + 2H_2O - \Delta G$$

In addition to the oxygen consumption by heterotrophic and nitrifying bacteria there are some other processes that contribute to the respiration of activated sludge. Like nitrifiers, the autotrophic *sulphur bacteria* and *iron bacteria* utilize inorganic compounds instead of organic matter to obtain energy and use carbon dioxide or carbonate as a carbon source. Sulfur bacteria are able to oxidize hydrogen sulfide (or other reduced sulfer compounds) to sulfuric acid:

$$H_2S + O_2 \rightarrow H_2SO_4 - \Delta G$$

Iron bacteria oxidize inorganic ferrous iron to the ferric form to obtain energy:

$$Fe^{2+} + O_2 \rightarrow F^{3+} - \Delta G$$

In addition, *protozoa* and other higher organisms contribute to the overall biological oxygen consumption in activated sludge. These organisms occur in varying numbers in the activated sludge depending on the loading of the treatment plant. Finally, some inorganic electron donors such as ferrous iron and sulfide can be chemically oxidized utilizing oxygen and contribute to the respiration of activated sludge.

## MODELING RESPIRATION

Modeling is important in the design and control of the activated sludge process, as well as for understanding the basics of respiration. In the *traditional* modeling approach oxygen consumption is associated with growth and decay of microorganisms (Grady and Lim, 1980). In the *death-regeneration* approach, adopted in the Activated Sludge Model No. 1 (Henze et al., 1987), oxygen consumption is associated only with aerobic growth of heterotrophic and nitrifying biomass (Dold et al., 1980). Figure 4.1 shows schematically the main processes for heterotrophic growth and biodegradation for the two approaches.

Both approaches describe growth of biomass ($X_H$) as a process where oxygen is consumed. However, the traditional approach considers biomass decay as an additional oxygen consuming process in which decaying biomass is oxidized whereas inert matter ($X_P$) is formed. The model implies that when the activated sludge has run out of readily biodegradable substrate ($S_S$) and slowly biodegradable matter ($X_S$) from the wastewater, the remaining oxygen consumption is associated with biomass decay only.

According to the death-regeneration approach decaying biomass is split

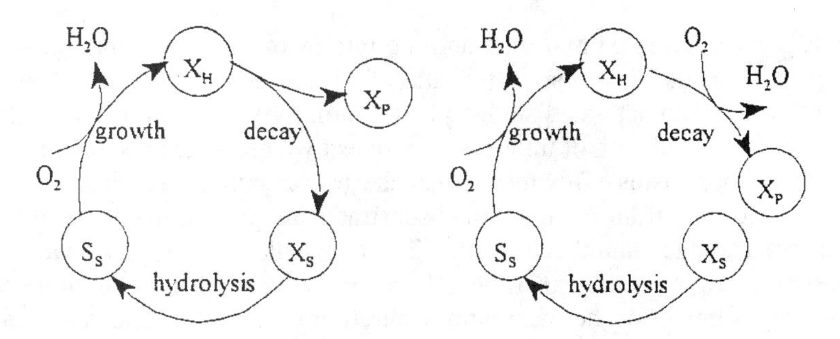


**Figure 4.1** Two modeling approaches for the activated sludge process: dead regeneration (left) and traditional (right).

into two fractions: inert matter and slowly biodegradable matter where the latter is hydrolyzed into readily biodegradable substrate. This process does not involve consumption of oxygen (or another electron acceptor). The death-regeneration model implies that even when all the substrate originating from the wastewater has been oxidized, there remains an oxygen consumption associated with the growth on substrate released from decay and hydrolysis, respectively. The amount of new biomass from released substrate is always less than the amount of biomass lost.

Both models imply that if biomass is left on its own without input of biodegradable matter from wastewater the respiration rate will gradually decrease until all the biomass has decayed. The respiration rate during this process is called *endogenous respiration rate.* The endogenous respiration rate of activated sludge can be defined in operational terms as the oxygen consumption rate in the absence of substrate from external sources. According to this definition, endogenous respiration includes not only growth and decay of bacteria but also oxygen consumption of protozoa. Note that in the microbiological literature endogenous respiration is defined in terms of maintenance only. The endogenous respiration rate is almost independent of the substrate concentration and is therefore indicative for the concentration of active biomass.

Usually in the activated sludge process a continuous input of biodegradable material exist from the influent. This results in a net growth of biomass and an associated respiration rate which is higher than the endogenous rate. This *actual respiration rate* is a function of the concentration of biodegradable matter in the aeration tank, which is in turn the net result of three processes: input from the influent, loss via the effluent and biodegradation. It is obvious that this balance is disturbed when activated sludge is sampled from an aeration tank and that the respiration rate measured in the sample is likely to be biased.

If the concentration of biodegradable matter is very high, the biomass will grow at its maximum rate and the rate of oxygen consumption will approximate its maximum value: the *maximum respiration rate.* In the real world of an activated sludge plant treating sewage, respiration is the result of the oxidation of multiple substrates by a heterogenous population of micro-organisms. This means that the true maximum respiration rate is reached only if all the individual substrates are present in excess. In an activated sludge plant this condition is not very likely to happen. In a well designed respirometric experiment, however, the appropriate condition for the measurement of the maximum respiration rate can be created and the measurement be used for model identification. Like endogenous respiration rate, maximum respiration rate is almost independent of substrate concentration and is therefore indicative of active biomass concentration.

Using the model concepts depicted in Figure 4.1, respiration rate can be expressed mathematically in terms of growth and decay. Traditional model (growth and decay of nitrifiers included):

$$r = \frac{1 - Y_H}{Y_H} \frac{\mu_{mH} S_S X_H}{K_S + S_S} + \frac{4.57 - Y_A}{Y_A} \frac{\mu_{mA} S_{NH} X_A}{K_{NH} + S_{NH}} + (1 - f_p)(b_H X_H + b_A X_A) \tag{4.1}$$

and the dead-regeneration model:

$$r = \frac{1 - Y_H}{Y_H} \frac{\mu_{mH} S_S X_H}{K_S + S_S} + \frac{4.57 - Y_A}{Y_A} \frac{\mu_{mA} S_{NH} X_A}{K_{NH} + S_{NH}} \tag{4.2}$$

It is seen from (4.1) that respiration rate is governed by the growth of both heterotrophs and nitrifiers, and the decay of both bacteria. At high substrate concentrations ($S_S$ and $S_{NH}$) the first and second term approach saturation and so does the respiration rate (maximum rate). If all the substrate is oxidized the first and second terms become zero and the lower—endogenous—respiration rate is governed by the biomass concentration. In the dead-regeneration model (4.2), even when external substrate is absent, the concentration will never approach zero since substrate is regenerated from decay and subsequent hydrolysis. It should be noted that during endogenous respiration ammonium released from decay will hardly be oxidized because it will be used for growth of biomass.

Note that in the description above, in accordance with the ASM No. 1, heterotrophic growth is associated with only one substrate ($S_S$). In the ASM No. 2 heterotrops can grow on two substrates with concomitant oxygen consumption. In addition, nitrification can be modeled as a two-step process: oxidation of ammonium to nitrite and subsequent oxidation of nitrite to nitrate; see also section ''Respirometry in Modeling''.

## MEASUREMENT OF RESPIRATION RATE

Common to all types of respirometers are (1) a reactor in which activated sludge from the treatment plant and, optionally, wastewater or a specific substrate are brought together and (2) a measuring device for assessing the rate at which the biomass takes up oxygen. Respirometers can be classified according to two criteria: (1) the phase where oxygen is measured (gas or liquid) and (2) whether or not there is an exchange of these phases with the environment (flowing or static). Figure 4.2 shows a scheme which is applicable to all types of respirometers. Generally, there are two phases, a gas phase and a liquid phase. The gas phase (including bubbles) contains oxygen which is transferred to the liquid phase. The latter contains biomass and dissolved oxygen that is transported to the micro-organisms. The biomass can be either in suspension or attached to a carrier. Oxygen can be measured in either the gas phase of the liquid phase or both. There can be an input and output of either gas or liquid or both. Based on this general scheme we can describe all respirometric measuring techniques in terms of oxygen mass balances.

### TECHNIQUES BASED ON OXYGEN MEASUREMENT IN THE LIQUID PHASE

The majority of the techniques based on the measurement of oxygen in the liquid phase use an electrochemical DO-sensor. The DO-sensor generally consists of two electrodes in an internal electrolyte solution separated from the liquid by a semi-permeable membrane (Hitchman, 1978). Dissolved oxygen molecules diffuse from the liquid through the membrane into the internal solution. The molecules are reduced on the cathode, which generates an electrical current. This current is proportional to the diffusion rate of the oxygen molecules through the membrane which, in turn, is proportional to the DO concentration in the solution. The relationship between electrical current and DO concentration is established by calibrating the DO sensor. It can be shown that for a DO sensor, water

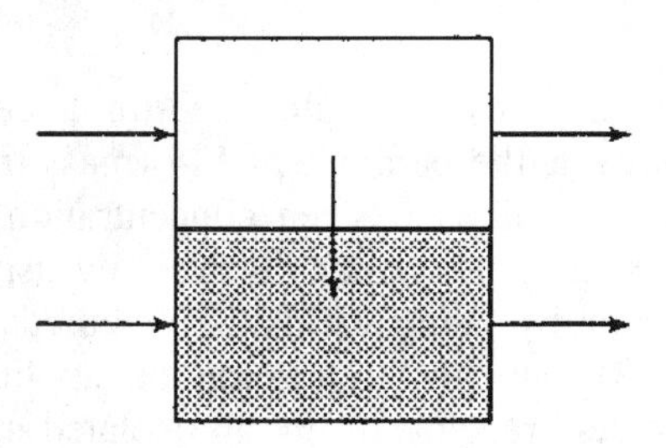

**Figure 4.2** Scheme for all respirometers.

saturated air is equivalent to oxygen saturated water. This is used to calibrate the DO-meter in water saturated air at 100% DO.

The simplest technique for measuring respiration rate is based on a static liquid phase (no input and output) and no gas phase. The DO-mass balance of this system will not contain terms for convective flow of DO and mass transfer from gas to liquid phase:

$$\frac{dC_L}{dt} = -r \tag{4.3}$$

Hence, to obtain respiration rate $r$ only the differential term has to be determined. This can be done by measuring the decrease of DO concentration $C_L$—due to respiration—as a function of time. Inherent to this technique is that DO becomes exhausted after some time due to the limited amount of DO in the liquid phase. For continuing measurement of respiration rate the liquid needs to be reaerated in order to supply new DO. The procedure for determining $r$ according to "Standard Methods" (APHA, 1995) is an example of this technique: A sample of activated sludge is brought into a bottle ("BOD-bottle"), aerated and sealed from the air (no gas phase, static liquid phase). Then, using a DO-sensor the decrease of DO concentration in the well mixed liquid is measured and recorded. For continuing measurement the liquid may be reaerated and sealed again. Respiration rate is calculated from the slope of the DO-decline. The technique is used in various commercial respirometers which automatically sample activated sludge, perform repetitive aerations and measurements of the DO-decline, and calculate respiration rate.

The need for reaerations can be eliminated by continuously aerating the biomass. Formally, this means that there is a gas phase with input and output. The simple mass balance (4.3) must be expanded with an oxygen mass transfer term:

$$\frac{dC_L}{dt} = K_L a(C_L^* - C_L) - r \tag{4.4}$$

Respiration rate can now be calculated in a continuous way by measuring $C_L$. However, the parameters $K_L a$ (mass transfer coefficient) and $C_L^*$ (saturation dissolved oxygen concentration) must be known. These may be determined at regular intervals by using reaeration tests and look-up tables or by using estimation methods (Lindberg, 1997; Vanrolleghem, 1993). The technique can be applied in a full scale activated sludge batch reactor or in an isolated mini-reactor i.e. a respirometer.

Another technique with a continuous supply of DO is based on a sole liquid phase with continuous input and output. Here the DO in the input liquid guarantees continuous presence of oxygen in the respirometer. The mass balance for this technique includes flow terms:

$$\frac{dC_L}{dt} = \frac{Q_L}{V_L}(C_{L,\text{in}} - C_L) - r \tag{4.5}$$

The practical implementation consists of a closed reactor through which liquid with a sufficient high DO concentration is pumped continuously. DO concentrations must be measured at two different locations: in the input stream and in the output stream. The liquid flow $Q_L$ and the cell volume $V_L$ must be known in order to calculate $r$. In contrast to the previously described technique, this technique allows the establishment of steady state conditions in the respirometer with respect to substrate concentration: substrate or wastewater can be added continuously to the input stream.

When both liquid phase and gas phase have an input and output the following DO mass balance needs to be solved in order to calculate $r$.

$$\frac{dC_L}{dt} = \frac{Q_L}{V_L}(C_{L,\text{in}} - C_L) + K_La(C_L^* - C_L) - r \tag{4.6}$$

Obviously, a respirometer based on such a technique requires measurement of DO at two locations and knowledge of four parameters.

## TECHNIQUES BASED ON OXYGEN MEASUREMENT IN THE GAS PHASE

Respirometric techniques based on measuring gaseous oxygen always deal with two phases: a liquid phase containing the respiring biomass and a gas phase in which the oxygen measurement takes place. The main reason for measuring in the gas phase is to overcome difficulties associated with interfering contaminants common in the liquid phase. Gaseous oxygen is measured by physical methods such as the paramagnetic method, or gasometric methods. Oxygen is one of the few gases that show paramagnetic characteristics and so it can be measured quantitatively in a gaseous mixture by using the paramagnetic method. The method is based on the change in a magnetic field as a result of the presence of oxygen, and this change is proportional to the concentration of gaseous oxygen.

*Gasometric methods* measure changes in the concentration of gaseous oxygen. According to the gas law, these can be derived from changes in the pressure (if volume is kept constant, *manometric method*) or changes in the volume (if pressure is kept constant, *volumetric method*). Because gasometric methods rely on the gas law these can only be applied to closed measuring systems (no input and output streams). This provokes the need for reaerations and thus temporary interruption of the measurements limiting the possibility for continuing monitoring of the respiration rate (see also liquid phase techniques). However reaerations are not needed if the oxygen used is replenished, eg. by supplying pure oxygen from a reservoir or by using electrolysis. The rate at which oxygen is supplied is equivalent to the biological respiration rate. Because carbon dioxide is released from the liquid phase as a result of the biological activity (see section Fundamentals of Respiration) this gas has to be removed from the gas phase in order to avoid interference with the oxygen measurement. In practice this is done by using alkali to chemically absorb the carbon dioxide produced.

The simplest gas phase technique for measuring respiration rate is based on a static liquid phase and static gas phase (no input and output). See also Figure 4.2. In addition to the DO mass balance on the liquid phase, an oxygen mass balance on the gas phase must be considered:

$$\frac{dC_L}{dt} = K_La(C_L^* - C_L) - r \tag{4.7}$$

$$\frac{d(V_GC_G)}{dt} = -V_LK_La(C_L^* - C_L) \tag{4.8}$$

Hence, in order to calculate $r$, the change in the oxygen concentration in the gas phase, $dC_G/dt$, must be measured and also knowledge of $dC_L/dt$ is required. If a gasometric method is used $dC_G/dt$ is related to the change in volume or the change in pressure (see above).

Another technique is based on a flowing gas phase, i.e. the biomass is continuously aerated with air (or pure oxygen) so that the presence of sufficient oxygen is ensured. In comparison to Eqs. (4.7) and (4.8) two transport terms must be included in the mass balance on the gas phase:

$$\frac{dC_L}{dt} = K_La(C_L^* - C_L) - r \tag{4.9}$$

$$\frac{d(V_GC_G)}{dt} = F_{\text{in}}C_{G,\text{in}} - F_{\text{out}}C_G - V_LK_La(C_L^* - C_L) \tag{4.10}$$

In order to allow calculation of $r$, the gas flow rates, $F_{in}$ and $F_{out}$, and the oxygen concentrations in the input and output streams, $C_{G,in}$ and $C_G$ must be known in addition to the variables of the previous technique. Of these, $C_G$ has usually to be measured and the others are set or known. A gasometric method can not be used here, and the measurement of $C_G$ is done for example by the paramagnetic method. The technique can be applied to either an isolated mini-reactor or a full scale aeration tank. In the latter case, when there are input and output streams on the tank, transport terms must be added to the mass balance on the liquid phase. Additional measurement of dissolved oxygen will then be inevitable for a correct assessment of respiration rate.

## MEASUREMENT OF RELATED VARIABLES

In addition to the different types of respiration rate mentioned in a previous section (Modeling Respiration) a number of other variables can be measured using the respirometric techniques described above. In some cases, however, the technique is specifically implemented in an instrument to measure that particular variable and not the respiration rate. Examples are *BOD-meters* and *toxicity-meters* which do not necessarily provide measurements of respiration rate.

A BOD-meter measures the *Biochemical Oxygen Demand* (BOD) of wastewater by integrating respiration rate over a certain time interval. An alternative is the BOD-probe, which consists of immobilized living cells, a membrane and a DO-sensor. The signal of that sensor is a measure of the substrate concentration of the sample in which the BOD-probe is submerged.

A *toxicity-meter* that is based on respirometry measures the effect of inhibitory wastewater or individual components on the oxygen uptake rate, eg. as a percentage of the rate, in absence of the inhibitor. The actual respiration rate might be used to assess toxicity. However, since this rate also depends on the loading, it is difficult to discriminate between effects from loading changes and toxic inputs. Therefore, a respiration variable should be used that is less sensitive to loading of the activated sludge, such as maximum respiration rate. If the biomass in the respirometer is loaded with an excess of wastewater or substrate the maximum respiration rate can be approached. This rate serves as a reference level in the absence of toxicity. At the same time the exposure of the biomass to a possible toxicant will be higher than in the plant so that toxic wastewater is detected before it inhibits the plant. In general, for a respirometer to be used as an early warning system, the wastewater must be sampled upstream of the treatment plant.

Other related variables are model states and parameters: since oxygen consumption is associated with substrate utilization and biomass growth, respirometric techniques are useful for activated sludge model identification. A respirometer can be constructed with the purpose of monitoring model states and parameters such as substrate concentrations and growth rates, respectively. See section ''Respirometry in Modeling.''

## GENERAL REMARKS TO THE ABOVE TECHNIQUES

The DO mass balances given above only apply to completely mixed reactors with a fixed volume. In a respirometer as an isolated mini-reactor the first condition can easily be met. If necessary a non-constant volume can be accounted for by including a differential term for volume in the mass balance.

Techniques with flowing gas phase can be applied to both isolated mini-reactors and full scale aeration tanks. However, the condition of a completely mixed reactor may not easily be satisfied in full scale respirometer. Application of the techniques described with Eqs. (4.3) and (4.5) to full scale tanks are likely to be unsuccessful because of practical constraints.

One of the applications of a respirometer is to measure actual respiration rate in a full scale aeration tank. To this end it may be practical to use the aeration tank as a respirometer. An isolated respirometer using the technique according to Eq. (4.5) is capable of measuring actual respiration rate if wastewater is added to the input stream in such a manner that the mini-reactor is equally loaded as the full scale tank. Isolated respirometers using techniques described with Eqs. (4.3) and (4.4) are only reliable if the time between sampling and measurement is very short, a condition that is not easily met. Another application of a respirometer is to do dynamic response experiments under well defined conditions, i.e. in order to extract information on wastewater and activated sludge characteristics. It is obvious that an isolated mini-reactor type respirometer is best suited for this application.

Isolated respirometers may contain activated sludge suspension sampled from the full scale plant but also fixed biomass. In the latter case the biomass is not as representative of that in the tank and the respirometer will most likely be used for measurement of wastewater characteristics.

All commercial respirometers are based on one of the above respirometric techniques. These respirometers exist in different degrees of automation ranging from simple equipment to fully self-operating instruments that automatically perform sampling, calibration, fault diagnosis, calculation and processing of the respirometric data.

## RESPIROMETRY IN MODELING

As previously described (section Modeling Respiration) oxygen consumption is directly associated with both biomass growth and substrate removal, and respiration rate is therefore a useful variable for identification of models describing these processes. *Model identification* is the process of determining the best mathematical model structure and accompanying parameter values by using measured data (see also chapter 2). One important step in model identification is *model calibration* in which numerical values of model parameters such as stoichiometric and kinetic coefficients are determined. Another aspect of model identification is *state estimation* where model variables such as concentrations of components are assessed. Various respirometric procedures to determine model variables and parameters of various biokinetic models are proposed in the literature (Ekama et al., 1986; Henze and Gujer, 1992; Henze and Gujer, 1995; Rozich and Gaudy, 1992). There are two possible approaches to the assessment of model parameters and variables: *direct methods* focus on isolated parameters which are directly related to the measured variables, whereas *optimization methods* use a more or less simplified complete model that is fitted to the measured data.

## DIRECT METHODS

Respirometric determination of specific model parameters and variables traditionally has been carried out through special controlled measurements on bench, pilot and full scale. For example the influent concentration *readily biodegradable substrate* ($S_S$) can be obtained from the response of respiration rate in a single reactor activated sludge plant upon feed termination (Ekama et al., 1986). The observed precipitous decrease in respiration rate is proportional to the $S_S$ concentration in the influent. A recently proposed method is based on the measurement of respiration rate under different loading conditions of a continuously operated respirometer (Witteborg et al., 1996). The influent $S_S$ is obtained by numerically solving a set of mass balances each of which pertaining to a different loading condition. Another method is the batch test where a small amount of wastewater is mixed with a larger amount of activated sludge grown on that wastewater. The test consists of the addition of a wastewater sample to endogenous activated sludge in the respirometer, and monitoring the respiration rate as a function of time. Once the biodegradable matter from the wastewater is depleted, the respiration rate drops to the endogenous level, which is—in modeling terms—the rate associated with the oxidation of $S_S$ generated from decay or associated with biomass decay alone (see Fundamentals of Respiration). The extra cumulative oxygen consumption in addition to the endogenous consumption may be related to the influent $S_S$ that is oxidized.

The above methods do not directly provide the influent concentration $S_S$: only a portion of the $S_S$ is oxidized and the remainder (the yield

$Y_H$) is reorganized into new cell material (see section Fundamentals of Respiration). Hence knowledge of $Y_H$ is required in order to obtain $S_S$ from respirometric measurements. Furthermore, since nitrification may contribute significantly to respiration, oxidation of ammonia must be accounted for, e.g., by inhibition of nitrifiers or by accomplishing constant maximum nitrification.

Figure 4.3 shows a respirogram from which $S_S$ of the added wastewater can be assessed. The respirometric response consists of an initial high respiration rate due to the oxidation of $S_S$, a lower rate due to nitrification, and finally a tailing indicating oxidation of slowly degradable matter. In this figure the endogenous respiration rate is already subtracted from the measured rate. Comparison with the ammonium respirogram confirms that the plateau after the initial peak is associated with nitrification. The cumulative oxygen consumption in addition to the nitrogenous consumption is 23 mg per liter wastewater added (shaded area). The wastewater $S_S$ concentration is, therefore (assuming $Y_H$ = 0.67): 23/(1 – 0.67) = 70 mg $l^{-1}$. Physical methods to assess $S_S$ have also been proposed, based on the difference in COD values of influent and effluent filtrates (Wentzel et al., 1995). For the example in Figure 4.3, the 0.45 μm filtrate COD's of influent and effluent (i.e., the endogenous sludge) were 210 and 51 mg/l, respectively, and therefore the influent $S_S$ can be calculated: 210 – 50 = 159 mg/l. This value is higher than that from the biological method, as is usually the case. When using biokinetic activated sludge models, it is obvious that $S_S$ in terms of biological response is preferred rather than in terms of physical properties.

An alternative of the above method is to inhibit nitrification. This is particularly recommended when nitrification is less explicit. Figure 4.4 shows respirograms from two consecutive wastewater additions. Prior to the second addition the nitrification in the sludge was inhibited by adding 10 mg allylthiourea per liter final mixture. The latter respirogram allows a more convenient assessment of the concentration readily biodegradable matter in the wastewater.

Figure 4.3 shows that respirograms may exhibit a number of degradable components. By using the inflection points, sufficiently detailed respirograms may be decomposed in several components, e.g., $S_S$, ammonium and nitrite. Hence, at each time instant the remaining BOD of each individual component can be obtained and related to the concentration of that component. By relating concentration and respiration rate, and applying the Monod model the half-saturation coefficient and maximum conversion rate of each component can then be assessed. This technique is used in the GPS-X module RespEval™ described in Chapter 7. This software automatically filters and decomposes respirograms, and calculates a number of model parameters.

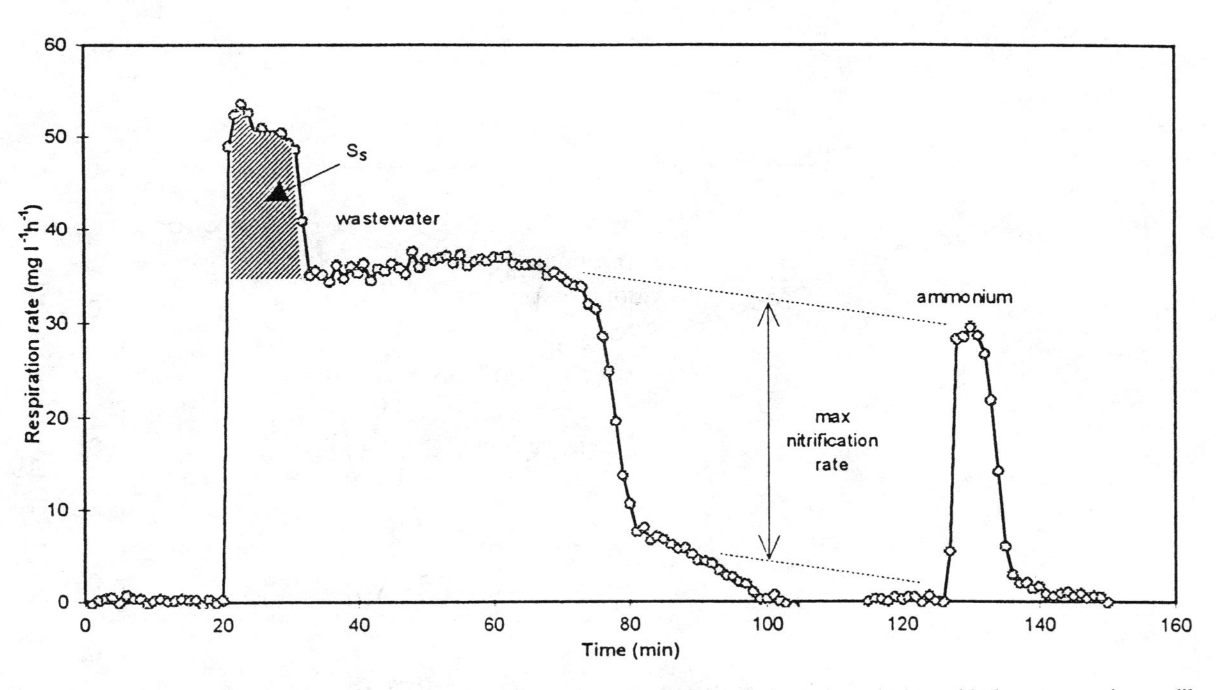


**Figure 4.3** Respirograms of wastewater and ammonium (Temmink, 1994). The shaded area is equivalent with the concentration readily biodegradable mater ($S_s$).

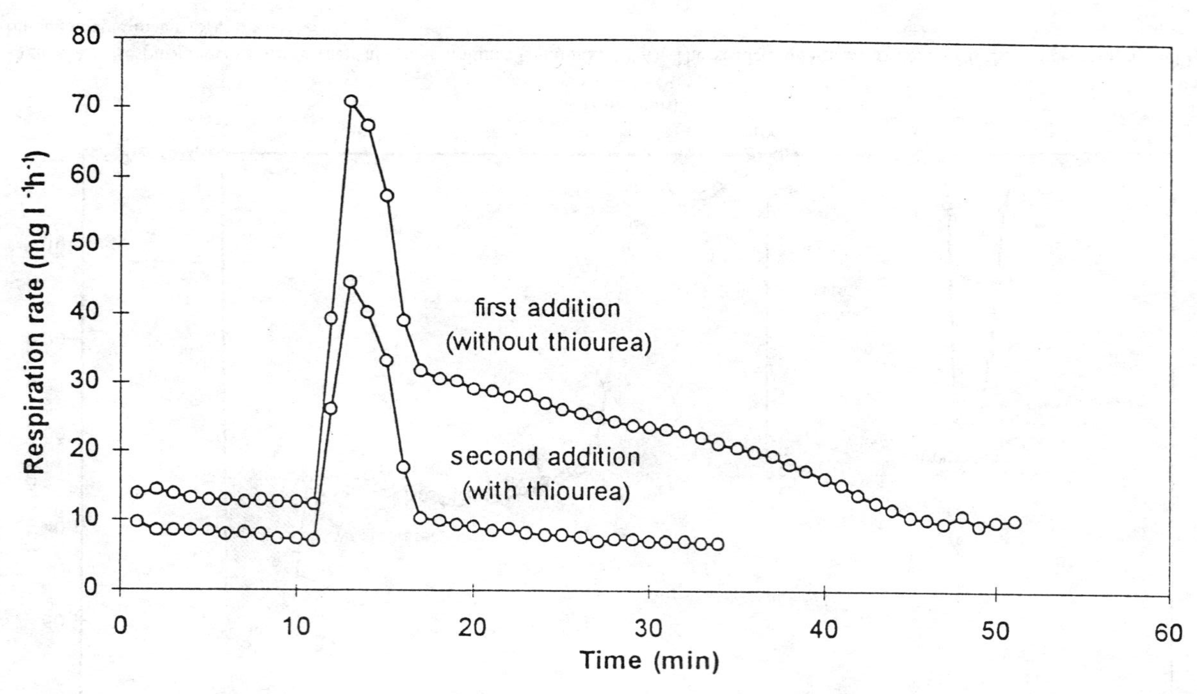


**Figure 4.4** Respirogram of wastewater, with and without inhibition of nitrification (adapted from Spanjers and Vanrolleghem, 1995).

The respirogram in Figure 4.3 shows a tailing which can be attributed to the oxidation of readily biodegradable matter released from hydrolysis of *slowly biodegradable matter*. It is difficult to assess the concentration slowly degradable matter $X_S$ from a respirogram by calculating the cumulative oxygen consumption because of uncertainty about the contribution of $X_S$ to the total respiration rate: the measured rate is for a significant part of the respirogram the result of simultaneously occurring processes and the rate due to hydrolysis alone cannot be identified. For the same reason it may be difficult to assess readily degradable matter and nitrifiable nitrogen from a respirogram. Therefore, it may be useful to apply model optimization techniques in order to assess model parameters and variables (see next section).

Other examples of model parameters that can be obtained from respirometric measurements are the *decay rates* for heterotrophic biomass and for nitrifiers, $b_H$ and $b_A$ respectively, and the yield coefficient for nitrifiers $Y_A$. One method for determining $b_H$ is by measuring respiration rate of endogenous sludge, in the presence of nitrification inhibitor, over a period of several days. The natural logarithm of the endogenous rate is a linear function of time of which the slope equals $b_H$. Another method allows simultaneous determination of both $b_H$ and $b_A$. The method is based on the measurement of the maximum respiration rates obtained after addition of a mixture of acetate and ammonium. This is repeated several times over a long enough period. Figure 4.5 shows the result of two such additions with a time interval of 7 days. The first plateau represents the maximum rate of heterotrophs oxidizing acetate and the second plateau that of nitrifiers. In terms of the ASM1 these rates follow from Eq. (4.2) by assuming high substrate concentrations:

$$r_{mH} = \frac{1 - Y_H}{Y_H} \mu_{mH} X_H \tag{4.11}$$

$$r_{mA} = \frac{4.57 - Y_A}{Y_A} \mu_{mA} X_A \tag{4.12}$$

If $Y$ and $\mu$ are assumed constant, $r_m$ only is a function of the active biomass concentration $X$. It can be shown that the natural logarithms of $r_{mH}$ and $r_{mA}$ are a linear function of time of which the slopes equal $b_H$ and $b_A$, respectively. Here it is assumed that the decay rate of heterotrophs oxidizing acetate is representative of all heterotrophic organisms.

It should be noted that traditional methods not always directly lead to parameters in terms of the ASM1 and require some conversion in order

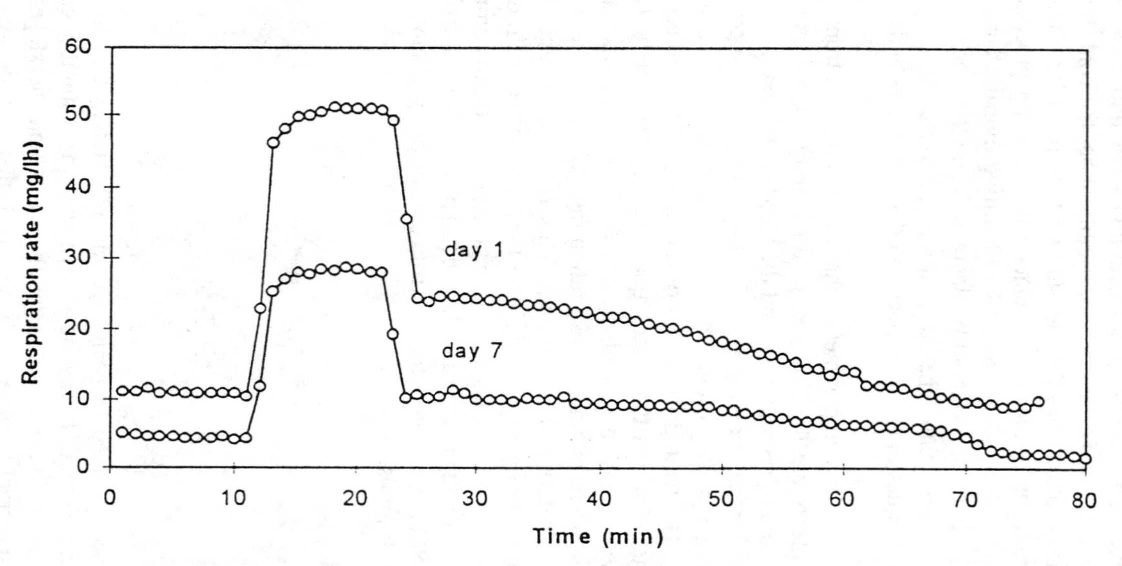


**Figure 4.5** Respirograms from a mixture of acetate and ammonium, spaced 7 days in time, for the simultaneous determination of $b_H$ and $b_A$ (Spanjers and Vanrolleghem, 1995).

to be applicable to that model. For example the decay coefficient $b_H$ obtained according to the above methods should be converted as follows:

$$b_H(\text{ASM1}) = \frac{b_H}{1 - Y_H(1 - f_P)} \tag{4.13}$$

where $f_P$ is the fraction of $X_H$ leading to $X_P$ (see also Figure 4.1).

The yield coefficient for autotrophs $Y_A$ can be assessed from the cumulative oxygen consumption measured after addition of a known amount of ammonium to activated sludge:

$$Y_A = \frac{4.57 - A}{1 + Ai_{XB}} \tag{4.14}$$

where $A$ is the cumulative amount of oxygen (in $mgO_2$) per mg ammonium nitrogen added and $i_{XB}$ is the mass of nitrogen per mass of COD in the biomass.

The yield coefficient for heterotrophs $Y_H$, however, cannot easily be assessed using respirometry alone. Sometimes a well defined substrate such as acetate is used to estimate the yield of the heterotrophic biomass by measuring the amount of oxygen associated with the oxidation of a known amount of acetate. Assuming complete utilization of the acetate the yield coefficient can be calculated from the amount of oxygen utilized per mass of oxygen theoretically required to oxidize the acetate completely, which is equivalent with $(1 - Y)$. The yield coefficient obtained is then assumed to be representative for the yield coefficient of all heterotrophs. However, there is a risk associated with the use of acetate: the yield for acetate can vary according to the history of the sludge. This is illustrated in Figure 4.6 which shows respirograms resulting from two consecutive additions of acetate to one and the same sludge sample. The first respirogram shows a slow response upon the addition of acetate, suggesting poor adaptation, and a tailing, suggesting oxidation of slowly degradable (stored) material. Upon the second addition the respiration rate raises rapidly and there is no tailing. From the total extra oxygen consumption and the known amount of acetate the yield coefficients from the first and second addition can be estimated as 0.61 and 0.69 g cell COD per g acetate COD, respectively. However, if the tail end is not included in the calculation of the oxygen consumption from the first addition both respirograms provide the same yield: 0.69 gCOD/gCOD. The experiment demonstrates that adaptation and storage play an important role in the evaluation of acetate respirograms and that care must be taken when assessing $Y_H$ from acetate oxidation measurements.

In the literature it is also suggested to estimate $Y_H$ by measuring, in continuous or batch experiments, the biomass formed per mass of soluble

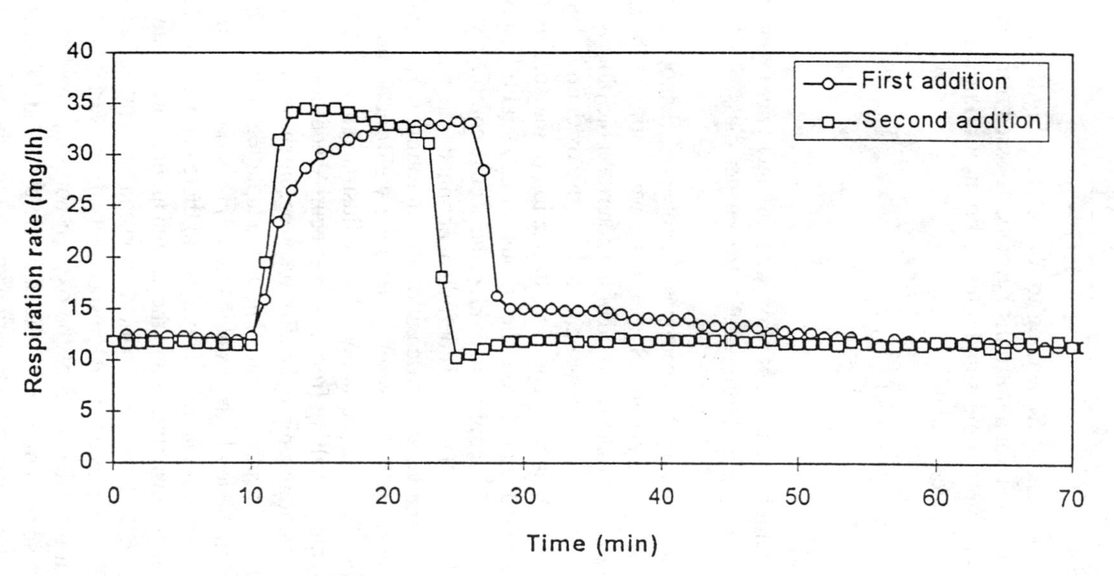


**Figure 4.6** Two consecutive additions of acetate to activated sludge (Spanjers, 1996).

substrate removed, both in COD units (Ekama et al., 1986; Henze et al., 1987; Rozich and Gaudy, 1992)

## MODEL OPTIMIZATION

In the section about Modeling Respiration, it was shown that respiration rate can be mathematically expressed in terms of biomass growth and decay. In activated sludge models such as the ASM1 these processes are, in order, related to hydrolysis and ammonification. Therefore, respiration rate is a useful variable for determination of model variables and parameters of activated sludge models. In the process of model optimization measuring data from optimal dynamic experiments are used in combination with mathematical curve fitting techniques to match the response of the model to the measured respirometric data. Obviously, it is not possible to obtain all model variables and parameters uniquely from respirometric data only. For example from batch experiments values for *parameter combinations* such as $S_S(1 - Y_H)$ or $\mu_{mH}X_H(1 - Y_H)/Y_H$ may be obtained but not constituting individual variables and parameters. These would require additional (respirometric) experiments.

Batch experiments are generally considered useful to assess model variables and parameters. They allow accurate adjustment of critical process conditions such as the substrate to biomass ratio, pH and temperature. Moreover, repetitive additions of the same or various samples to biomass can easily be done. For example additions of a well defined substrate such as ammonium prior to the addition of wastewater allows the determination of nitrification parameters separately. This is illustrated with the experiments depicted in Figure 4.7, where first the respirometric responses to additions of nitrite and ammonium are used to assess the nitrification parameters for a two-step nitrification model. Subsequently, these parameters are employed to obtain parameters for heterotrophic degradation from the overall respirometric response to the addition of wastewater. Using an appropriate optimization technique the concentrations and parameters shown in Table 4.1 were obtained from the respirograms.

It should be noted that the optimized model here was a modified version of the ASM1 with the most important modifications: two-step nitrification, first order hydrolysis, no ammonification and constant endogenous respiration.

The sequence of respirometry and model optimization as depicted in Figure 4.7 can be automated with the purpose of monitoring concentrations and model parameters at a treatment plant. Figures 4.8 and 4.9 illustrate this with the results of a respirometric device that automatically samples activated sludge, adds the various substates, monitors the respirograms, and calculates the concentrations and parameters. The figures show for

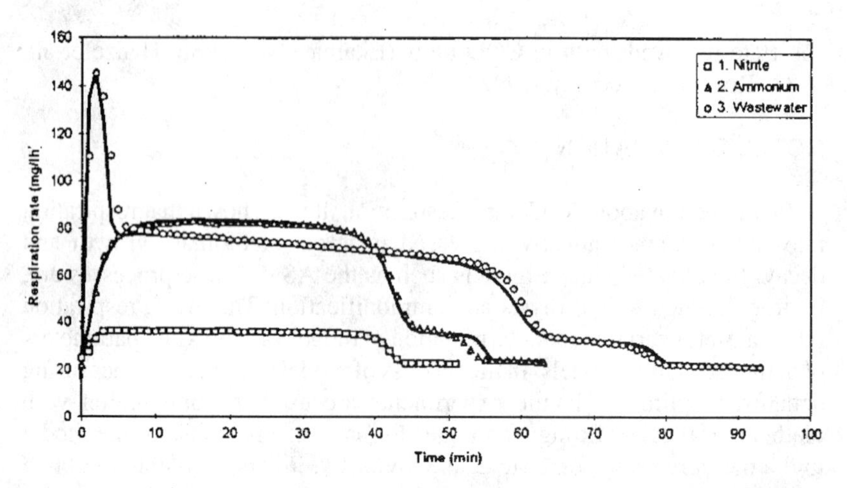


**Figure 4.7** Respirometric response from the addition of, successively, (1) nitrite, (2) ammonium, and (3) wastewater to activated sludge (Brouwer, 1997).

two different periods of several days the maximum nitrification rates $\mu_{mNH}X_{NH}$ and $\mu_{mNO}X_{NO}$, and the ammonium concentration $S_{NH}$, obtained from model optimization. For comparison, in Figure 4.9 the analytical ammonium and Kjeldahl wastewater concentrations are plotted along with the respirometric $S_{NH}$ to demonstrate the good agreement between the two different methods. The sharp drop of the nitrification rate on January 11th was a result of an increased waste flow (Figure 4.8) and the sudden decrease of $S_{NH}$ on the 27th coincided with a rain event (Figure 4.9).

TABLE 4.1.

| Wastewater Concentrations | |
|---|---|
| $S_S(1 - Y_H)$ | readily biodegradable substrate |
| $S_{NH}$ | ammonium |
| $X_S(1 - Y_H)$ | slowly biodegradable matter |
| **Model Parameters** | |
| $Y_{NH}$ | yield coefficient ammonium oxidizers |
| $Y_{NO}$ | yield coefficient nitrite oxidizers |
| $k_H$ | first order hydrolysis coefficient |
| $K_S(1 - Y_H)$ | half saturation coefficient $S_S$ |
| $K_{NH}$ | half saturation coefficient ammonium |
| $K_{NO}$ | half saturation coefficient nitrite |
| $\mu_{mH}X_H(1 - Y_H)/Y_H$ | maximum heterotrophic oxidation rate |
| $\mu_{mNH}X_{NH}$ | maximum rate ammonium oxidizers |
| $\mu_{mNO}X_{NO}$ | maximum rate nitrite oxidizers |

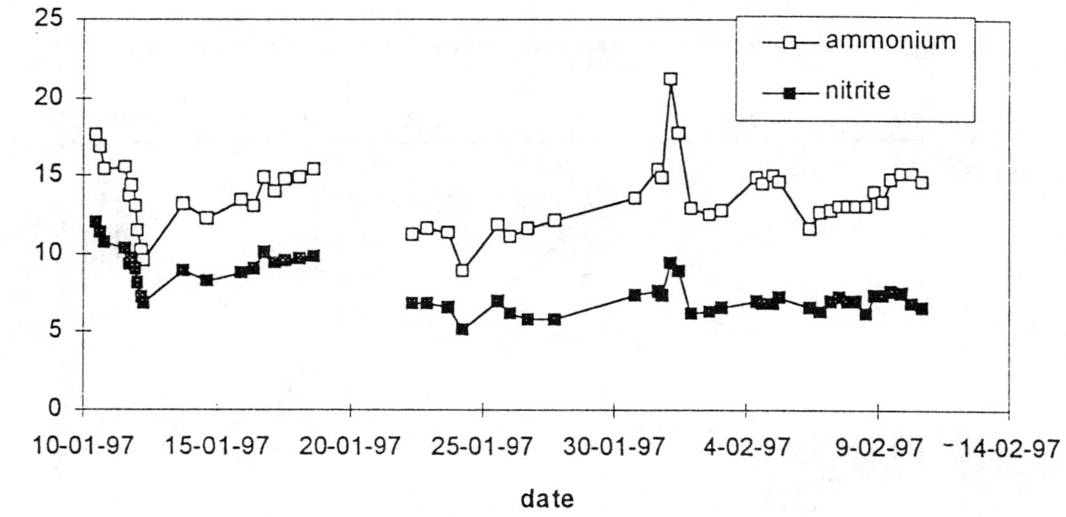


**Figure 4.8** Maximum nitrification rate from repeated respirometry and model optimization (Brouwer, 1997).

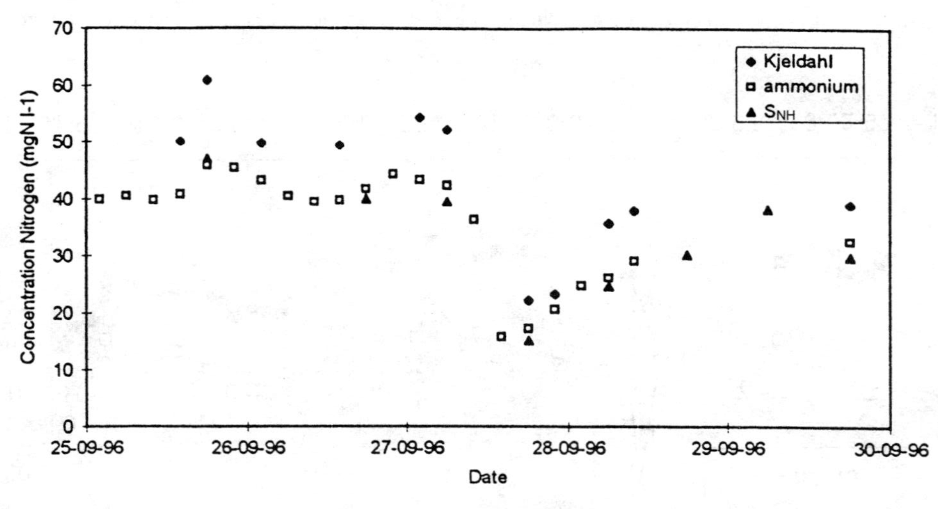


**Figure 4.9** $S_{NH}$ from repeated respirometry and model optimization, in comparison with analytical values form ammonium and Kjeldahl.

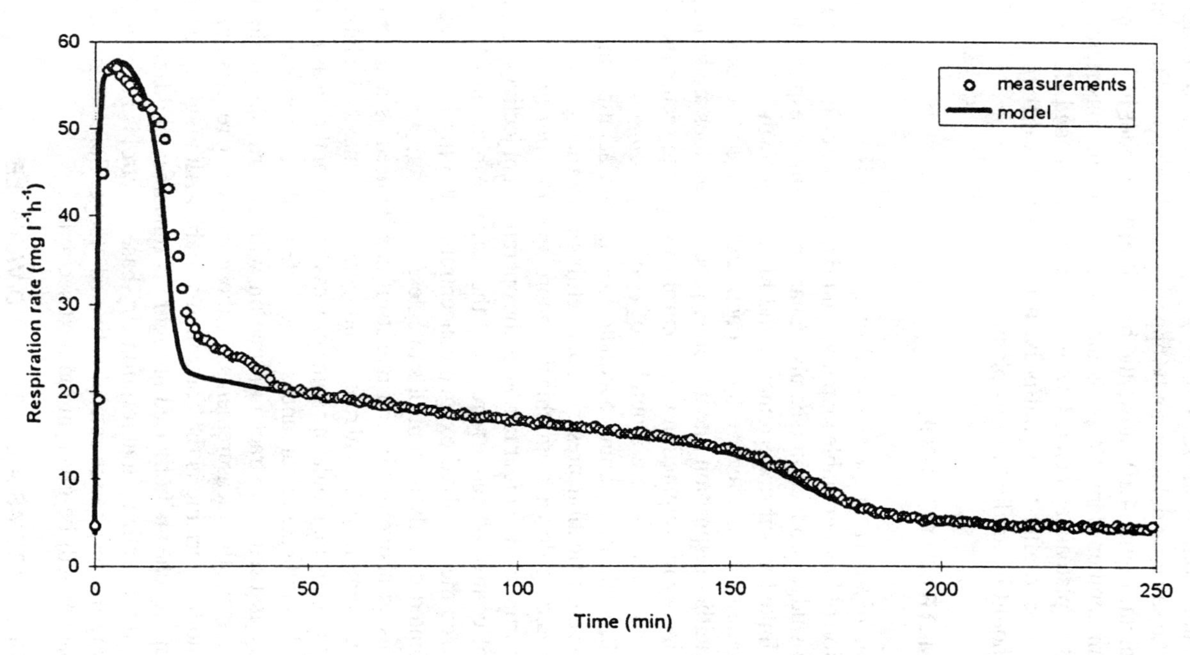


**Figure 4.10** Respirometric response from the addition of wastewater to activated sludge. Illustration of an inappropriate model structure (Brouwer, 1997).

The results shown in Figures 4.8 and 4.9 are all based on the same model structure. In these particular cases the model appeared to be appropriate for the wastewater-biomass interaction, as was proved by a good model fit. However, the same model may not be appropriate when applied to another treatment plant. Figure 4.10 shows the best fit of the model in Figure 4.7 to data from another plant. Respirograms from this plant consistently did not exhibit a plateau associated with nitrite oxidation and do show a shoulder occurred after the initial peak. It is clear that a different model structure should be applied to this plant.

## RESPIROMETRY IN CONTROL

The *basic objective* of the activated sludge process is to achieve a low concentration of biodegradable matter and nutrients in the effluent together with a low sludge production (see also Chapter 3). Since respiration rate is directly linked to substrate removal and biomass growth it is obvious that respirometry is an effective tool in activated sludge process control. However, neither respiration rate nor other process variables are generally deployed directly in conjunction with control strategies to secure the basic objectives. To achieve these a number of *operational objectives* have to be specified, such as: maintain adequate loading and aeration intensity, accomplish denitrification, preserve good settling properties, etc. Likewise, further proceeding down hierarchically, more specific *control objectives* have to be formulated in order to realize the operational objectives. Typical examples of control objectives are: keep the actual respiration rate at 42 mg $l^{-1}h^{-1}$, keep the dissolved oxygen concentration at 3 mg $l^{-1}$. These can be implemented directly in control strategies.

As suggested above respiration rate itself can be used as a controlled variable. For example the actual rate or endogenous rate may be maintained at a certain value by manipulating some process handle, or the measurement may be used to indicate disturbances or activate an alarm. However, often respirometry is used to extract information with a particular biological significance from the measurements. In these cases not respiration rate itself but another variable is the controlled variable. Although specifying these two approaches is not based on rigorous fundamental differences it may be helpful in realizing that respirometry-based control is not restricted to controlling the respiration rate of activated sludge. Some examples of both approaches will be given in the next sections.

### RESPIRATION RATE AS A CONTROLLED VARIABLE

One example of using respiration rate as a controlled variable is a strategy in which wastage flow is manipulated to maintain the endogenous respiration rate in the aeration tank at a set point value. The underlying idea is that endogenous respiration rate is a measure of active biomass concentration, and that an optimum endogenous rate (cq. active biomass concentration) exists such that an adequate process performance is guaranteed. Figure 4.11 shows an example of this control strategy for a completely mixed aeration tank. To ensure that control was based on measurement of endogenous respiration rate, the controller was only active during the low loading period (3:00–8:30 A.M.) where concentration readily biodegradable substrate was assumed negligible. After 3:00 A.M., as wasting commenced, the respiration rate started to decrease as a result of the decreasing active biomass concentration. On the day shown the desired set point of 22 mg $l^{-1}h^{-1}$ was almost but not entirely achieved in the allocated time of wasting. Apparently the controller was not tuned appropriately here.

In another strategy the influent flow rate is manipulated based on the measurement of the actual respiration rate in the aeration tank. It is obvious that this strategy requires an influent storage tank. Figure 4.12 shows an example of such a strategy at an industrial wastewater treatment plant where, by manipulating the influent flow rate, the actual respiration rate was controlled to maintain a stable organic loading.

## CONTROLLED VARIABLE DEDUCED FROM RESPIRATION RATE

If control is based on information extracted from respirometric measurements, not respiration rate itself but some other variable is the controlled variable. A common example of a deduced variable is the specific respiration rate that has been used as a measure of the loading condition of activated sludge. This is obtained by dividing actual respiration rate by biomass concentration in terms of volatile suspended solids. Strategies have been proposed to maintain the specific respiration rate at a set point by manipulating the influent distribution or the return activated sludge flow (Spanjers et al., 1998).

The percentage inhibition of the biomass due to a specific compound or to the wastewater is another typical deduced variable. This variable may be obtained by dividing the respiration rate in the presence of inhibitor, by the rate of sludge where the inhibitor is known to be absent. Possible bias from changes in substrate concentration may be avoided by measuring maximum respiration rate in the presence of a well defined substrate (see section Modeling Respiration).

The Biochemical Oxygen Demand, obtained from batch or continuous respirometric measurements, is another example of a deduced variable. A number of proposals exist in which one of the plants' handles is manipulated on the basis of the BOD. Figure 4.13 shows an example of a control strategy in which the aeration intensity is manipulated in concordance

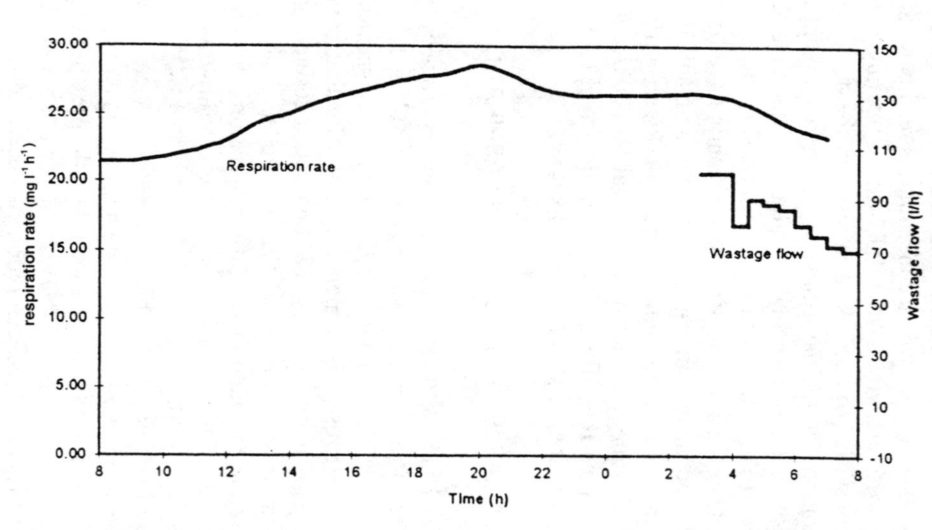


**Figure 4.11** Control of endogenous respiration rate by manipulating wastage flow. Set point respiration rate: 22 mg $l^{-1}h^{-1}$ (Stephenson et al., 1983).

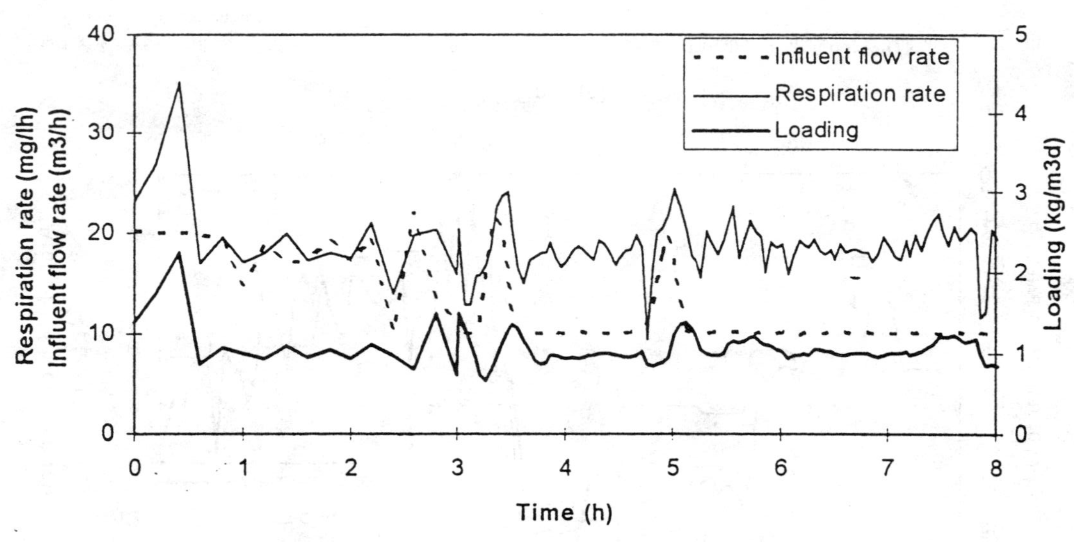


**Figure 4.12** Control of actual respiration rate by manipulating influent flow (Kim, 1997).

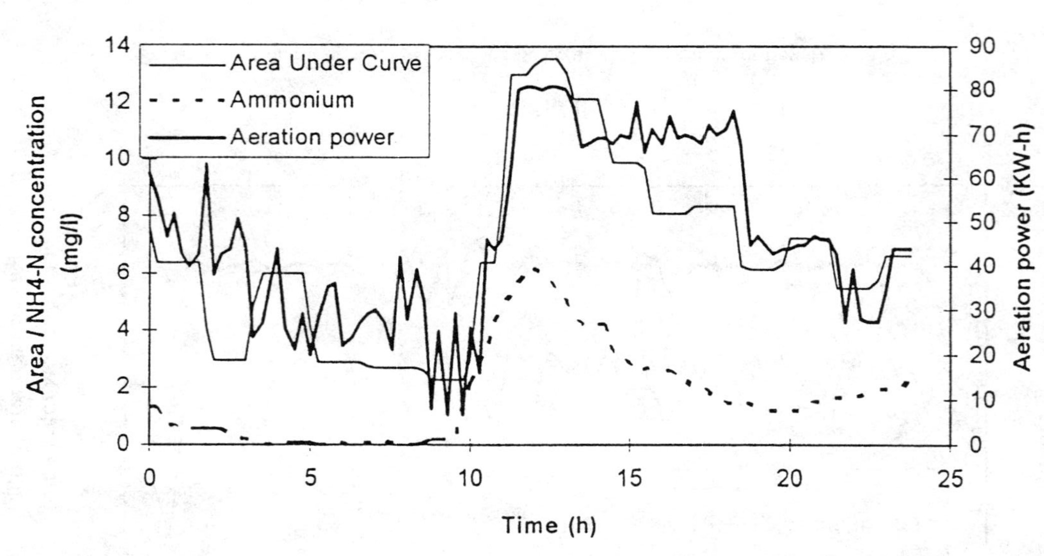


**Figure 4.13** Control of effluent BOD by manipulating aeration intensity (Draaier et al., 1997). The area under the respirogram is representative of the effluent BOD.

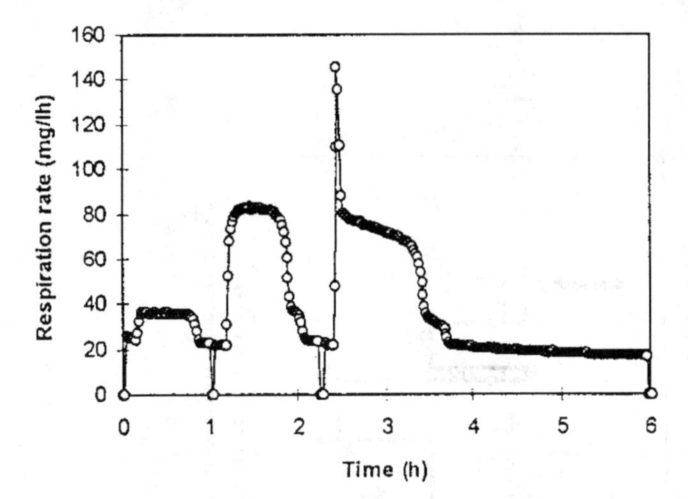


**Figure 4.14** Series of automatically generated respirograms for model based control of the aerobic volume (Brouwer, 1997).

with the BOD. The latter is derived from the area under the curve of batch respirograms and is identified as equivalent to the ammonium concentration in the effluent.

In a number of cases respirometry involves experiments in which wastewater, or substrate, and sludge are brought together, and in which specific process conditions are established. Respirometric information is then usually interpreted in a modeling context and deduced variables are model variables. An example is shown in Figures 4.14 and 4.15 where data obtained from batch respirometric experiments are used to manipulate the aerobic volume in a nitrifying/denitrifying activated sludge plant with the purpose of optimizing the denitrification via nitrite in the anoxic volume. A respirometer automatically samples activated sludge from the aeration tank and performs a number of additions. The respirograms from nitrite and ammonium additions are used to establish kinetic parameters for nitrification, i.e. calibrate a model for two-step nitrification. These are used in a model optimization on a respirogram from a wastewater addition to calculate, among others, the concentration of nitrifiable nitrogen $S_{NH}$ in the wastewater (see also section Model Optimization). This value together with the influent flow, and using the calibrated nitrification model allows calculation of the required aerobic volume to convert all the ammonium into nitrite. The calculated and the imposed aerobic volumes of the reactor for a period of about one week are depicted in Figure 4.15. The imposed volume is subject to discrete changes of 110 liter and is bounded by lower and upper constraints of 220 liter and 550 liter, respectively.

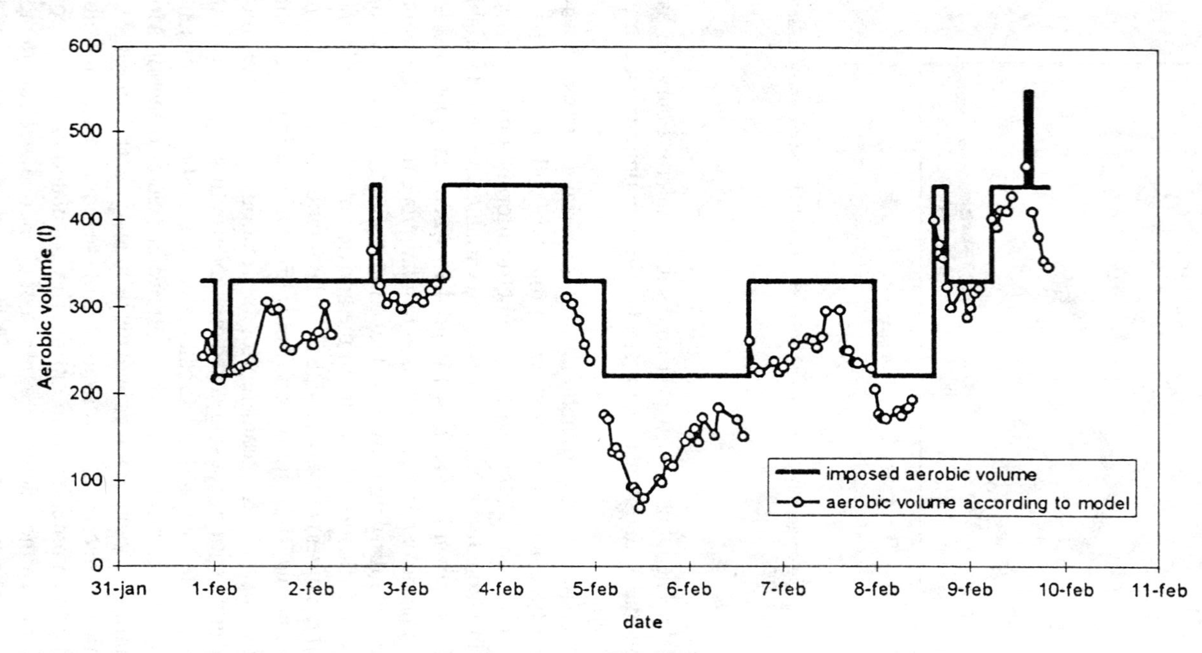


**Figure 4.15** Model based control of aerobic volume. The calculated volume is based on information obtained from series of respirograms as depicted in Figure 4.14. The imposed volume is the real volume bounded by practical constraints.

## EPILOGUE

This chapter deals with respirometry in modeling and control of the activated sludge process. Some fundamentals are provided in order to explain the significance of respiration in the biochemical reactions of the activated sludge. It is shown that respiration rate is directly linked to substrate removal and biomass growth and that respirometry is, therefore, a useful tool in activated sludge process model identification and control.

Because good model identification and control hangs on thorough knowledge of measurement techniques, the basic principles of measuring respiration rate are discussed. It is shown that respirometers can be classified in terms of the phase where oxygen is measured (gas or liquid) and whether these phases are flowing or static. It is explained that respiration rate is obtained by solving oxygen mass balances on these phases.

The application of respirometry in modeling is illustrated with a number of examples of obtaining model variables and parameters from respirometric experiments. It is rationalized that two approaches can be distinguished: direct methods focus on isolated parameters which are directly related to the measured variables, whereas optimization methods use a more or less simplified complete model that is fitted to the measured data.

It is illustrated with some examples that respirometry can be used as an effective tool in activated sludge process control. Although not justified by fundamental reasons, a distinction can be made between two approaches: respiration rate as a controlled variable, and respirometry as a basis for more specific biological information, e.g. model parameters or state variables, for use in control.

Recent publications show an increasing interest in the use of respirometry in modeling and control of the activated sludge process. It must be noted, however, that not all model variables and parameters can be assessed, and that no process can be controlled by using respirometry alone. This does not change the fact that respiration rate is generally considered as one of the most sensitive variables on the basis of which the activated sludge theory can be validated. In addition, the IAWQ has established a task group with the mission to write a Scientific and Technical Report on respirometry in control. Also, it is typical that the respirometric procedure for BOD determination after being disappeared from the Standard Methods, is included again in the 19th edition. Clearly, the new boom can be attributed to the progress in the development of reliable commercial respirometers in combination with evolution of advanced software.

## REFERENCES

APHA (1995) American Public Health Association, *Standard Methods for Examination of Water and Wastewater,* 19th edition, Washington DC.

Brouwer H. (1997) To be published. Wageningen Agricultural University, The Netherlands.

Clarke L. C. (1959) Electrochemical device for chemical analysis. US Patent 2913386, November 17, 1959.

Dold P. L., Ekama G. A., Marais G.v.R. (1980) A general model for the activated sludge process. *Prog. Water Technol. 12,* 47–77.

Draaier H, Buunen A. H. M. and van Dijk J. W. (1997) Full-scale respirometric control of an oxidation ditch. *7th IAWQ Workshop on ICA,* Brighton, UK, 6–9 July 1997.

Ekama G. A., Dold P. L., Marais G.v.R. (1986) Procedures for determining influent COD fractions and the maximum specific growth rate of heterotrophs in activated sludge systems. *Wat. Sci. Technol. 18* (6), 91–114.

Grady C. P. L. and Lim H. C. (1980) *Biological wastewater treatment: Theory and applications,* Marcel Dekker Inc., New York.

Henze M., Grady C. P. L., Gujer W., Marais G.v.R. and Matsuo T. (1987) *Activated Sludge Model No. 1,* IAWPRC Scientific and Technical Report No. 1, IAWPRC, London.

Henze M. and Gujer W. (eds.) (1992) Interactions of Wastewater, Biomass and Reactor Configurations in Biological Treatment Plants. *Proceedings of the IAWPRC Specialised Seminar,* Copenhagen, Denmark, 21–23 August 1991. *Wat. Sci. Technol. 25* (6).

Henze M. and Gujer W. (eds.) (1995) Modelling and Control of Activated Sludge Processes. Selected *Proceedings of the IAWQ Specialised Seminar,* Copenhagen, Denmark, 22–24 August 1994. *Wat. Sci. Technol. 31* (2).

Hitchman M. (1978) *Measurement of dissolved oxygen.* Wiley & Orbisphere.

Kim C. W. (1997) Stable loading control in a field-scale activated sludge plant using an on-line respiration meter. *Environmental Engineering, Res. 2,* No. 1, 1–8.

Lindberg C.-F. (1997) Control and estimation strategies applied to the activated sludge process. Ph.D. thesis Uppsala University, Sweden.

Montgomery H. A. C. (1967) Review Paper: The determination of biochemical oxygen demand by respirometric methods. *Wat. Res. 1,* 631–662.

Rozich A. F. and Gaudy A. F. (1992) *Design and operation of activated sludge processes using respirometry.* Lewis Publishers, Chelsea, Michigan.

Spanjers H. and Vanrolleghem P. (1995) Respirometry as a tool for rapid characterization of wastewater and activated sludge. *Wat. Sci. Technol. 31* (2), 105–114.

Spanjers H., Vanrolleghem P., Olsson G. and Dold P. (1996) Respirometry in control of the activated sludge process. *Wat. Sci. Technol. 34* (3–4), 117–126.

Spanjers H. (1996) Unpublished data. University of Ottawa, Canada.

Spanjers H., Vanrolleghem P., Olsson G. and Dold P. (1998) Respirometry in control of the activated sludge process; Principles. Scientific and Technical Report No. 1, ISBN 1 900222043 IAWQ, London, UK.

Stephenson J. P., Monaghan B. A. and Yust L. Y. (1983) Pilot scale investigation of computerized control of the activated sludge process. Canada Mortgage and Housing Corporation, Report SCAT-12.

Temmink H. (1994) Unpublished data. Wageningen Agricultural University, The Netherlands.

Vanrolleghem P. (1993) On-line Modelling of Activated Sludge Processes: Development of an Adaptive Sensor. Ph.D. Thesis University of Gent, Belgium.

Witteborg A., van der Last A., Hamming R. and Hemmers I. (1996) Respirometry for determination of the influent Ss-concentration. *Wat. Sci. Technol. 33* (1) 311–323.

CHAPTER 5

# Intelligent Systems for Control and Decision Support in Wastewater Engineering

## INTRODUCTION

IN the previous edition of this text the goal of a knowledge-based system was described as representation of expertise using new tools from the field of artificial intelligence. This notion has weathered the intervening years well and with experience in the development and deployment of knowledge-based systems, we now have a better understanding of their strengths and weaknesses. More importantly, we have learned that knowledge-based systems are a collage of various representational techniques, each leveraging the other to produce an application that is much more than the sum of its parts. There have been disappointments, as may have been expected, some surprises, and successes. In this edition, we upgrade the designation for this type of system from knowledge-based to intelligent, reflecting our expanding ability to represent—in more varied ways—deep, expert knowledge and implement applications that meet our expectations for a truly smart system. There is great opportunity for intelligent systems in wastewater engineering because the problems that arise in wastewater applications parallel those in related industries where these techniques have been used successfully. For example, in manufacturing, intelligent systems techniques now round out the standard suite of tools on the automation practitioner's workbench.

In wastewater engineering the system of interest is composed of many different types of processes. The *onion model* shown in Figure 5.1, describes the relationship between these different processes. Processes in-

Michael W. Barnett, Gensym Corporation, 125 Cambridge Park Drive, Cambridge, MA 02140.

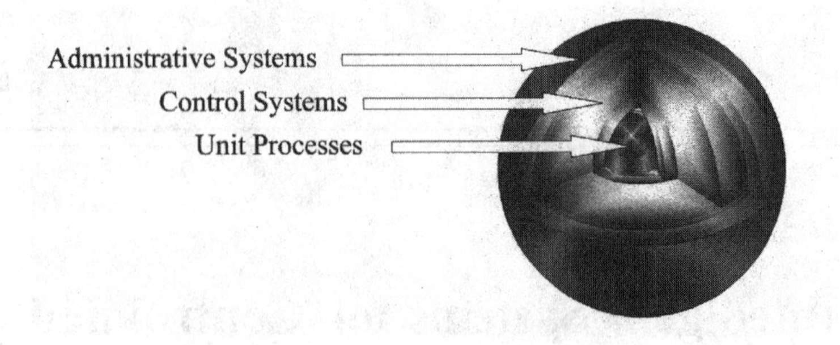


**Figure 5.1** The onion model of dependence between systems in an enterprise.

clude engineered processes such as valves, pumps, treatment units and supervisory control systems, as well as business processes such as those for managing human resources, equipment maintenance, documentation, administrative and financial sub-systems. This perspective highlights the interactions and inter-dependencies between processes and the importance of the enterprise in achieving the goal of intelligent control of water quality.

Intelligent systems have broad applicability for process, production and enterprise optimization. The *pyramid model* shown in Figure 5.2 demonstrates the architecture we strive to achieve. At the lowest level is the process equipment. Ideally the design of this equipment takes into account the dynamics encountered in actual operation. Monitoring and control systems lie above the lowest level and provide the information infrastructure necessary for higher levels in the pyramid. Many plants are now either installing or upgrading to new monitoring and control systems, for example, distributed control systems (DCS) and supervisory control and data acquisition (SCADA) systems. Advanced control and optimization systems rely on data supplied by lower levels and enable tighter control and coordination between unit processes as well as implementation of higher-level strategies for control.

The two highest levels of the pyramid have historically been the domains of non-real-time, transaction-based software that manage the ill-structured

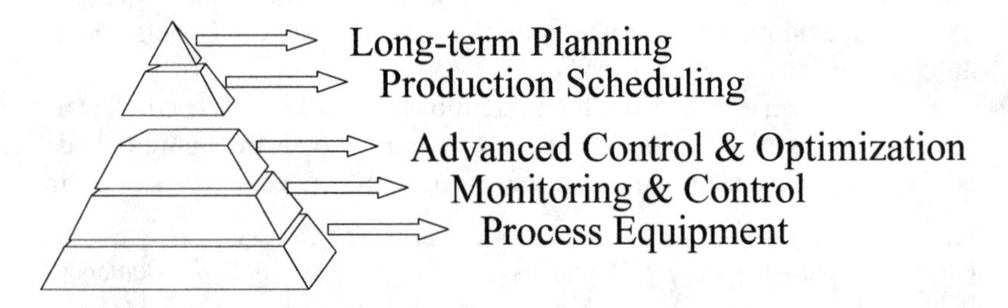


**Figure 5.2** The pyramid design model of hierarchial information systems.

process of taking production needs and translating these into a production schedule that can be implemented in an efficient way. Long-term planning is difficult, requiring intelligence on future trends and planning to meet these needs in a cost-effective way. These two levels are shown separated from the lower levels, because communication between the upper and lower levels has always been a problem. This issue is slowly being addressed in order to achieve the goal of a fully integrated enterprise resource management system. In wastewater systems, production scheduling and long-term planning don't always receive as much attention as other levels, however, these activities are an important component of water resource management. Cities, municipalities, industries and contract operations companies often struggle with the best way to manage water resources to meet environmental, safety and cost constraints. Intelligent systems have a significant role to play in optimization of these higher levels in the pyramid.

The goal of this chapter is to present enough background material to understand important issues and sufficient technical detail to get you started in the right direction. Pointers to references with greater detail on each topic are provided in all sections. Material is included to expand the view of intelligent systems to encompass all processes in the ''enterprise'' that address the key business issue of delivering effluent water of a desired quality at the lowest cost. The chapter is divided into four sections. The first section (Representation) covers the tools and techniques that are often associated with the field of artificial intelligence (AI) as it developed from the larger area of computer science. In this section, we discuss objects, rules and knowledge quality or certainty—all within the context of representation. The second section (Modeling) concerns traditional ways of building models. Modeling is another way of describing the task of representation but in practice has a definition that is more restricted. In this section, we discuss different types of models and concentrate on two major types, that is, mechanistic and empirical. The third section (Optimization) develops the idea of ''if-what'' (in contrast to ''what-if'') analysis, that is, using optimization techniques with models to determine the inputs needed to obtain a desired output. This is different from modeling and representation, as it has to do with applying models in a special way. Topics in the last section (Control) can be viewed as techniques for optimization, but control is unique as it concerns closing the loop between sensors and actuators. Olsson covers this topic in detail in an earlier chapter of this text. The purpose of this section is to stress the importance of extending conventional control systems and to present a framework for integrating many small pieces to realize an intelligent system.

## REPRESENTATION

Every intelligent system is a model of something. A model, in turn, employs a representation to capture the knowledge that it contains. Figure

5.3 shows some of the continua along which models lay. Models can represent all kinds of knowledge, for example, expertise about how to operate a unit process in an optimum way, physical, chemical and biological mechanisms, the meanings of patterns in data, and strategies for planning. Math is a powerful ways to represent knowledge. Development of new representations for knowledge as well as tools and techniques to do so efficiently is the contribution artificial intelligence is making to automation of the enterprise. This section describes some important representational techniques that were developed in the computer science and artificial intelligence fields.

## OBJECTS

Object-oriented design and development techniques started as a way to simplify computer programming languages, software application development and software maintenance. The central concept of object-oriented software is an object. An object represents just that, for example, a car, a bank account, or a recipe. Objects can also be abstract, for example, a flying-object or an elliptical-transformation-object. Objects are organized into classes that are related in various ways as shown in Figure 4. Class definitions include a number of variables, or attributes, that have fixed values and sub-routines, or methods, that define the behavior of the class. For example, a bank-account object may have an ending-balance attribute and a method used to determine how the bank-account object calculates its ending balance given a deposit or withdrawal amount. In Figure 5.4, the Reactors class has length, width and depth attributes and a method for calculating the detention time [calcDT ()].

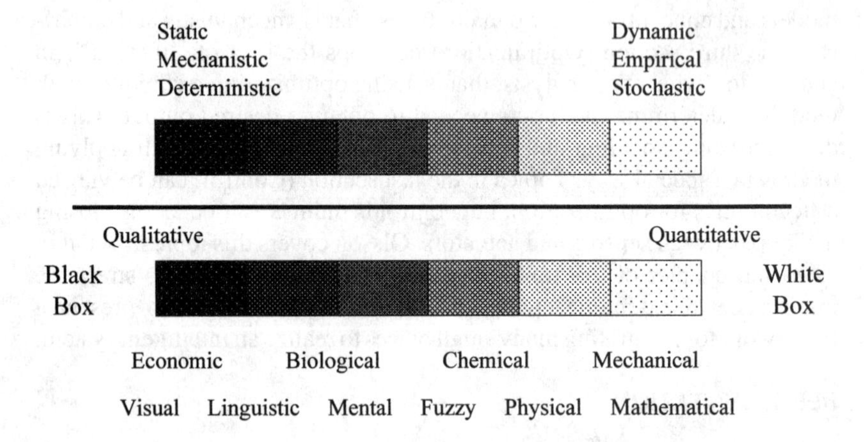


**Figure 5.3** Types of models along two descriptive continua.

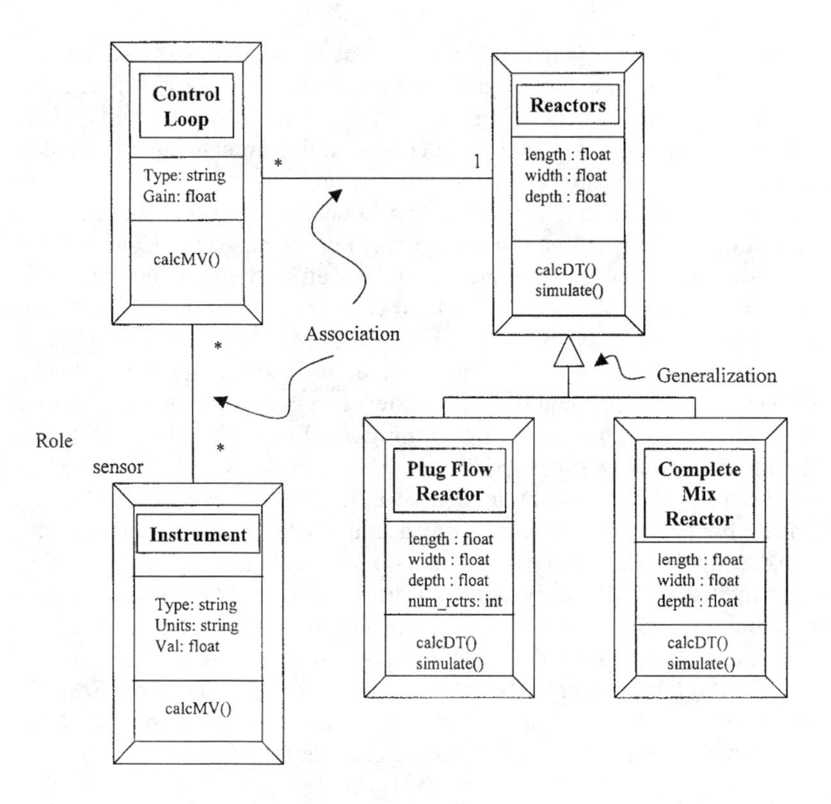


**Figure 5.4** An object oriented class diagram for common treatment process components.

The two most important types of relation that a class can participate in are *generalization* (or inversely, *specialization*) and *association,* also shown in Figure 5.4. For example, Plug-Flow and Complete Mix reactors are both a kind of Reactors, which is a more general class. A Reactor may also be associated with one or more control loops. Associations impose responsibilities on a class. For example, in Figure 5.4, the line from Reactor to Control Loop implies that a Reactor knows something about the Control Loop associated with it (what kind it is, what its attribute values are, etc.). Generalization relations are convenient if the child classes are permitted to inherit the attributes and behavior of the parent class. Thus, Plug Flow and Complete Mix reactors inherit the length, width and depth attributes from Reactor as well as the calcDT () method. Although the attribute and method names are the same, each class has the option of redefining the attribute value or the type of behavior implemented in the method as appropriate for that object.

In non object oriented languages, data and code are separate. In contrast, in an object-oriented language data and code are a single indivisible thing. For example, in the language C, segments of code are organized into functions or subroutines, while units of data are individual variables or collections of variables called structures. Indiscriminate mixing of data and code results in messy, hard-to-understand, and ad-hoc programs. In an object-oriented language data and code are combined in the class definition as demonstrated in Figure 5.4. For example, the Control Loop class defines the data (Type and Gain) as well as the code that results in behavior (calc MV). This notion of data encapsulation has important consequences in software development because it is a key to simplifying the program design and achieving re-usable software components. Objects conveniently localize and organize knowledge in a single place by storing data related to the object and by defining the things the object can do and the interface between that object and the rest of the world. This interface can be thought of as a contract that specifies the services provided by the object. For example, to obtain a bank account balance in an object-oriented banking application, you would need to know that the bank account object has a method called ''get-balance.'' There are numerous ways to implement this function properly, depending on the way the bank does it's accounting, where the data are located, etc. This implementation may also change over time as the bank changes its databases or accounting methods. Still, the exact manner in which the calculation is done is not of interest to the user, who is only concerned about the final result.

### Object Concepts in Practice

Early object-oriented languages in general did not deliver on the promise of dramatic increases in productivity through greater software re-use. However, recent object-oriented, visual development environments have achieved this goal due to the availability of many re-usable software components. Maturity of the language constructs and the tools available for development have had a huge impact on object-oriented software development. Languages for the Internet show promise for delivering reusable components that simplify many programming tasks and enable rapid development of net-wide distributed applications.

Object-oriented design (OOD) is a guiding force in model development. Programming languages and other kinds of software tools are beginning to grow OOD modules (Quatrani, 1998). Since any modeled system can be described in OOD terms—now using a standard modeling language (Fowler and Scott, 1997)—several workers have begun to define and organize useful sets of classes for building models. For example, work has been done on defining standard classes for process control in batch manufacturing (Instrument Society of America, 1995) and the utilities business (Taylor, 1995) to name just two. This area is likely to grow with the development of various classes and re-usable modules, including those that would be useful for building water quality management applications.

## RULES

A rule-based computer program is any program that uses rules as its primary form of knowledge representation. Expert systems, as the term is widely used, are often rule based. Rules describe knowledge in a very explicit way, using a simple IF-THEN structure. Models in which chunks of knowledge are best described in this way, and in which these chunks are logically chained together to achieve the desired representation, are good candidates for a rule-based system. Rules are also convenient as a way of reacting to real-time or other events, thus they are convenient for monitoring the passage of time.

Expert systems as a field evolved as a result of a shifting emphasis within the AI community from the development of general problem solving methods to specific, knowledge-intensive methods for solution of narrowly defined problems. Work in this field has resulted in the development of many tools and techniques that have been successfully used to build computer systems capable of expert level problem solving.

The often-cited example of an expert system is MYCIN, a program developed as part of the Stanford Heuristic Programming Project during the 1970s and early 1980s (Buchanan & Shortliffe 1984). The application consists of a knowledge base that contains the rules, an inference engine that processes rules, and a user interface for entering rules and interacting with the user. The inference engine reasons primarily by backward chaining through the rules as described below. Development of the knowledge base occurs beforehand during a process referred to as knowledge engineering—essentially extracting the appropriate rules from an expert. Commercial expert system development tools incorporate these features and reduce the programming burden on the developer.

The knowledge representation used in a rule-based system is the conventional rule. The conventional rule form is as follows;

IF A, THEN B

where A is called the antecedent of the rule and B is called the consequent. The expressions A and B can be complex, consisting of smaller expressions combined using the logical connectives ‘and’, ‘or’ and ‘not’. The antecedents and consequents consist of object-attribute-value triples, e.g., ammonia (object) concentration (attribute) is high (value). An antecedent or consequent can have the value TRUE or FALSE.

There are four ways in which rules most commonly are invoked (or 'fired'). These are (1) scanning, (2) event detection, (3) forward chaining, and (4) backward chaining.

Scanning at a selected frequency and event detection can initiate a rule invocation, which then proceeds according to either a forward or backward chaining strategy. This is the way that rules are chained together to develop a line of inference.

Backward chaining or goal-directed inference begins from a particular state (the goal) and moves backward through the rules from the THEN side to the IF side to find evidence to substantiate the goal:

Given:

| | | |
|---|---|---|
| Goal: | Determine if C is TRUE | |
| Assume: | A is TRUE | |
| Rules: | IF B THEN C | (6.1) |
| | IF A THEN B | (6.2) |

Conclude: C is TRUE

Given the goal of proving that C is TRUE, the inference engine searches for rules that can conclude that C is TRUE. Rule 6.1 can conclude this but Rule 6.1 requires that B be TRUE. The inference engine thus sets up B as a sub-goal and searches for rules that can conclude that B is TRUE. Rule 6.2 can be used for this purpose but again Rule 2 requires that A be TRUE. By looking in the database (where the assumed data are stored) the inference engine finds that A is TRUE thus B can be inferred and since B is TRUE, C can be inferred.

Backward chaining is often used for analysis problems such as diagnosis. In diagnosis problems a set of goals are established beforehand. Rules are then written in terms of these goals. In addition, goal-directed inference mimics the focusing ability of experts in diagnosis. The program MYCIN described above used a backward chaining strategy.

Forward chaining is often called data-directed inference since the data drive the inference from IF side to THEN side in the rules:

Given:

| | | |
|---|---|---|
| Assume: | A is TRUE | |
| Rules: | IF A THEN B | (6.3) |

IF B THEN C (6.4)

Conclude: C is TRUE

In this case, given the fact that A is TRUE, it can be inferred using Rule 6.3 that B is TRUE. If B is TRUE then, by Rule 6.4, C is TRUE. The conclusion is the same as the goal in the backward chaining example above. The difference is the manner in which the inference is made. Forward chaining is often used for synthesis problems, i.e., combining separate elements into a coherent whole. Forward chaining rule-based systems that perform configuration tasks are common, for example, creating a list of resources needed to construct a pipeline across three states, or deciding on the type and number of add-on components needed to prepare a computer system for a customer.

### Rules in Practice

Expert systems that are primarily rule-based are a small fraction of the total number of intelligent applications currently deployed in business and industry even though many of these applications use rules in their implementation. Diagnostic systems comprised of large numbers of rules are effective, however, these are always hand-crafted and tailored to a specific domain of expertise. Large rule bases are hard to specify, debug and maintain and alternative techniques can sometimes achieve similar ends. For example, causal digraphs (Stanley and Vaidhyanathan, 1998) represent reasoning chains well, and are easier to specify and maintain. Rule-based programming as the sole method of knowledge representation has not become as prominent as was originally thought. In many cases, knowledge expressed in several rules can be contained in procedures and methods, thus the observation that rules can be conveniently incorporated into existing procedural code. The most effective applications and application development tools integrate rule-based constructs with the other representational tools and techniques described in this chapter.

## METHODS

The traditional way of expressing knowledge that consists of a sequence of operations that execute sequentially and/or in parallel is to write a procedure. In object-oriented languages, procedures are called methods and are associated with a particular class as described above. Methods are the core repository for knowledge in most applications. Methods perform any task that can be programmed with the available language libraries and packages or other software components coded in the language's constructs.

### Combining Objects, Rules and Methods

Methods, rules and objects work together to implement intelligent tasks. Rules are useful for reacting to, or initiating an event when some behavior reacts to or causes the event can be expressed most conveniently in a chunk. Other rules may be invoked or procedures may take over, perhaps causing other events to occur. This chain of events is the same as forward and backward chaining described above and can lead to very complex behavior. Figure 5.5 shows an example of how rules, procedures and objects work together in monitoring a number of tanks. In this figure, a single rule monitors the tanks, firing the rule consequent when any tank pH rises above a certain value over a certain time. When the rule fires, it sets an alarm status for the tank and starts an alarm procedure that could lead to other procedures or the firing of more rules.

Note how the rule is expressed in a general way to refer to any tank (an object) that is currently being monitored, thus taking advantage of an object-oriented class hierarchy like that shown in Figure 5.4. In fact, each

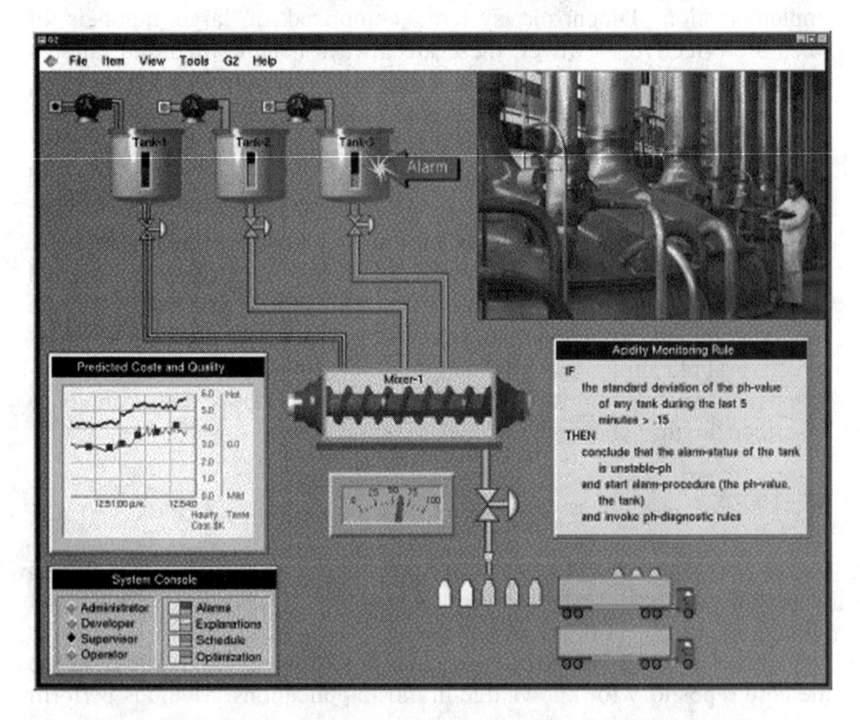


**Figure 5.5** Combining rules, objects and procedures in an intelligent application.

item shown in this figure is an object with its own individual attributes and behaviors. For example, the small valves at the tank outlets are objects that have attributes such as flow rate and valve position, where the flow is calculated based on the valve position. This computation is expressed as a method that is invoked for each valve object every time a signal is received at its input port—a pipe that is also a defined object.

Note the key features of this simple example:

- Rules conveniently express chunks of knowledge that react to events in time.
- Objects organize and localize data and behavior.
- Procedures contain complex algorithms.
- The combination results in complex behavior.

Implementation of intelligent tasks always leverages the strengths of individual representational tools.

## CERTAINTY

An important aspect of intelligent system development concerns management of knowledge quality, or knowledge certainty. There is no comprehensive theory to handle this problem thus many ad hoc procedures have been developed. Two methods will be introduced. These are certainty factors and fuzzy logic. Both find application in real-world problems and are widely used in industrial applications. The text by Buchanan & Shortliffe (1984) includes a separate section devoted to reasoning under uncertainty and discusses in detail the use of certainty factors. A general discussion can be found in Winston (1984).

### Certainty Factors

A logical first step in developing a method to handle uncertainty would be to borrow methods from probability theory. This was the approach taken in the past by researchers in the medical field with some success (Gorry & Barnett 1968) but strict adherence to the procedures can require huge amounts of statistical data. Such data usually cannot be obtained in practice so a simplified, heuristic approach is taken. One example is the certainty factors approach. This method, or a modification of it, is commonly used to deal with knowledge quality considerations in rule-based system, including MYCIN.

The certainty factors approach takes a practical view and relates the procedures for calculating certainties directly to the action of firing rules. In the certainty factors approach it is assumed that there is certainty in knowledge input to the rule, i.e., the degree to which the antecedent is true, as well as in the rules themselves, i.e., how true is the rule itself. Three issues must be considered;

(1) How to combine certainties associated with multiple-antecedent rules to form the rule input certainty
(2) How to transform the rule's input certainty into the output certainty
(3) How to combine rule output certainties when many rules have the same consequent

A certainty factor (CF) is a number between −1 and +1 indicating the degree of belief that a statement is TRUE. When CF = 1, the statement is definitely TRUE. When CF = 0, there is no evidence to support either the truth or falsehood of the statement. When CF = −1 the statement is definitely FALSE. A rule may have several antecedents, each with a different degree of certainty.

One way to deal with the first issue above is to take the minimum of all the antecedent certainties. The reasoning here is that the highest certainty that can be expected for the combination of antecedents is that associated with the evidence that is least certain; like assuming that the strength of a chain is that due to its weakest link. This method is simple and will suffice for many problems. The second issue above is handled by taking the product of the antecedent certainty and the rule certainty to obtain an updated rule certainty.

As inferences are made in the system, certainty factors are updated. As an example, assume that it is asserted, with a certainty factor $\alpha$, that the influent to an activated sludge process is high in carbohydrates and fats, and that a rule exists stating that (Geselbracht et al. 1988);

if the influent to an activated sludge process is high in carbohydrates and fats,

then sludge bulking will be a problem at the plant with a rule certainty factor of $CF_1$

Here $CF_1$ is the rule certainty factor. Using the product rule described above, the new certainty factor for the rule is

$$CF_{1,new} = \alpha * CF_1 \tag{6.5}$$

So if the value of $\alpha$ is 0.5, and the value of $CF_1$ is 0.9, then $CF_{1,new}$ for this case would be 0.45. The new certainty, $CF_{1,new}$, associated with the consequent of this rule (the assertion that sludge bulking will be a problem) is given the value 0.45.

Finally, to deal with the third issue, that is, combining certainties from many rules, special formulas are used. For example, imagine that the assertion above, that sludge bulking will be a problem at the plant, has an associated certainty factor $\beta$. If there is another rule (Geselbracht et al. 1988);

if the BOD/N ratio or the BOD/P ratio is too high

then sludge bulking will be a problem at the plant.
with rule certainty factor $CF_2$

which has the same consequent, the following formulas are used to update $CF_2$ (Dym & Levitt 1990);

$$\text{For } \beta \text{ and } CF_2 > 0,\ CF_{2,new} = \beta + CF_2 - \beta * CF_2 \tag{6.6}$$

$$\text{For } \beta \text{ and } CF_2 < 0,\ CF_{2,new} = \beta + CF_2 + \beta * CF_2 \tag{6.7}$$

$$\text{Otherwise, } CF_{2,new} = (\beta + CF_2)/[1 - \min(|\beta| + |CF_2|)] \tag{6.8}$$

In the example above $\beta = 0.45$ and let $CF_2 = 0.9$, thus, the new certainty factor $CF_{2,new}$ is;

$$CF_{2,new} = 0.45 + 0.9 - 0.45 * 0.9 = 0.945 \tag{6.9}$$

The belief in the assertion ''sludge bulking will be a problem at the plant'' increases substantially as a result of the additional strong evidence provided by the second rule.

### Fuzzy Sets

In the section above, uncertainty about an assertion is passed around with the assertion and the certainty for a rule is changed as evidence is combined and inferences are made. Inference is controlled by the inference engine, which may use forward or backward chaining, or some combination of these. Fuzzy set theory provides methods for making inferences using fuzzy variables. In this type of system control of inference is different from conventional rule based systems as described above. The methods of fuzzy set theory have been widely used in the area of process control including research conducted 20 years ago on control of the activated sludge process (Beck 1978). In this section, the basic concepts are presented followed by a discussion of how the methods can be used in a straight forward way to implement an 'expert' control algorithm.

Fuzzy rules can be defined just as conventional rules, however, fuzzy sets or fuzzy membership functions define the value portion of the attribute-value pairs. As an example Figure 5.6(A) shows the ordinary set representation of young, middle and old ages, whereas, Figure 5.6(B) shows the fuzzy set representation of each of these age groups. The fuzzy set young has the membership function $\mu A(y)$ which has values ranging from 0.0 to 1.0 depending upon age. The membership functions shown in Figure 5.6(B) are exponential functions, but they could have any shape.

The phrase used to describe a fuzzy membership function is comprised of a fuzzy variable and a linguistic value. For example, the phrase 'age is young' refers to a fuzzy variable (or attribute) 'age' and a linguistic value 'young'. These attribute-value pairs are defined for both antecedents and consequents in fuzzy rules such as (Beck 1978);

IF MLSS concentration is low

THEN make the SOLIDS WASTE RATE low

where the fuzzy variables are in bold and their linguistic values are underlined. Many fuzzy rules like this, which reflect the experience of plant personnel can be obtained from experts much like they would be in development of a conventional rule-based system.

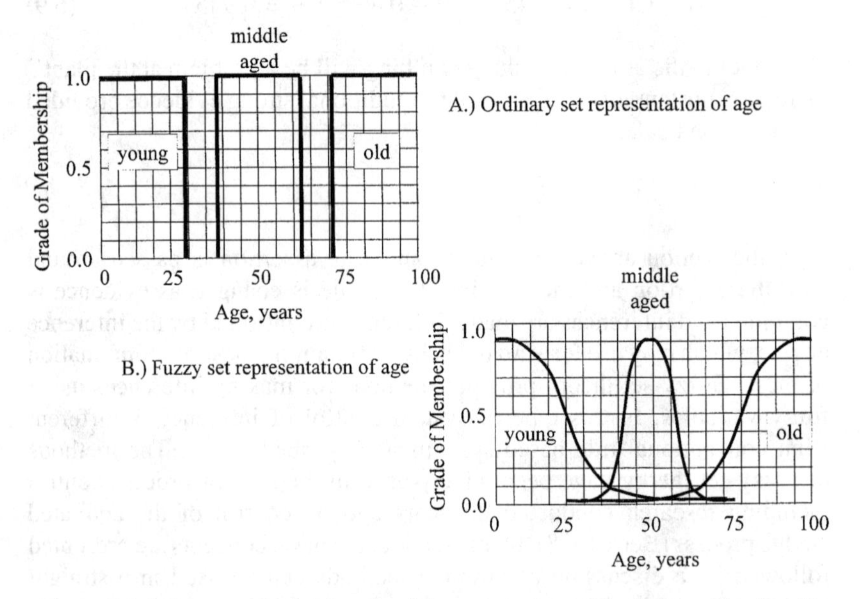


**Figure 5.6** Ordinary set (A) and fuzzy set (B) representations of the fuzzy variable "age."

A simple, but powerful manner for developing a fuzzy control system applies the concepts of fuzzy sets and fuzzy rules as described above with fuzzy relations and fuzzy algorithms. A fuzzy relation determines the relationship between linguistic variables in the antecedent and consequent of a fuzzy rule. It is defined as follows;

$$u_{Ri}(x,y) = \min[u_x(x),\ u_y(y)] \qquad (6.10)$$

where $u_{Ri}(x,y)$ is the ith fuzzy relation defining the relationship between $x$ and $y$ based on the ith fuzzy rule, $u_x(x)$ is the fuzzy membership function defined for the linquistic variable $x$, $u_y(y)$ is the fuzzy membership function defined for the linquistic variable y and min is the minimum operator.

This equation provides a simple way to calculate, given a fuzzy rule, the degree of relationship between two variables. For example, given $x$ and $y$ equal to the quantitative values of the MLSS concentration and SOLIDS WASTE RATE (SWR) respectively, we can imagine that if $x =$ 500 mg/L then the degree of membership of this value in the class ''low'' would be close to 1, say 0.9, that is, $u_{\text{LOW-MLSS}}(500) = 0.9$. The membership function for $u_{\text{LOW-SWR}}(y)$ is calculated in a similar manner for each value of $y$. Construction of the relation defined by the equation above can be visualized as show in Figure 5.7. To do this, translate each membership function along the axis of the other membership function. Notice that at each point in the $x$-$y$ plane, two values of the degree of membership $z$ are defined—one for each of the two membership functions. If we take the minimum of these two, we create a three-dimensional surface. The points on the surface reflect the strength of the relation between the two linguistic statements (the antecedent and consequent) at any point corresponding to quantitative values for $x$ and $y$. The surface quantifies the expert rule expressed linguistically above.

A fuzzy algorithm is constructed by combining several fuzzy relations (rules). This is defined as;

$$u_A(x,\ y) = \max[u_{Ri}(x,\ y)] \qquad (6.11)$$

where $u_A$ is the fuzzy algorithm defining the relationship between $x$ and $y$ based on all the fuzzy relations, and max is the maximum operator. Continuing with our visualization, a fuzzy algorithm is obtained by dropping all the fuzzy relation surfaces on top of one another and deriving a new surface by taking the maximum membership function value corresponding to each $x$-$y$ value. The resulting fuzzy algorithm surface would be like that shown in Figure 5.8. This figure shows, based on the expert fuzzy rules, how SOLIDS WASTE RATE should be changed given the current value of the MLSS. The best value for the SWR at a given MLSS

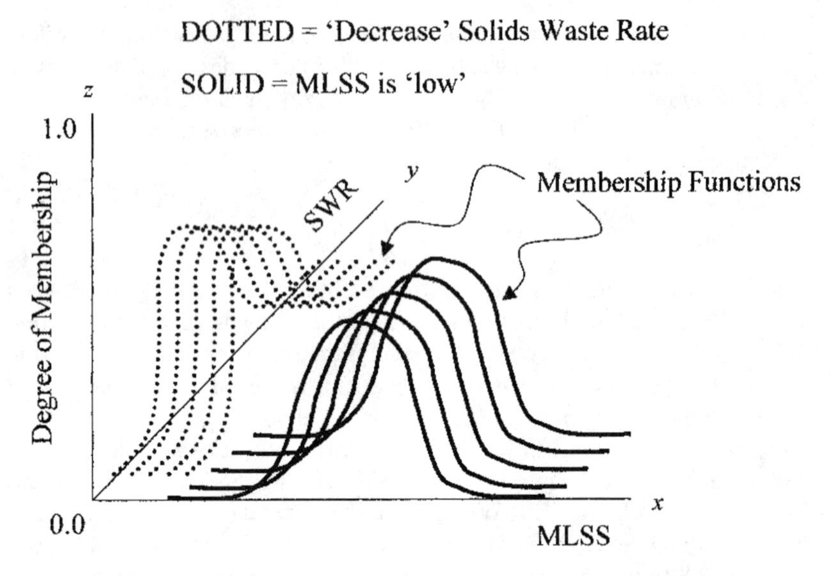


**Figure 5.7** Combining two fuzzy rules to develop a fuzzy algorithm.

is the SWR with the highest degree of relation, thus the *ridge* of this surface is of greatest interest. Looking at the surface directly from above, and drawing a single curve along the ridge of the surface results in a two-dimensional curve relating MLSS and SWR as specified by combining all the fuzzy rules that relate MLSS to SWR.

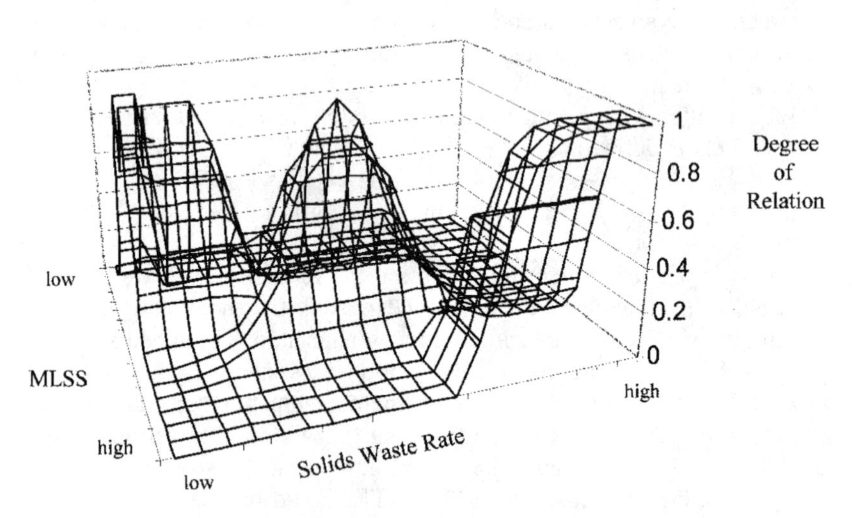


**Figure 5.8** Fuzzy algorithm relating MLSS concentration and solids waste rate.

Fuzzy rules can be manipulated in this way to define a relationship between any number of observed and manipulated variables. The relationships can be arbitrarily complex, depending only on the number of fuzzy variables and the definitions of the fuzzy membership functions. Of course, the effectiveness of the resulting fuzzy algorithm is dependent upon the correctness of the original rules. In contrast to the certainty factors method, fuzzy set methods tightly link linguistic and quantitative expressions of knowledge. To get a sense of the difference, consider how the certainty factor approach would handle the MLSS-SWR rule. In the CF approach, the rule CF would be combined with the belief that the antecedent is true to arrive at a CF that the SWR should be low. One would then have to decide how to adjust the SWR based on the level of certainty.

### Fuzzy Sets in Practice

Fuzzy set theory has caught on well in many industries. Fuzzy technology has been used in home appliances, elevator control systems and control of many other complex processes. Debates about the appropriateness of fuzzy set theory compared to probability methods continue, however, there are practical reasons for using the fuzzy approach. These include the simplicity of specification, the intuitive nature of the tools and the apparent success of fuzzy systems in practice, particularly in automatic control. Some vendors position fuzzy control as a simpler, more robust alternative to conventional PID control.

There are numerous references available on the topic of fuzzy logic, particularly in the areas of control, decision support and pattern recognition. Zadeh (1973) presents a clear description of the techniques in one of the original papers on fuzzy set theory and an early review can be found in (Tong, 1984). A number of applications are described in Mamdani & Gaines (1981) and more detailed discussions of theory are given in Kaufmann (1975) and Kosko (1992). A recent tutorial textbook presents the basics, several practical applications and the necessary software for studying simple implementations (Cox, 1994). Some recent research papers include papers on fuzzy control strategies for the activated sludge process by Yin and Stenstrom (1996), Tsai et al. (1996), and Spall & Cristion (1997) as well as a fuzzy pH controller presented by Menzl et al. (1996).

## MODELING

There are many different kinds of models as shown in Figure 5.3 and these are categorized in different ways. Static, or steady-state models are contrasted to dynamic models whose variables change with respect to some other dimension (usually space or time). Mechanistic models based on first principles are compared to empirical models that are derived instead from the information in raw data. Deterministic models that give predictable output for a set of inputs differ from stochastic models that have a probabilistic component in their output. Another useful perspective is shown at the bottom of Figure 5.3 in which models lie along a continuum from qualitative to quantitative. Several names given to these models at various points along the continuum are shown in the figure. This section stresses the mechanistic-empirical classification as a basis for description of math models because this feature often dictates the approach taken in automated analysis and control of engineered processes. One model technique of each type is described. Both can be central components of an intelligent system.

## MECHANISTIC

Dynamic mathematical models are ubiquitous in the literature. These are sets of differential equations that describe variation over time and space of important variables in physical, chemical or biological processes. Models of this type are dynamic, expressing the change of variables in time and space, but they can also be simplified to the steady state by setting the derivatives equal to zero. This was the approach taken in the past because until a few years ago it was difficult to solve differential equations. That is no longer the case and now an entire segment of the commercial environmental market is devoted to building and applying dynamic mathematical models. Using commercial tools, we can now build large math models quickly and then investigate the characteristics of these models through simulation. Advanced computing has been a great enabling tool for modelers. Only a few years ago it was argued that computing power was limiting what we could do with models, however this is no longer the case.

Figure 5.9 shows a simple math model, expressed in a convenient, standardized format recommended by the IAWQ's Task Group on Mathematical Modelling for Design and Operation of Biological Wastewater Treatment (Henze et al., 1987). This matrix is used to express the most important, or state, variables, and their relationships to other variables and model parameters. The first column shows the important processes, the last column presents the rate equations for each of these processes and the intervening columns show the stoichiometric coefficients for each of the state variables which are affected by the listed processes. Units are shown at the bottom and stoichiometric and kinetic parameters are defined in the lower left- and right-hand corners. To write the dynamic equation

Continuity

| Component → i / j Process ↓ | 1 Xb | 2 Ss | 3 So | process rate, $\rho_j$ $[ML^{-3}T^{-1}]$ |
|---|---|---|---|---|
| 1 Growth | 1 | -1/Y | $\frac{-(1-Y)}{Y}$ | $\frac{\mu Ss}{k+Ss} Xb$ |
| 2 Decay | -1 | | -1 | b Xb |
| Observed Rates | $\rho_j = \sum_j r_{ij} = \sum_j v_{ij}\rho_j$ | | | |
| | Biomass $[M(COD)L^{-3}]$ | Substrate $[M(COD)L^{-3}]$ | Oxygen $[M(-COD)L^{-3}]$ | |

mass balance

Stoichiometric Parameters:

Y = true growth yield

Kinetic Parameters

μ = maximum specific growth

K = half saturation constant

b = decay rate

$$\frac{dS_s}{dt} = input - output + \hat{\mu}\frac{S_s}{Y(K_s + S_s)}X_b$$

$$\frac{dX_b}{dt} = input - output + \hat{\mu}\frac{S_s}{K_s + S_s}X_b - bX_b$$

$$\frac{dS_0}{dt} = input - output + \left(\frac{-(1-Y)}{Y}\right)\left(\hat{\mu}\frac{S_s}{K_s + S_s}X_b\right) - bX_b$$

**Figure 5.9** Standard presentation for mathematical process models (Henze et al, 1987).

that describes the variation of a state variable, simply start at the top of the column for that state variable and, moving downward, multiply each stoichiometric coefficient (blank cells are assumed to be zero) by the corresponding process rate and sum these terms. Set this sum equal to the derivative as shown in Figure 5.9.

Many models developed for biological wastewater treatment are presented in this format. There is useful qualitative information in these matrices. For example, adding the coefficients in a given row of the table results in a sum that can be used to check continuity. This sum must be equal to zero for mass to be conserved. Also, it is clear in this example that the maximum growth rate that can be achieved is bounded at the value of $\hat{\mu}$.

### Mechanistic Models in Practice

Math models are powerful tools for analysis, design and operation of wastewater systems because they can closely represent real system dynamics. This type of model has been used to intelligently re-design plants and as a discovery tool for inventing new modifications that improve wastewater treatment. Mechanistic models are also useful in operations where they can be used in an open-loop forecasting or advisory mode and for closed-loop automatic control. The remaining chapters of this text examine applications that use mechanistic models in this way. In the case of automatic control, simplified models must be used. If the model is too complex, then it is not easy to ensure the representational correctness and maintain calibration of the model. The distinction between linear and non-linear model becomes important in addressing the calibration issue since there are very robust ways of handling calibration for non-linear models, for example, by using Kalman filters (Grum 1997).

## EMPIRICAL

Non-mechanistic approaches are often used in practice because they are ideally suited to situations where large quantities of data must be analyzed or when mechanistic models are not available or are difficult to calibrate on-line. A stochastic modeling approach was discussed in the previous edition of this text (Hiraoka & Fujiwara, 1992). These techniques can be extended through the use of methods that handle general non-linear models. One flexible, widely used empirical technique is described here in the context of modeling, namely the application of artificial neural networks (ANN).

## Neural Networks

A neural network is an information processing structure that can be thought of as a type of input-output model. Inspired by biological neural nets, they are composed of many interconnected units or neurons as shown in Figure 5.10. Each unit (the squares in Figure 5.10) consists of a summation function and an activation function as shown in Figure 5.11. A network can be configured for any number of inputs or outputs. Networks with the architecture shown in Figures 5.10 and 5.11, called layered feed-forward networks, are among the most commonly used in practice.

Inputs, either discrete or continuous values, enter the neural network and values calculated at each unit are propagated forward to the network outputs. Neural networks are non-linear systems, thus they are suitable for general modeling problems and have wider applicability than linear statistical models. As with other types of models, neural networks are data intensive, requiring sufficient data for a proper calibration.

Neural network training algorithms enable the network to learn input/output relationships by adaptive changes to network connections, unit characteristics and the weights (W in Figure 5.11), or coefficients between units which quantify the effect of one unit on another. One widely used method for training layered feed-forward networks is back propagation. In this method network outputs are compared to target values (see Figure 5.10) and the error is used to make changes in network weights. Back propagation is based on the generalized delta rule, which defines an optimization method for determining the values of the weights. Since the first description of this rule in 1986 (Rumelhart et al. 1986), there have been many speed and efficiency improvements in training algorithms. Commer-

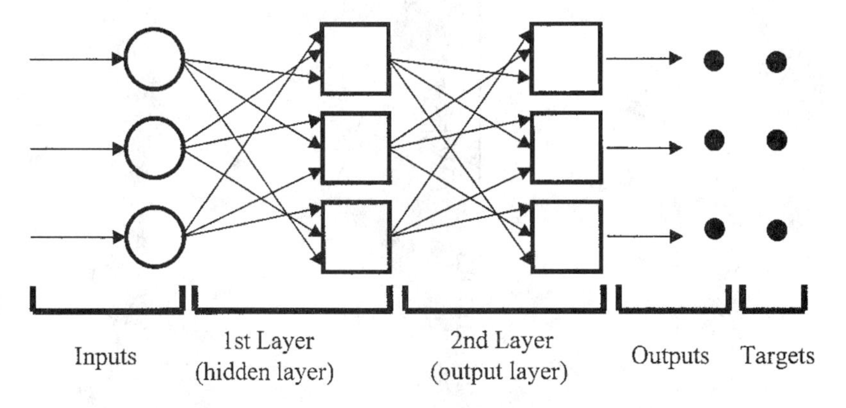


**Figure 5.10** A two-layered feed-forward neural network (3 inputs and 3 outputs).

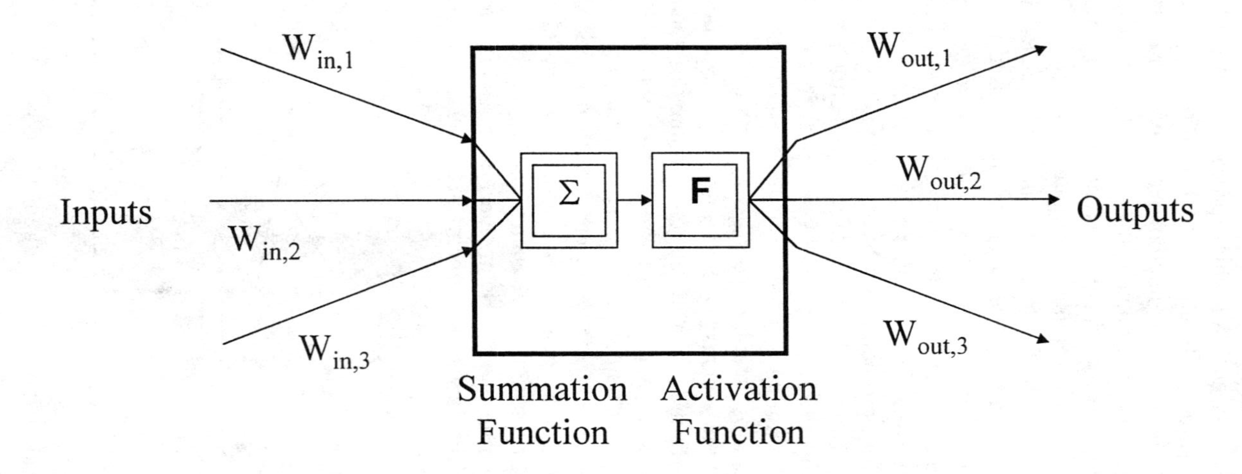


**Figure 5.11** Components of a unit (neuron).

cial packages implement these algorithms and provide necessary auxiliary tools to help in configuring and deploying neural networks.

In wastewater systems modeling, most neural networks described in the literature are function mapping networks, however, other types of neural networks also are used. The mapping problem is to implement an input-output functional relationship by presenting the network with inputs and target outputs and, by applying a learning algorithm, teach the net the relationship. Inputs and outputs can be continuous variables, e.g., air flow rate, temperature, etc., or discrete binary, e.g., ''the scrubber is on,'' ''component concentration is high,'' etc. Back propagation networks, like that shown in Figure 5.10 are often used for function mapping problems.

Modifying the learning procedures and the network architecture shown in Figures 5.10 and 5.11 results in other types of neural network. Rho networks are good for learning multivariate probability density functions from process data. This makes Rho nets ideal for single-class membership problems, that is, detecting whether a specific data pattern describes a state that is a member of a certain class. For example, a Rho net can be trained with normal operating data and then used to detect when a process is not in a normal state. Radial basis function networks (RBFN) extend these techniques to solution of multiple-class membership problems such as distinguishing between toxic, organic or hydraulic disturbance events. Auto-associative networks (AAN) operate as a type of signal processor that is effective for learning correlations between variables. In training this type of network, the inputs and outputs are the same data sets and the net learns correlations between data. An architectural restriction prevents over-learning during training, which results in minimization of the sum of the squared differences between targets and networks predictions. Once trained, an AAN can be used to filter the noisy data or—since the net knows the correlations between input data—to substitute for one of the missing input variable values.

## Neural Networks in Practice

Neural networks have been used extensively in industry for modeling and control of dynamic processes. For example, oil and gas companies use neural networks to model compositions of streams from a crude oil flash tower unit, for model-based sensor validation and to control pipeline operations. Neural networks are flexible, adaptable and capable of representing any kind of relationship between inputs and outputs. As a general-purpose modeling tool, this technique finds application in many practical situations. Neural networks have been used to

- recognize faults by applying learned fault models

- forecast disturbances using learned dynamic models
- combine data from redundant sensors to give a more reliable sensor reading
- perform statistical quality control
- adaptively control non-linear processes
- adaptively tune other kinds of process controllers

As suggested by this list, many applications that apply neural networks also make use of other tools. Hybrid techniques that apply combinations of fuzzy or stochastic reasoning, mechanistic process models as well as neural networks are common.

## OPTIMIZATION

Optimization in this chapter, refers to the situation where one already has a model of a system and is interested in using the model to answer the question; What model inputs will result in a desired model output or behavior? In contrast to investigatory, or 'what-if', simulations, this kind of analysis suggests a need to 'run the model backward' to find out what inputs would be needed to obtain some desired output behavior. The desired behavior is obtained when important model indicator variables are maximized or minimized. These variables are represented in an equation called an objective function. For example, we might ask; what combination of influent liquid flow rate, airflow rate and solids residence time is needed to achieve an effluent BOD of 5 mg/L? When the model has no simple solution and many model inputs are permitted to vary simultaneously, this amounts to more than solution of a set of algebraic equations.

An 'if-what' analysis requires special tools—tools which can be difficult to develop and are certainly more time-consuming to run. The most useful tools are those designed to handle constrained optimization problems. These are problems in which one requires both a desired output and imposition of restrictions on the range of inputs that can be tried or on the way the solution is developed.

This section will describe two relatively new techniques developed for complex constrained optimization problems, namely, genetic algorithms and agent-based optimization. Genetic algorithms are widely used in optimization because they are robust and flexible. They are also interesting since they are inspired by the mechanisms of natural selection and natural genetics, which is much different from the approach taken in traditional mathematical optimization. In this section, genetic algorithms are discussed from the perspective of dynamic scheduling problems followed by an example demonstrating important concepts. Genetic algorithms work well for many problems, however, in some cases, the model and the expression of the objective function become unwieldy such that a single program can not be prepared to find the optimum. In these cases it becomes necessary to consider distributing the knowledge among several collaborating entities called agents. The method of collaboration causes an optimum to emerge from the interaction of agents. Agent-based optimization has shown considerable promise for complicated optimization problems that can not be handled with either tradition methods or genetic algorithms.

## GENETIC ALGORITHMS

Genetic algorithms (GA) are a relatively new approach to optimization (Goldberg, 1989). First described by John Holland in the 1970's, genetic algorithms have been applied in many kinds of optimization problems including dynamic scheduling of industrial processes to meet product specifications under resource constraints and control of gas pipelines to maintain pressures within a pre-determined range while minimizing the total power consumed.

Genetic algorithms are search algorithms that are inspired by the mechanisms that control natural selection. The search is directed through selection of candidate solutions that are the 'best' as defined by an objective function. The most powerful aspect of genetic algorithms is the method of building candidate solutions through a combination of structured information exchange that is also somewhat random.

To understand this more clearly we will examine how genetic algorithms could be used to solve a common problem that occurs in a production facility, including a wastewater treatment plant. The problem is development of a production schedule. This typically involves the sequencing of production activities over time to meet certain requirements. For example, in wastewater systems it may be necessary to decide which treatment trains to operate in which sequence based on available manpower and expectations for incoming flow and load. In a collection systems it would be useful to know how to route flows to take advantage of pipe volume and reduce loads on the treatment plant. The difficulty solving this type of problem varies widely, depending on the particulars of the application but these can be divided into two classes:

- expert scheduling applications: In this type of application, a scheduling expert (someone on staff) is able to generate a schedule, but it requires a lot of the expert's time to prepare the schedule, so only a few alternatives can be considered. In these cases it may be difficult for operators to modify the schedule when conditions change and the expert is not available. This type of application can use rules to express the way the expert prepares the schedule.

- complex constraint applications: In this type of application, complex relationships exist between the sequence of activities and the resources required to execute the sequence. In principal these aspects are mutually dependent. We can't decide which is the best sequence, or even if a sequence is feasible, without knowing the loading of resources that the sequence will induce. Conversely, we can't say anything useful about resource loading unless we know the sequence of activities that will consume those resources.

In both cases, the key to finding a solution is having a way to express the pertinent objects (unit processes, etc.), variables and constraints, and a method for creating alternative combinations of object-variable values that will lead to an optimum. The ideal method would search among these different possibilities, generating new, candidate solutions and direct the search towards an optimum. The optimum solution is the combination of object-variable values that gives the desired model behavior and output.

Figure 5.12 shows how an initial solution to a scheduling problem is converted into a better solution using a genetic search algorithm. The procedure starts with a string, roughly analogous to a chromosome, which encodes one candidate solution to the problem. For example, in the case of a production schedule, a single chromosome represents operation of

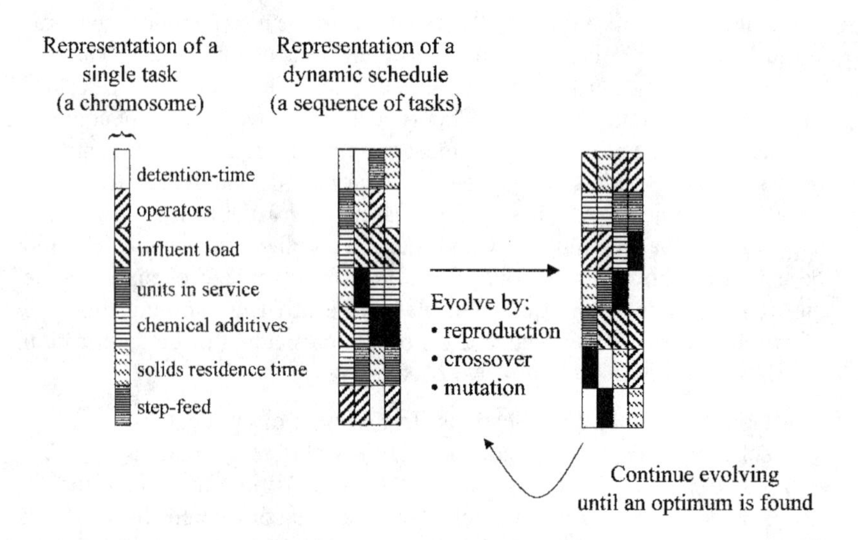


**Figure 5.12** Finding an optimum schedule with a genetic algorithm.

the facility with a certain load, number of reactors on-line, hydraulic detention time, etc. If the schedule is dynamic, then each string represents a single task and the dynamic schedule—defined as a sequence of tasks—is represented by a sequence of chromosomes. Each candidate solution has an associated fitness that is determined by how close these conditions come to the desired result as measured by an objective function. A typical objective function might include the error between desired and predicted treatment efficiencies, the cost of energy (pumping, aerations, etc.) and the cost of chemicals.

Encoding new candidate solutions creates new generations. New candidates are constructed in three ways. Reproduction is the process of copying chromosomes in proportion to their objective function values. Strings that have high (i.e., better) objective function values are copied more than strings with low values. This process implements a kind of 'artificial' natural selection. Strings that have been successful at meeting the goal have a better chance of being propagated across generations. Crossover is a mating of two strings resulting in new generations that have some variable-values from one string and the balance from another string. Crossover implements a process of combining ideas that tends to result, in the long run, in improved candidates. As you can imagine, there are various modifications to these simple ideas that can be implemented, however, much of the power of genetic algorithms derives from these two simple processes for generating new candidate solutions.

The third process is mutation. Mutation is the process of modifying strings in a random way, for example, by inserting an un-tried variable-value or combination. As in natural systems, mutation is not always good and often results in very bad strings, but it does ensure diversity in the population of strings and can make a difference when existing strings and their combinations are not contributing to a better solution. Mutation in genetic algorithms is used sparingly to avoid premature loss of good strings.

A simple example, demonstrating development of a single generation in a genetic algorithm, will serve to clarify the important concepts.[1] Consider a problem that has four different variables and that each of these variables can have one of two different values. Then assume that the problem is to discover which combination of values is the 'best' based on some criteria that we also define. For example, in a wastewater treatment plant, imagine that the airflow to each of four different locations (''Quads'') along an aeration basin from inlet to exit can be set on either low or high. What is the best combination of these airflows that results in optimum treatment?

[1] This example is based on one described in greater detail in the textbook by Goldberg (1989).

For this problem, we can represent each possibility, or candidate solution, with a simple string;

> ''airflow in Quad 1 (inlet), airflow in Quad 2, airflow in Quad 3, airflow in Quad 4 (outlet)''

and code the values for each airflow using either a one or zero (since there are only two possibilities). A one (1) means that the airflow is set on high and a zero (0) means that the air flow is set on low. Thus, one possibility would be;

1010

Corresponding to the situation with the airflow on high in Quads 1 and 3, and on low in Quads 2 and 4. Many other possibilities can be defined and in fact we know that there are exactly 16 different possibilities ($2^4$). Notice also that we can use this 'binary code' to represent integers from 0 to 15.

Next, define what will be the 'best' solution from among the 16 different possibilities. Imagine that the best solution is coded by '1000', where the Quad 1 airflow is set on high and the remaining Quad airflows are set on low. This is reasonable as we would expect that the highest airflow rate is needed where the load is highest, that is, near the influent end of the aeration basin. Note that if we interpret the string above as a binary number, its value in base 10 is '8' ($1 \times 2^3 + 0 \times 2^2 + 0 \times 2^1 + 0 \times 2^0$). Define the objective function as;

$$F(x) = 8 + (x - 8), \qquad \text{for} \quad x < 8 \tag{6.12}$$

$$F(x) = 8 - (x - 8), \qquad \text{for} \quad x \geq 8 \tag{6.13}$$

where $x$ = the Base 10 value of the binary encoded string.
This objective function definition results in the highest value being assigned to the string ''1000'' with successively lower values as the Base 10 value moves away from this value as shown in Figure 5.13. In an actual situation we would want to define an objective function in terms of the treatment that is obtained, for example, using the values of effluent BOD resulting from treatment with a given aeration pattern. This completes our definition of the representation for this problem.

Now we can use the methods of genetic algorithms to find the optimum solution. First, start by creating five candidate solutions at random (objective function values are shown in parentheses):

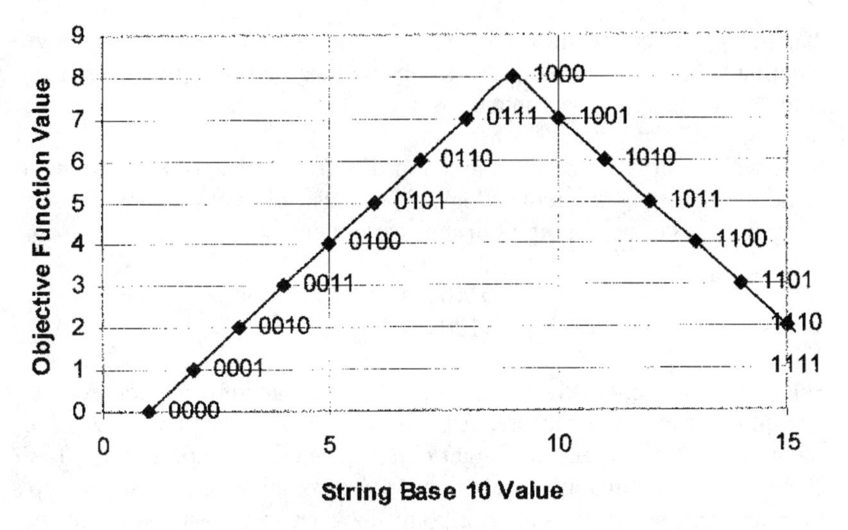


**Figure 5.13** Objective function values for each candidate string.

0110 (6)
1100 (4)
0010 (2)
0001 (1)

The total of all the objective function values is 13 and the percentages of this total are 46%, 31%, 15% and 8%, respectively for each string. First, we allow each of these strings an opportunity to reproduce. If we allow those with the highest objective value the biggest opportunity to reproduce, then we would expect that the ''0110'' string—being the 'fittest' of the group—would have the greatest chance of participating in reproduction. If we select from this initial population of strings based on these probabilities we might get;

0110
0110
1100
0010

You can experiment with this on your own using two decks of playing cards. Mark 46 cards with a certain color, 31 cards with a different color, 15 cards with yet another color and 8 cards with a fourth color (100 cards total). Shuffle the cards and then select one from the deck to choose a string. Return the card to the deck, shuffle and select another card. Note

that in the selections shown here, the string with the lowest objective function value is no longer in the group and that there are now two strings with the high objective function value. This new set of strings is the 'mating pool' for the next generation.
Next we use crossover to combine strings from the new population to make the next generation of strings. To do this, select at random two sets of 'mates' from the mating pool. For example:

$$0110 \times 0110$$
$$1100 \times 0010$$

Selecting a string and then selecting a mate at random from one of the remaining strings created this list. Of course, there are many ways to accomplish this step and we might obtain much more variety if we used 20 strings in our population rather than 4. To complete the crossover step, for each pair, select at random a point between two positions along the string and swap the values on either side of the point. Using the strings above (objective function values for the new generation are shown in parentheses);

$$0|110 \times 0|110 \quad \Rightarrow \quad 0110(6) \text{ and } 0110(6)$$
$$1|100 \times 0|010 \quad \Rightarrow \quad 1010(6) \text{ and } 0010(2)$$

Note that the total objective function value for the new generation is now 20, compared to the original total of 13. Also, a string now exists with a '1' in the first position, which may eventually lead to the solution that we know is the best, that is, 1000. The last step is mutation. Typically, this is imposed on only a small fraction of the total population, so we will allow mutation to only one string at one position, both selected at random. For example, mutating the third position of the second string that resulted from the first mating pair shown above results in:

$$0110 \Rightarrow 0100(4)$$

This mutation actually lowers the objective function value for this string from 6 to 4, but notice that it also creates a string in the new generation that has a '0' in the third position. Before the mutation, there was no such string. This new, mutated string could have a positive impact on future generations, perhaps hastening our progress to finding the best string, which also has a '0' in the third position, that is 1000.

This simple example could be easily solved using other optimization techniques. Still, it highlights the key strengths of genetic algorithms, including generality and robustness. Any constrained optimization can be represented in such a way as to allow solution with a genetic algorithm. Once the representation is set, a simple set of procedures can be prepared that are effective for a wide range of optimizations. Moreover, genetic algorithms are very effective at dealing with difficulties that arise in non-linear optimization due to by 'poorly behaved' objective functions that have deep valleys, numerous peaks, ridges, steps and random variations.

### Genetic Algorithms in Practice

The robustness and flexibility properties of genetic algorithms have made them a popular optimization method. In industry, these techniques are often used in facility-wide or enterprise-wide resource planning. In water distribution, genetic algorithms have been used very successfully to reconfigure a pipe network resulting in savings of 50% over the original design (Frey et al. 1996). One author has reported on the use of a genetic algorithm to find an optimal solution to the problem of allocating financial resources in a utility (Miller 1997).

The problems that arise in application of genetic algorithms are the same as those which plague the type of complex, constrained optimization described in the introduction to this section. As you have noticed from the example, it may take many generations to obtain a good solution and for large problems it may take thousands of generations. For problems in which the number and size of 'strings' is large, the computational load is immense, prohibiting the practical application of this approach. In these cases the problem is subdivided in a way that permits the use of a genetic algorithm approach.

## AGENTS

An agent is anyone or anything that acts as a representative for another party for the purpose of performing tasks that are requested by, or beneficial to the represented party. A software agent is a computer program that performs these tasks within a computing environment. However, with this definition any program we currently use qualifies as an agent, so in addition to these characteristics, we add three more, namely, intelligent, autonomous and mobile.

Agents can perform many tasks, but in the area of intelligent process control, one very important task is optimization. A key motivation for using agents is reduction of application complexity. Eventually, optimization problems become so large that it is impractical to employ a single, mono-lithic routine to evaluate the optimum. In wastewater engineering we expect that agent technology will find use with networks of individually automated treatment plants, pumping stations and collection systems. An example of this type of system is given in Chapter 8 of this text.

The following material summarizes the work of Nissen (1995) published on the Internet. Additional information on this topic can be found in Maes (1995) and a Special Issue of *Communications of the ACM* (1994). Nissen lists five key characteristics that differentiate agents from other types of software applications. These are

(1) Autonomy: Agents take actions that lead to completion of a task without being requested to do so. Agents act on their own and are motivated to move across a network in an effort to seek out the information needed to complete a task.
(2) Communication: Agents must be able to obtain data and 'talk with' other programs in order to get the information needed to complete a task. This assumes the existence of standard ways of communicating with external databases and other sources of information.
(3) Cooperation: Communication can be one-way, intrusive, obstructive or even destructive. Agents have the ability to work in consort with other agents and programs to achieve goals in a way that is mutually beneficial.
(4) Intelligence: Agents have their own complex behaviors. Rules, objects and procedures represent knowledge that impart qualities to agents that we regard as intelligent.
(5) Adaptive: Agents can examine the external environment and then adapt their actions to improve the probability of successfully achieving their goals.

Agents make use of a distributed model for computer applications as shown in Figure 5.14. In this model the agent is a specialized program that is initiated by a client who has requested a service. Prior to the development of agents, this action resulted in a remote procedure call (RPC) to a procedure or method running on a machine somewhere else on the network. Remote procedure calls typically involve a significant overhead in order to complete a transaction with the server. In contrast, an agent program does not initially exist on the distant server machine. In the agent *remote programming* paradigm, the client computer is permitted both to call procedures on the server and to supply procedures to be executed, that is, instructions about how to satisfy the client's needs. Combine this technique with the features listed above and the result is a very flexible technique for distributing intelligence across a network. One important capability that results is the ability to set up negotiation channels among agents with the purpose of creating small *economies* that can be directed towards optimization.

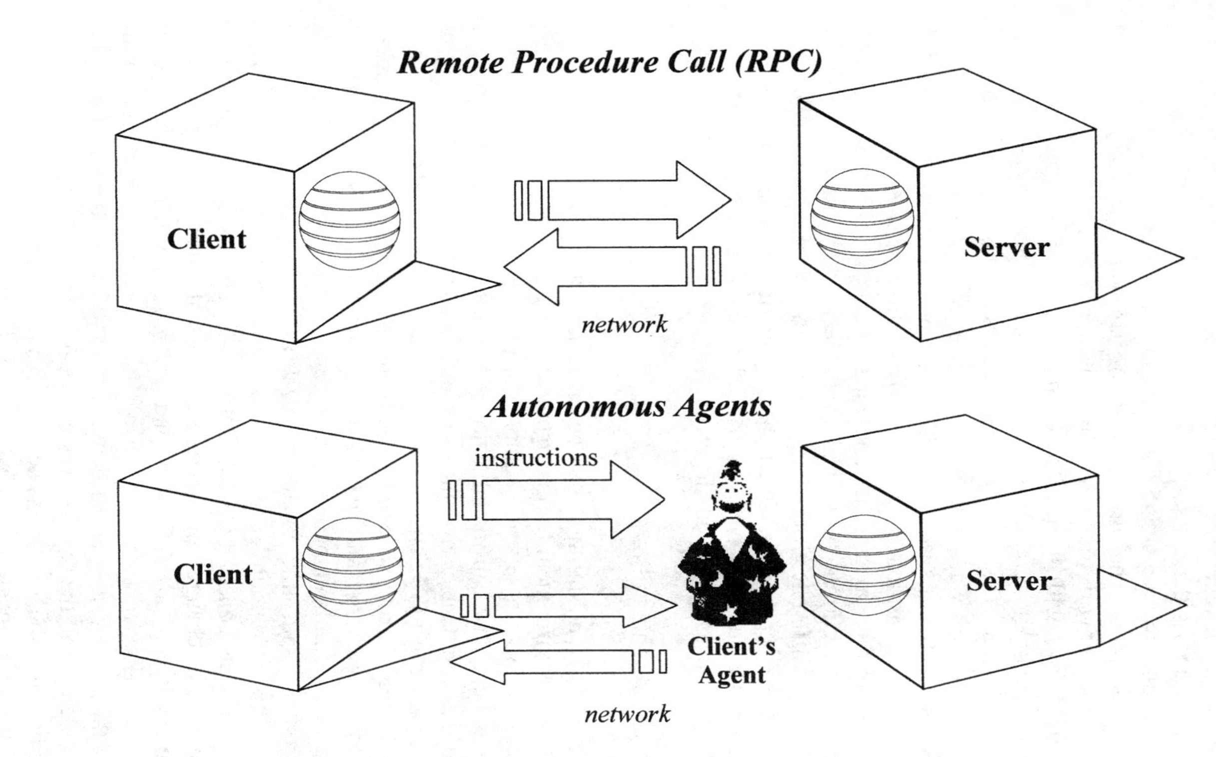


**Figure 5.14** Agents distribute processing for higher performance.

The remote programming paradigm using agents offers two key advantages, namely, performance and customization. The need for greater performance in large optimization problems has motivated the development of agent technology. Since the work of an agent is distributed across the network, it operates more efficiently than a single program running on a single client machine collecting data from various server machines. The proliferation of software applications in any information network, including those used in automation of wastewater systems, brings with it an increase in the need to consider interactions between applications and the way in which resources are allocated to each application. Distributed agents deal with this type of problem in a very efficient way. For example, we can envision a wastewater collection-treatment system in which treatment operations are scheduled in an optimal fashion based on a negotiation between unit processes that indicate their availability and capacity and a collection system that reports on its current flows and available hold-up capacity. This may not be too different from what is done using conventional methods, however the manner in which the negotiating is conducted, the way in which the optimum *emerges from* the negotiating and the efficiency of whole process is much different.

### Intelligent Agents in Practice

Ohtsuki, et al. in this text describe a system that implements an agent-type environment for management of package wastewater treatment systems. There are several examples of prototype and pilot agent-based systems being used in manufacturing. Several companies market agent-based development environments. Work funded by the National Center for Manufacturing Sciences involves a pilot test of a complex, multi-plant manufacturing scheduling problem that is much like the problem of allocating energy and manpower resources among several operating wastewater treatment plants in a regional utility. Solution to this type of problem with traditional optimization approaches is not feasible, because the ''solution'' can not be obtained in real time or is sensitive to a particular constraint, such that a small change in one plant could cause radical changes in the schedules of all plants. The agent-based approach is not attempting to find an optimal schedule. Instead, the plants are allowed to operate in a ''free enterprise'' fashion, learning about the buying interest (or demand for service) of the assembly plant, and bidding on the work. Agents located in each plant make local decisions and the overall system of agents operates like an economy. With the right incentives, it works well.

## CONTROL

The goal of conventional regulatory closed-loop control is the maintenance of operating variables at a set point. Existing control systems do this quite well. But, keeping in mind the components of a simple loop from sensor to controller to actuator, consider the various reasons why this might not be adequate. What happens to the system under control if

- The sensor readings are not correct because of failure such as drift or deterioration.
- The sensors don't provide enough information about changes in important process variables.
- The model dictating how to maintain the setpoint is not valid under all conditions or is hard to tune or calibrate.
- The setpoints have to change based on the controlled-system inputs or due to interactions with other systems.
- One control loop interferes with the operation of another loop because of interactions and interdependencies.
- An actuator fails or changes too fast.

Moreover, what is the appropriate way to respond to each of these problems? A persistent problem that occurs in regulatory control is that as the amount of process automation increases, the operator is faced with the need to make increasingly difficult decisions, especially when an unusual or abnormal event occurs.

Real-world control systems are always wrapped with a layer of specialized rules and procedures designed to address these issues. For example, levels of automated control are often defined, with procedures for moving between automatic and manual control when a sensor fails. Other problems arise even if all control loops and fail-safe measures are operating properly. For example, distributed control systems typically include the ability to generate alarms when a component fails, however, multiple alarms are generated in cases where several components monitored along a causal chain from the root cause to the ultimate effect. These sympathetic alarms result in operator confusion and more problems in identifying the actual cause of a process fault. To avoid this confusion, it is necessary to implement an intelligent alarming facility—often based on specialized algorithms or rules.

Plants that want to extend conventional control and sequencing operations to a higher level need to apply the knowledge of experts to predict, control and optimize processes. Because of the interdependence of unit processes in a plant, this all needs to take place in a distributed, networked environment where computers communicate information about each of the unit operations. The task of representing and distributing intelligence across an enterprise is difficult and this explains some of the early disappointments with intelligent systems implementations.

Optimized management is the ultimate goal of automating the enterprise. To achieve this, the technologies discussed in the sections above must be combined to realize implementation of intelligent systems. Figure 5.15 is a re-drawing of Figure 5.2 that shows important information and decision flows. Information flows from the left upward towards the center while decisions flow downward and to the right. Knowledge of the plant system and how to operate it is incorporated in the various modules of this figure. Control can be open loop, in which case a human operator or manager interrupts the information or decision flows. Control can also be closed loop with a tight link from the sensing instrument through the various blocks to the mechanical actuators that manipulate important plant system inputs. Each of the systems described in the remaining chapters of this text follows this framework for construction of an intelligent system.

Conventional control as commonly used today is designed primarily to regulate process variables, not implement business strategies. Defining and implementing this strategy is a difficult problem that requires the tools and techniques described throughout this text.

## SUMMARY

This chapter reviews commonly used techniques for implementing intelligent decision support systems for real-time process control. Representation of knowledge is a key theme as each of the techniques has as its goal the implementation of knowledge in a computing machine so that it can be reused for the purposes of optimizing management of a complex operation. Models are used to formalize this knowledge and practitioners now have available a number of tools for creating these models. They range from the highly quantitative, deterministic models to qualitative, fuzzy models that behave a lot like human experts. Adaptive models are an important contribution of artificial intelligence. Neural networks help to extract knowledge from data and learn about current plant conditions based on this knowledge. Finally, the representational richness that these tools provide, enable development of systems for optimization. Once we know how processes respond to inputs and define strategies to operate processes, we can then search for an optimum operating strategy, that is, we can develop automated operations management systems. Optimization techniques like genetic algorithms permit this to be done in a robust way. One interesting technique that uses software agents doesn't specifically search for an optimum. Instead, an optimum emerges from a collaboration/negotiation process that acts much like a socioeconomic system.

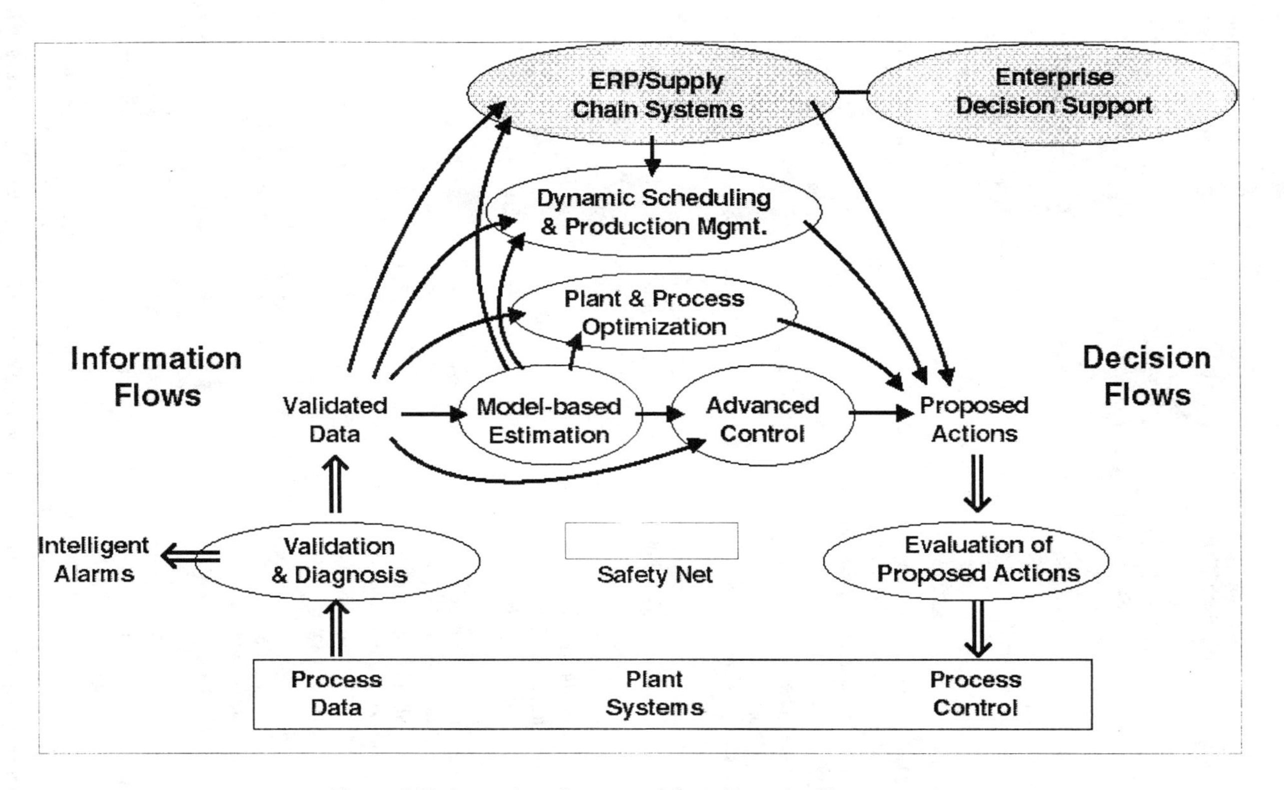


**Figure 5.15** Integration of many modules achieves intelligent control.

## REFERENCES

Beck, M. B., "Modelling and Operational Control of the Activated Sludge Process in Wastewater Treatment," International Institute for Applied Systems Analysis, Laxenburg, Austria, November 1978.

Buchanan, B. G. and E. H. Shortliffe. 1984. *Rule-Based Expert Systems, the MYCIN Experiments of the Stanford Heuristic Programming Project.* Addison-Wesley Publishing Company, Reading Massachusetts.

Charniak, Eugene, McDermott, D. V., *Introduction to Artificial Intelligence,* Addison-Wesley, 1987.

Clocksin, W. F., Mellish, C. S., *Programming in PROLOG,* Springer-Verlag, 1984.

Communications of the ACM. 1994, "Intelligent Agents," Special issue of *Communications of the ACM,* Vol. 37, No. 7; July.

Cox, Earl. 1994, *The Fuzzy Systems Handbook: A Practitioner's Guide to Building, Using, and Maintaining Fuzzy Systems,* Academic Press, Boston, MA.

Davis R., "Knowledge Based Systems," *Science,* 23:957–963, 1986.

deKleer, J., & J. S. Brown, "A Qualitative Physics Based on Confluences," in *Qualitative Reasoning about Physical Systems,* D. G. Bobrow, Ed., MIT Press, Cambridge Mass., 1985.

Dym, C. L. and R. E. Levitt. 1991. *Knowledge Based Systems in Engineering.* McGraw-Hill, Inc.

Fowler, Martin & Scott, Kendall. 1997, *UML Distilled: Applying the Standard Object Modeling Language,* Addison-Wesley, Reading, Massachusetts.

Frey, J. P., Simpson, A. R., Dandy, G. C., Murphy, L. J. and Farrill, T. W. 1996. "Genetic algorithm pipe network optimization: the next generation in distribution system analysis," *Public Works,* 127(7):39–42.

Geselbracht, J. J., E. D. Brill, Jr. and J. T. Pfeffer. 1988. "Rule-Based Model of Design Judgement About Sludge Bulking," *Journal of Environmental Engineering,* Vol. 114, No. 1.

Goldberg, D. E. 1989, *Genetic Algorithms in Search, Optimization, and Machine Learning,* Addison-Wesley Publishing Company, Inc., Reading, Massachusetts.

Gorry, G. A. and G. O. Barnett. 1968. "Experience with a Model of Sequential Diagnosis," *Computers and Biomedical Research,* Vol. 1, pp. 490–507.

Grum, M. 1997, *WATERMATEX '97: A Symposium Review,* Water Quality International, September/October, International Association on Water Quality, London, England.

Henze, M. Grady, C. P. L. Jr., Gujer, W. Marais, G. V. R. & Matsuo, T. 1987, *Scientific and Technical Reports No. 1—Activated Sludge Model No. 1,* International Association on Water Quality, London, England.

Hiraoka, M. & Fujiwara, M. E. 1992, "The Use of Time Series Analysis in Hierarchical Control Systems," in *Dynamics and Control of the Activated Sludge Process,* J. F. Andrews, Ed., Technomic Publishing Co., Inc. Lancaster, PA.

Instrument Society of America. 1995, "S88.01: Batch Control Part 1: Models and Terminology—ANSI/ISA-1995," ISA.

Intelligent Agents: A Technology and Business Application Analysis Mark Nissen. November 30, 1995. See http://haas.berkeley.edu/~heilmann/agents

Kaufmann, A. 1975. *Introduction to the Theory of Fuzzy Subsets—Volume I: Fundamental Theoretical Elements.* Academic Press Inc., Orlando, FL.

Kosko, B. 1992, *Neural Networks and Fuzzy Systems,* Prentice Hall, Englewood Cliffs, NJ.

Maes, Pattie. 1995, "Intelligent Software," *Scientific American,* Vol. 273, No. 3, pp. 84–86.

Mandami, E. H. and B. R. Gaines, eds. 1981. *Fuzzy Reasoning and Its Application.* Academic Press, Orlando FL.

Menzl, S. Stuhler, M., & Benz, R. 1996, "A Self-Adaptive Computer-Based pH Measurement and Fuzzy-Control System," *Water Res.* (G.B.), 30:981.

Miller, C. R. 1997, "An Integrated Approach to Financial Resource Planning Using an Optimization Model," presented at *Proceedings for the Specialty Conference of the Water Environment Federation, Computer Technologies for the Competitive Utility,* Philadelphia, PA, June 15–18, Water Environment Federation, Alexandria, Virginia.

Nissen, Mark. 1995, "Agents to Increase the Intelligence of the Intelligent Hub"; presented at the December 1995 CommerceNet CALS Working Group Meeting, in Sunnyvale, CA.

Quatrani, Terry. 1998, *Visual Modeling with Rational Rose and UML,* Addison-Wesley, Reading, Massachusetts.

Rumelhart, D. E., Hinton, G. E. & Williams, R. J. (1986), "Learning Internal Representations by Error Propagation," in D. E. Rumelhart & J. L. McClelland, eds., *Parallel Distributed Processing: Explorations in the Microstructure of Cognition., Vol. 1., Foundations,* Bradford Books/MIT Press, Cambridge, Massachusetts.

Spall, J. C. & Cristion, J. A. 1997, "Neural Network Controller for Systems with Unmodeled Dynamics with Applicatoins to Wastewater Treatment," *IEEE Transactions on Systems, Man and Cybernetics—Part B. Cybernetics,* Vol. 27, No. 3, pp. 369–375.

Stanley, G. M. and Vaidhyanathan, R., "A Generic Fault Propagation Modeling Approach to On-Line Diagnosis and Event Correlation," 3rd IFAC Workshop on On-Line Fault Detection and Supervision in the Chemical Process Industries, Solaize, France, June 4–5, 1998.

Taylor, David A. 1995, *Business Engineering with Object Technology,* John Wiley & Sons, Inc. New York.

Tong, R. M. 1984, "Retrospective View of Fuzzy Control Systems," *Fuzzy Sets Sys.,* Vol. 14, pp. 199–210.

Tsai, Y.-P., Quyang, C.-F., Wu, M.-Y., & Chiang, W.-L. 1996, "Effluent Suspended Solid Control of Activated Sludge Process by Fuzzy Control Approach," *Water Environ. Res.* 68:1045.

Winston, Patrick H., *Artificial Intelligence,* 2nd Ed., Addison-Wesley, 1984.

Yin, M. & Stenstrom M. K. 1996, "Fuzzy Logic Process Control of HPO-AS Process," *J. Environ. Eng.,* 121:484.

Zadah, L. 1973. "Outline of a New Approach to the Analysis of Complex Systems and Decision Processes," *IEEE Transactions on Systems, Man and Cybernetics,* SMC-3, pp. 28–44.

# PART II: APPLICATIONS

CHAPTER 6

# Dynamics and Control of Urban Drainage Systems

## INTRODUCTION TO CONTROL OF SEWER NETWORKS

### BACKGROUND

WASTEWATER collection and conveyance systems represent a crucial part of the urban infrastructure. The cost and complexity of these systems is great, especially in highly urbanized areas. Managers, engineers, and operators of these systems are faced with difficult problems related to the operation and maintenance of their facilities. In addition to the issues related to the operation and upkeep of the system, many sewerage agencies are facing increasing public concern about the environmental impact of Combined Sewer Overflows (CSO's) and local flooding. In many instances, these operational challenges need to be faced in an atmosphere of limited resources and fiscal pressures to "achieve more with less". Often, fiscal pressures are accompanied by concurrent increase in performance requirements from the regulatory agencies.

The design practices for sewer networks are usually conservative and include significant safety factors. As sewerage network design does not normally include consideration of real time control, there are often opportunities to optimize the utilization of the existing system through operational strategies.

The problem of sewage spills and local flooding has traditionally been addressed by large scale capital improvement programs that focus on construction alternatives such as sewer separation or construction of storage

Z. Cello Vitasovic and Siping Zhou, Reid Crowther Consulting, Inc., 155 NE 100th, Suite 301, Seattle, WA 98125.

facilities. The cost of such projects is often high, especially in older communities where the population density and the value of land is high. In the last few years, Real Time Control (RTC) of conveyance systems has been emerging as an attractive alternative. Although there are still only a few documented implementations of RTC to large urban sewerage systems, this technology has been successfully implemented to large urban systems.

The basic concept behind RTC of sewer systems is fairly straightforward: the conveyance system is controlled in real time with the objective of optimizing the utilization of in-line storage available within the system. The cost of the control system is often a fraction of the cost required for alternatives that include construction of new storage facilities. Such savings were clearly demonstrated in the application of RTC to a large urban system in Seattle, Washington. This model-referenced automatic control system (Vitasovic et al., 1990) is presented as a case study later in this chapter.

## COMPONENTS OF URBAN DRAINAGE SYSTEMS

Urban drainage systems normally include the following components:

(1) *Drainage basins:* the overall urban drainage system is normally subdivided into a number of smaller geographic areas. These areas include open spaces and residential, commercial, and industrial developments. Drainage basins include many small conduits such as house connections. This part of the urban drainage system is usually modeled using hydrological models, which use rainfall as main input and produce flow hydrographs out of the basin as main output. These models take into account general features of the area, such as ground slope or permeability, but include only a simplified view of hydraulics.

(2) The *network* includes many pipes (conduits) which convey the wastewater to treatment facilities. The smaller, local systems normally feed into larger pipes which are often called interceptors. Older cities and communities often have combined systems, using a single network to convey both sanitary wastewater and stormwater. Most of the recently designed systems separate the wastewater network from the stormwater network.

(3) *Static control elements,* such as weirs, are fixed structures which regulate the flow through the network but are not adjustable in real time.

(4) *Dynamic control elements* can be manipulated in real time; these include gates, valves, inflatable dams, and pumps.

(5) *Instrumentation* in sewer networks normally includes rain gauges,

level sensors, and flow meters. Many urban drainage systems are also equipped with Supervisory Control and Data Acquisition (SCADA) systems, which provide connectivity between different points of the network and allow for remote monitoring and control. Use of radar for the purpose of measuring or predicting rainfall is emerging but is not yet widely used (refs).

The development of an automatic control system for a large urban drainage network includes several different aspects of engineering and requires not only knowledge of hydrology and hydraulics, but also an understanding of modeling, control theory, software issues, pumping stations, and SCADA products. Finally, for successful implementation of such systems we must consider the non-technical, organizational aspects (''people issues'') and ensure that the system will be accepted by the operations staff.

## LEVELS (MODES) OF RTC

Discussions about real time control often need further clarification and definition regarding the mode, or level, of control. One way to classify real time control systems is according to the extent of intervention required by the operator:

(1) Under *local manual* real time control, operators exercise control actions based on their direct observations and experience. Manual systems are usually reactive, and the control decisions are based on what the operator is able to see: which in most cases would be only the local conditions in each station rather than the overall state of the system. Especially for a system where the operators have limited access to on-line data, control actions are not intended or able to optimize the system, but are instead mostly geared towards avoiding disasters such as flooding or damage to facilities. A schematic of manual control is shown in Figure 6.1

(2) *Supervisory remote control* (Figure 6.2) provides a level of improvement from manual control, accomplished by implementing a level of automation immediately above the process layer. In case of sewer networks this is usually a SCADA system. Most SCADA systems are supplied by specialized vendors, and they often provide centralized control consoles that allow the operators to monitor the system from a centralized control center. In addition to receiving system-wide information, operators are sometimes allowed to exercise supervisory control and manipulate control elements remotely from the control

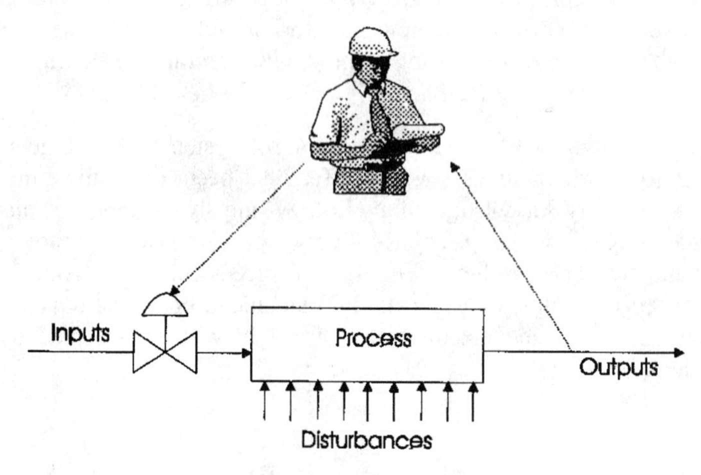


**Figure 6.1** Local manual control.

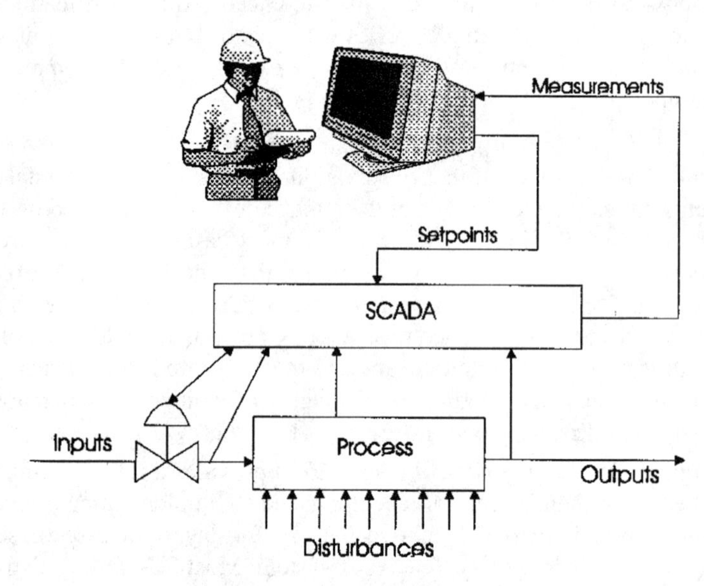


**Figure 6.2** Supervisory remote control.

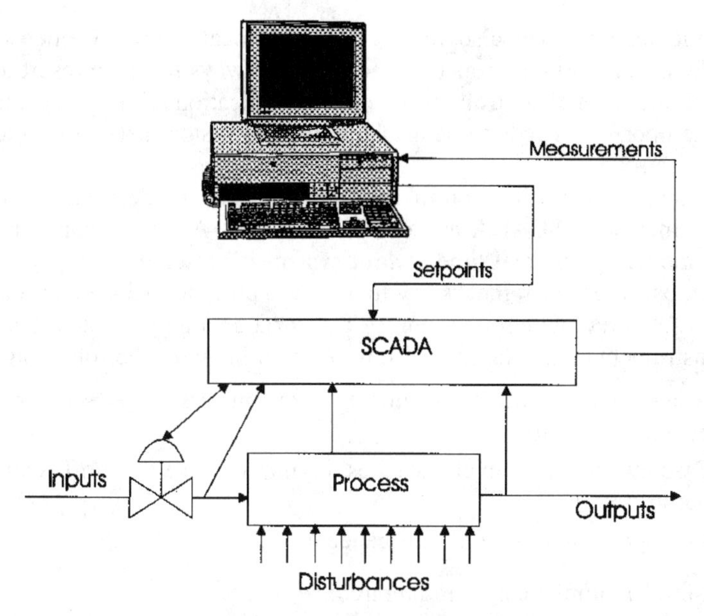


**Figure 6.3** Automatic remote control.

center. All the control decisions, although implemented through computers, are still made by the operators and based on their experience.

(3) The most advanced real time control systems exercise *automatic remote* control (Figure 6.3). Under automatic control, the control strategies and algorithms are implemented within computer programs. The sewer system is controlled without operator intervention. Examples of such systems include CATAD (Computer Augmented Treatment And Disposal) system implemented in Seattle, Washington and the system currently under design in Hamilton, Ontario.

Supervisory and automatic control levels (or: modes) of control are also classified as "remote" as they rely on SCADA system's ability to provide system-wide information to a central location (normally a control room) and also distribute the setpoints from a central location to all the remote sites. All the hardware and software required for local control resides within the remote site such as a pumping station or regulator station.

The three modes of control described above are not mutually exclusive, and they would normally appear as operational options from which operators could choose depending on the state of the system.

These three control modes are arranged hierarchically: both supervisory and automatic control require that all aspects of the SCADA system are operational, and they also require that the local control elements and functions are working properly. The system always needs to resort to the lowest safe level of control: in case of communication failure, for example, remote operation needs to be disabled and the system needs to switch to local control.

A completed real time control system will include modeling components intertwined with SCADA and control elements. A typical configuration for a modeling and real time control system is presented in Figure 6.4.

In most cases, the remote site will include a programmable logic controller (PLC), a remote terminal unit (RTU), or a data logger connected with the instrumentation. The input from the field includes the following:

(1) Analog signals from instruments (flow meters, level sensors, gate position indicators)
(2) Discrete signals (level switches in the wet well, on/off status of pumps, etc.)
(3) Counters (such as from rain gauges)

Control is implemented in hierarchical layers:

- At the bottom layer, there are hardware controllers which are wired to instruments on the input side and to the final control elements on the output side; for example, pump controllers will

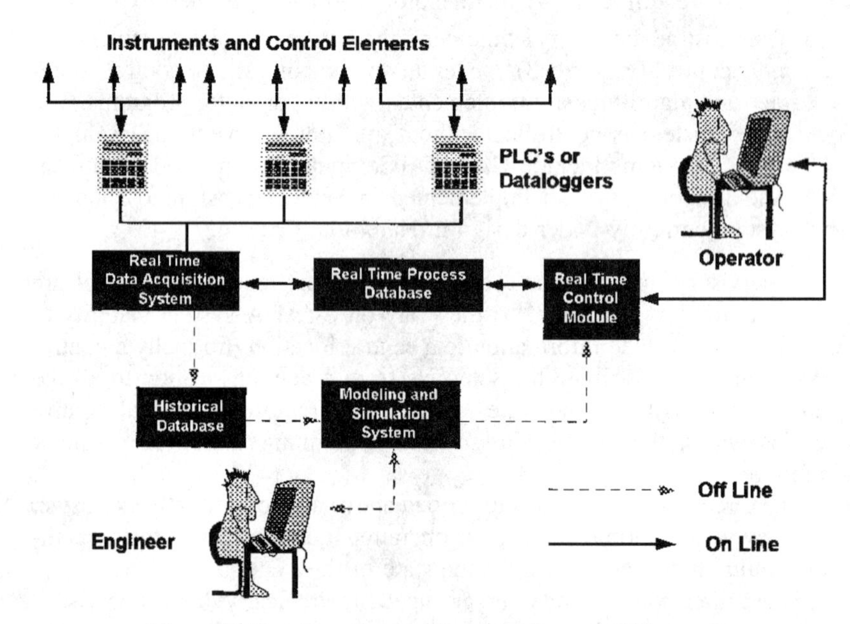


**Figure 6.4** Typical configuration for a modeling and RTC system.

normally be connected to wet well level sensors (for pump speed control) and level switches (for on/off control). They exercise control based on fixed setpoints which can normally only be changed by manipulating the hardware and the wiring.

- Individual stations can also be operated by the PLC's. When PLC's are introduced into a remote station, they normally intercept the connections between the field and the existing hardware controllers. PLC's provide greater functionality and flexibility since control actions can be defined through software such as ladder logic. The interface between the PLC and the hardware controllers is done so that if the PLC is disabled, the control automatically and immediately defaults down to the local hardware controller.
- Remote control is implemented from a central computer, usually situated close to the main control room. This is usually done by replacing the setpoint in the Proportional-Integral-Derivative (PID) control loop of the PLC. Thus, the PLC can always use the same standard algorithm but can point to different setpoints under different control modes. The remote setpoint can de determined by the operator (in supervisory mode) or derived by a computer program (in automatic mode).

The schematic presented in Figure 6.5 may help demonstrate these concepts.

## AUTOMATIC CONTROL

Automatic control resides in the central control computer, and it includes a number of different computer programs. The development of automatic control strategies includes design, testing, and implementation of automated algorithms for system operation. During the design stage, control strategies are superimposed on the mathematical model and evaluated through simulation, as shown in Figure 6.6.

Models enable the engineer to quickly and efficiently evaluate the operation of the system under different conditions including storm events of different sizes and different configurations of the network. Such testing would be impossible on-line for a number of reasons, including the fact that large storms occur infrequently.

Algorithms for controlling the overall network contain rules and/or procedures for operating many facilities in the system in a coordinated way. The input to the control algorithm are state variables such as flows, levels, or rainfall. The algorithms can fall into the following general categories:

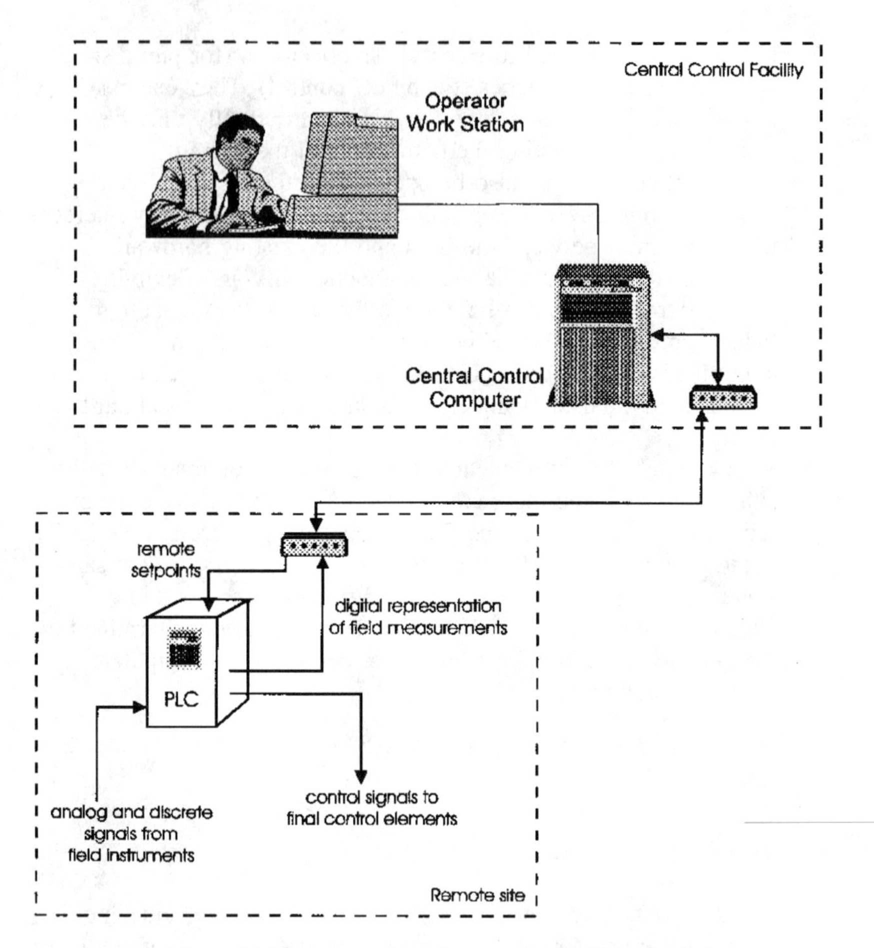


**Figure 6.5** Main components of an operational RTC system.

(1) Rule based systems define the response of the control system according to a predefined strategy. The rules can be derived from operational experience.
(2) Optimization algorithms use a mathematical description of the system to search for optimal strategy according to the objective function.

Control of sewer networks can be established for a number of different objectives, including the following:

(1) Reduction of overall CSO volumes
(2) Reduction of overall pollution loads from CSO's (includes quality aspects)
(3) Reduction/elimination of flooding

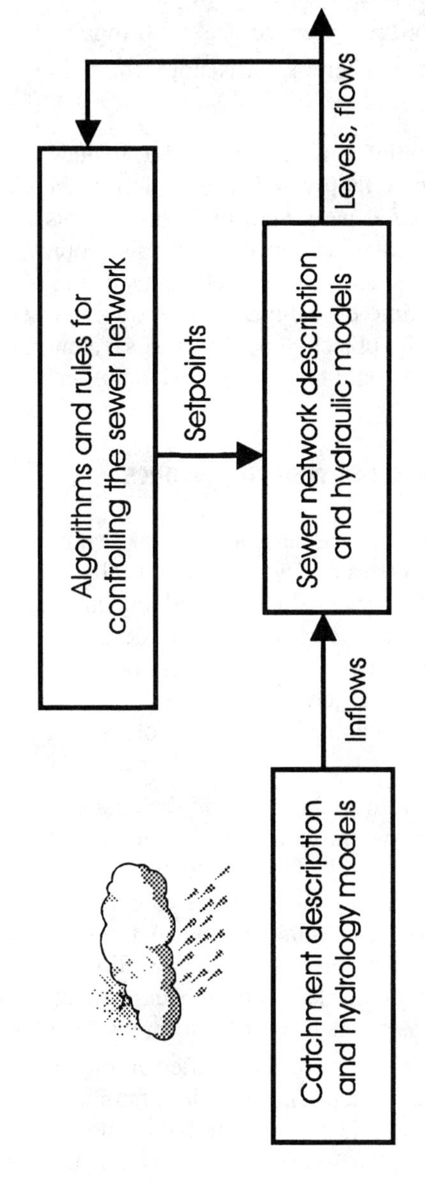


**Figure 6.6** Conceptual approach to RTC development.

(4) Elimination/reduction of CSO's in environmentally sensitive areas
(5) Reduction of energy required for pumping
(6) Equalizing the flow to the treatment plant
(7) Isolating parts of the system to deal with repairs or crisis situations
(8) Minimizing the number of starts/stops for pumps and movements by gates

The objective function can be structured to include one or more of the objectives listed above; the overall formulation of the objective function would consist of several parts with different weights indicating relative priority of each objective. The most appropriate objective function will depend not only on the specific characteristics of the sewer network, but will also change in time even for the same sewer system. For example, during dry weather the objective could be to save energy but during wet weather the objective could be to reduce flooding or CSO's.

## MODELING OF URBAN SEWER NETWORKS

Mathematical modeling has been accepted as a valuable tool in planning, design and analysis of sewer networks, and it is also is an important aspect of the methodology for the development and implementation of RTC. Most sewer system models include components for describing the hydrology and the hydraulics of the network. Main model components are presented in the schematic shown in Figure 6.7.

For the purpose of planning and design, computer models need to meet the following requirements:

(1) Models need to adequately represent the changes in flows and levels throughout the network, including the interactions between different elements of the network. In order to represent the dynamics of the system, most modern hydraulic sewer models contain numerical solutions to St. Venant equations of flow. A brief review of solutions to these equations will be summarized below. Since many sewer networks are affected by tidal influences or may experience flooding problems, it is important that models are able to predict the backwater effects.
(2) Models need to be calibrated and verified, as they are only a representation of the actual system and include a number of assumptions. Archived data on flows, levels, and rainfall are normally used to simulate storm events and the model parameters are adjusted according to the comparison of simulations with actual behavior of the system. Calibration and comparison against actual data is especially important for hydrologic models.

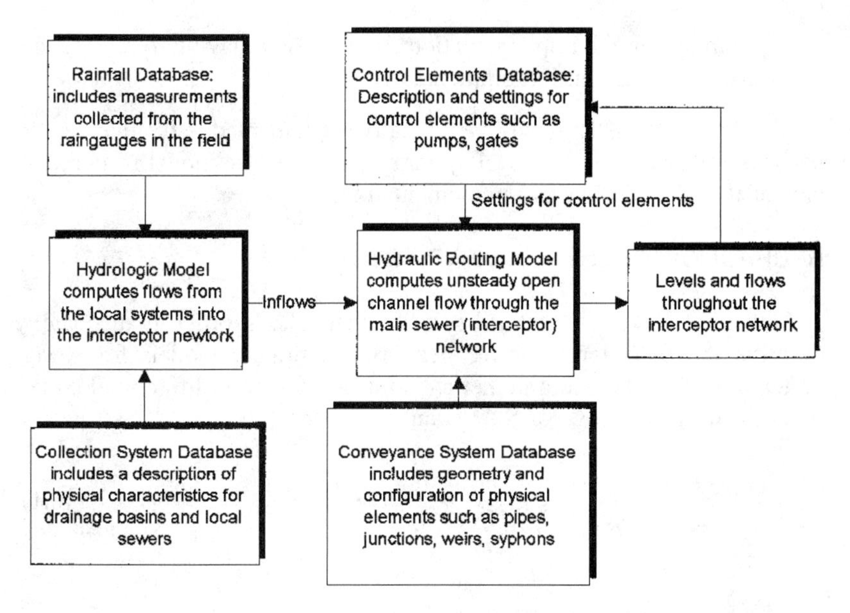


**Figure 6.7** Main components of hydrologic/hydraulic modeling system.

(3) Stability of the model is important for proper operation.
(4) Accuracy is essential for maintaining the credibility of the model.

The requirements listed above need to be met in order for the model to be successfully applied for planning studies and analysis. If the models are to be used for designing and implementing RTC, some additional requirements include the following:

(1) Speed is an important factor for RTC, especially if the model is to be used on-line as part of a model-referenced control strategy.
(2) In order to simulate operator intervention, it is necessary for the modeling software to allow the user full control over the execution of the program. The simulation needs to have an interactive connection with the user, so he can manipulate the settings for control elements during the simulation. This is not easily accomplished if the models run in a batch mode, as in that case the model will use the input files to define the simulation and the user will not have a mechanism to interact with the simulation. Interactive features of the model are also important if simulations will be used to train the operators.
(3) Integration of RTC functions into the model greatly facilitates the use of the model for RTC design and implementation. During the design stage, the user will test the control strategies on the model. Ideally, this mechanism of control could be directly transferred to the actual

system during the implementation stage without having to rewrite the software for the on-line implementation.

Since most of the RTC will be typically implemented in the interceptor network rather than in the local systems, the hydraulic model is a crucial part of the RTC design and implementation.

## HYDRAULIC MODELS FOR SEWER SYSTEMS

Sewer networks are modeled as open channels. In order to adequately describe the behavior of sewer networks, hydraulic models for sewer systems need to include a numerical solution to partial differential equations for unsteady flow, or St. Venant equations:

$$\frac{\partial Q}{\partial x} + \frac{\partial A}{\partial t} = q_l \tag{6.1}$$

$$\frac{\partial(\rho Q)}{\partial t} + \frac{\partial(\rho Q V)}{\partial x} + gA\,\frac{\partial(\rho y)}{\partial x} + \rho g A(S_f - S_0) = 0 \tag{6.2}$$

where: $Q$ = flow discharge; $A$ = cross sectional area $q_l$ = lateral inflow; $V$ = velocity; $\rho$ = density of water; $t$ = time; $x$ = length along the channel; $S_f$ = friction slope; $S_0$ = bed slope; $g$ = gravitational constant.

Equation (6.1) describes continuity, and Equation (6.2) includes the dynamic (momentum) equations which consider the force balance. The St. Venant equations are nonlinear hyperbolic partial differential equations which cannot be solved analytically. The assumptions inherent in these equations (Liggett, 1975) include the following:

(1) The surface of the water varies gradually.
(2) Friction losses in unsteady flow are the same as the friction losses in steady flow.
(3) Velocity distribution across the wetted area does not impact wave propagation.
(4) The wave movement can be considered two dimensional.
(5) The average slope of the channel bottom is small.

Although these equations have been in use for over a hundred years, they are normally found to be appropriate for describing gradually varied flow in open channels and have been confirmed by experiments (Colorado State University, 1970).

The numerical techniques commonly used for solving these equations include finite difference methods (Liggett, 1975). The two most commonly used finite difference methods on a fixed grid include the explicit and implicit approach:

(1) In the explicit approach, the flow through a particular link at time (step) t is expressed and solved explicitly by using only values for flows or levels at the previous time step $t - 1$. This scheme is easier to formulate, but suffers from limitations imposed by the Courant Criterion. The Courant Criterion describes the stability of the explicit solution in terms of the size of the simulation time step and the length of the conduit elements. In practice, this means that simulations need to be performed using a short time step in order to avoid instability of the model and erroneous results. Small computational time steps result in poor computational efficiency and slow simulations. Another way to avoid stability problems is to ''adjust'' the lengths of channel elements (pipes), but this way we are departing from the actual physical layout of the network when we describe the system for the purpose of modeling. A typical example of an explicit model is EXTRAN (Roesner et al., 1988).

(2) In the implicit approach, the solution for flow or level at a particular point in the network at time t is expressed in terms which include values for other elements in the network at time t: since the solution for our unknown variable includes other unknown variables, this approach requires that all the unknown variables are solved simultaneously. A sparse matrix is normally set up based on the unsteady flow equations and the sewer network connectivity. The time step used in the implicit approach is not affected by the Courant Condition, but it depends on the dynamics (rate and magnitude of change) of the inflows imposed on the system. The implicit approach will allow for longer time steps and thus faster execution times. Implicit approach is used by models incorporated into several advanced hydraulic system modeling packages including Mouse from Danish Hydraulic Institute, Hydroworks from Wallingford Software (Wixcey et al., 1985), and SewerCAT (Ji et al. 1995).

It is important to remember that although sewer system models have a long history and a number of successful applications, the model is only a perception of the actual system. Proper use of mathematical modeling and simulations requires information about the sewer network: the layout of the system, geometry of the conduits, rainfall records, flows, and levels. In addition to the availability of data, it is often important that the modeling staff has adequate experience and knowledge of modeling.

## ORGANIZATIONAL ASPECTS OF RTC IMPLEMENTATION

The urban drainage RTC system is typically designed by internal engineering staff or outside engineering consultants, and operated by the Operations and Maintenance staff. Some organizational considerations for the development of RTC system include the following:

(1) The development of an RTC system will require a broad range of technical skills and knowledge of several technical areas including hydraulics, control theory, instrumentation, and SCADA systems. The RTC project will therefore require a team approach and a well integrated project plan. Without a balanced approach to the project, a single "view" of the problem might come to dominate at the expense of others.
(2) Early and meaningful involvement of the Operations and Maintenance staff in the RTC development project is very important. The O&M staff should help define the functionality of the system, especially for the graphical user interface.
(3) The new capabilities provided by the RTC system will enable the O&M staff to accomplish more, but staff training will most likely be required in order to obtain the full benefit of the added functional capabilities.
(4) The RTC system will require "care and feeding" after it is implemented, as it will have to be periodically updated and adjusted. If the funding is planned and established in advance for such (mostly software) support activities, the risk of the system becoming unmanageable is much smaller.

It is important to remember that the most important components of an RTC system will be computer programs. The development of an RTC system will in many ways be a software development project. Unfortunately, many managers at large municipalities can recall "horror stories" regarding software projects which went out of control in terms of budget, scope, and schedule. In order to avoid the risks of project failure, the RTC system development needs to be carefully managed and organized.

## APPLICATIONS OF RTC TO URBAN COLLECTION SYSTEMS

Different levels of RTC have been considered for or implemented in a number of communities including Minneapolis (Callery, 1971), Detroit (Watt, 1975), Cleveland (Buczek and Chantrill, 1984), and Lima, Ohio (Brueck et al., 1981; Chantrill et al., 1996). A comprehensive review of RTC application to sewer networks was given by Schilling (1984). Although this reference is today somewhat dated, it includes a thorough description of issues related to RTC and discusses the status of RTC in many municipalities. Most of these control systems considered current values of local variables around each control facility, rather than the values predicted into the future for the duration of the event. The control decisions were made on the basis of observations from a single station rather than optimization on a system-wide basis.

## OPTIMIZATION OF SEWER NETWORKS

The problem of system-wide CSO minimization is conceptually similar to inventory-control problems in other industries, which have been successfully solved based on optimization theory. Research work on system-wide real-time control for CSO minimization can be traced back to the mid 70's. Several researchers, including Leiser (1974) and Trotta et al. (1977), developed off-line optimization models based on optimization theory and applied it to system-wide automatic computer control of CSO's. Dynamic programming was applied to solve the state equation set and related constraint conditions in these models. Due to heavy computational loads imposed by the solution algorithm, large systems with more controllable elements could not be addressed at that time. Most of this early work, thus, was focused on smaller systems with only a few inputs and outputs.

For the purpose of control, the sewerage network was reduced to a combination of simplified storage, transport/delay, and junction (converging and diverging) elements. The delay elements mimic the impact of long sewer trunks on the flows and levels throughout the system. In any optimization control system, there are always more (unknown) system variables than state equations and constraint conditions. Therefore, there are many solutions. The purpose of an optimization method is to select a solution which can best fit the objective function among many possible solutions of the state equations. It is important to note that this approach required that the direction of flow be predetermined and fixed for the duration of either the simulation or the operation of the control system. The control model had to assume that the direction of flow will remain the same throughout the simulation.

Gelormino and Ricker (1991) implemented an optimization algorithm to a large CSO control system in the Seattle Metropolitan area. The representation of this system included 23 storage reservoirs, and the key decision variables included gate flow rates at 23 locations throughout the sewer network. The model can be both linear or non-linear depending on which objective function is used. Gelormino's optimization model was solved by using MINOS 5.1, a package designed for large-scale linear and nonlinear programs. The control system did not consider delay elements, meaning that the influent flow variations at a trunk upstream would simultaneously affect both the downstream and the upstream gate flow. Although the system operated well and resulted in CSO reductions (Vitasovic et al., 1990), Gelormino and Ricker (1991) found that the overall match between the real measured data and the off-line simulated data was poor, questioning the validity of the approach.

The difficulty in achieving optimal operation in Gelormino's algorithm when the control model is fixed can be explained by looking at the principle of the optimization model. The optimization model produces the desired flows at each control station. The desired flows are selected from many gate flow combinations which only satisfy the flow continuity and do not necessarily satisfy the hydrodynamic conditions in the entire system. Therefore, those desired flows may not be realistic for the real physical system even if the desired flows theoretically give the minimized CSOs. Since the solution is not always hydrodynamically feasible, the desired flows generated by the optimization algorithm cannot be achieved.

### Rainfall Forecasting

Optimal control algorithms generate the best response of the system by considering the entire duration of the event. This requires that the algorithm utilizes forecasts of system inflows, which are the result of rainfall. In literature on optimal control of sewer systems this issue seldom receives significant attention, not because it is unimportant but because it is so difficult to deal with. In most cases, published studies and off-line simulations of optimization and control assume that the storm intensity is known to the optimization algorithm, meaning that accurate rainfall forecast would be available.

Rainfall forecasting is normally addressed using stochastic techniques and historical data. In some cases, forecast of rainfall implemented in actual operating RTC system is very simple (Gelormino and Ricker 1991). The neural network algorithm developed for the Hamilton, Ontario RTC system is showing some promising early results (see case study Nr. 2 below).

The difficulty of obtaining accurate rainfall forecasts will not invalidate the concept of using optimization to control sewer networks, but it will impact the efficiency of optimization. Inaccurate forecasts will lead to suboptimal control decisions, but the control decisions produced by the optimal control algorithm will still likely be a significant improvement over other modes of operation.

Rainfall information, while normally obtained through a network of rainfall gauges, can also be collected using radar technology. While a properly installed and well maintained rain gauge will give relatively good data for the discrete point at which it is located, rain gauge data often does not give an accurate measurement of rainfall over a wide area. Many storms are very dynamic, even chaotic in nature, with rainfall patterns and intensity in a state of constant and rapid change. Intense rain cells of relatively small size may cause the use of rain gauge data to lead to errors in estimating total accumulation on a drainage area. For every accurate measurements, a very dense network of rain gauges would be necessary, but the expense and logistical complexity of such a network is usually prohibitive.

Technology developed by RHEA, S.A., a French company, uses signals from Doppler weather radar to measure rain. The relatively ''rough'' measurements taken from the reflected energy of the radar beam are calibrated using contemporaneous data from a few strategically placed rain gauges. The combination of the local accuracy of rain gauges with the wide area coverage of radar provides predictions of rainfall over a wide area with an areal resolution of 1 sq. km. (0.4 sq. mi.) and a vendor-claimed accuracy of $\pm 10\%$. RHEA's system has been installed in Europe and used in real time for drainage system control or flood warning.

Another example use of radar for rainfall measurement is the research conducted by Prof. Ian Cluckey at Salford University in the U.K.

## COMBINING OPTIMIZATION WITH HYDRAULIC MODEL

Labadie et al. (1980) first developed a model combining optimization and dynamic unsteady flow routing for computation of RTC strategies. In this study the Marina branch of the planned North Shore Outfalls consolidation project in San Francisco was simplified to 3 reservoirs connected in series. No time lag elements were assumed among the reservoirs. A fully dynamic, unsteady flow model was explicitly included within a deterministic dynamic programming formulation of the control problem.

In order to ensure that the control action is feasible, the desired flows generated by the optimization model should not be used directly for sewer system control. The RTC system for The Region of Hamilton-Wentworth, Ontario (described below) combines a system-wide optimization with a hydrodynamic model, thus eliminating the problem introduced by the predetermined flow directions.

Recent RTC projects include the RTC system in Bordeaux, France and the European Union-sponsored Sprint project which included a pilot implementation of RTC in Bolton, England and pilot implementations of model-based control in Gothenburg, Sweden, and Copenhagen, Denmark. The Sprint project used programs from the Mouse line of software from Danish Hydraulic Institute (DHI).

## CASE STUDY 1: RTC IN SEATTLE, WASHINGTON

### Abstract

Municipality of Metropolitan Seattle[1] (hereafter referred to as Metro) is responsible for operating and maintaining the combined sewer system tributary to its West Point Treatment facility. As part of the agency's goal to reduce combined sewer overflow (CSO) volumes by 75% between 1985 and 2005, a real-time control system has been successfully implemented. A computer program automatically determines the settings for in-line regulator gates and pump speeds in order to maximize the use of in-line storage during storm events. A distributed network allows the control decisions to be implemented without operator intervention.

The new ''Predictive'' control program replaces a heuristic control algorithm that was developed early in the 1970's. The effort to develop an improved control algorithm began as a result of a study (refs) showing that extra capacity in the trunks could be utilized by improving the software, and as part of an overall hardware and software upgrade for off-site pump and regulator stations that was completed in 1990.

### Introduction

Metro has been using a real-time control system since the early 1970's. A 1985 upgrade to off-site hardware and software has enabled Metro to improve the control strategy to further reduce CSO volumes in its West Point System.

The West Point Treatment Plant conveyance system consists of large diameter trunks and interceptors, owned by Metro, with a total length of over 160 kilometers. These pipes serve an area of roughly 30,000 hectares, and are fed by pipes owned and maintained by local sewer agencies. A large portion of the system within the City of Seattle is considered a ''combined'' system, with both domestic wastewater and stormwater inflows. A significant portion of the service area (10,000 ha) area outside the City of Seattle is served by separated sewers.

Metro operates 13 pump stations in the West Point system, ranging in firm capacity from 0.2 to 5.7 $m^3/s$. In addition, several regulator stations utilize in-line sluice gates that allow wastewater to be stored in trunks during periods of high flow. A typical regulator station is shown in Figure 6.8.

The regulator gate is used to store excess flows in the trunk line. The outfall gate releases the excess flows to the receiving water if the capacity

[1] Now called the King County Department of Natural Resources.

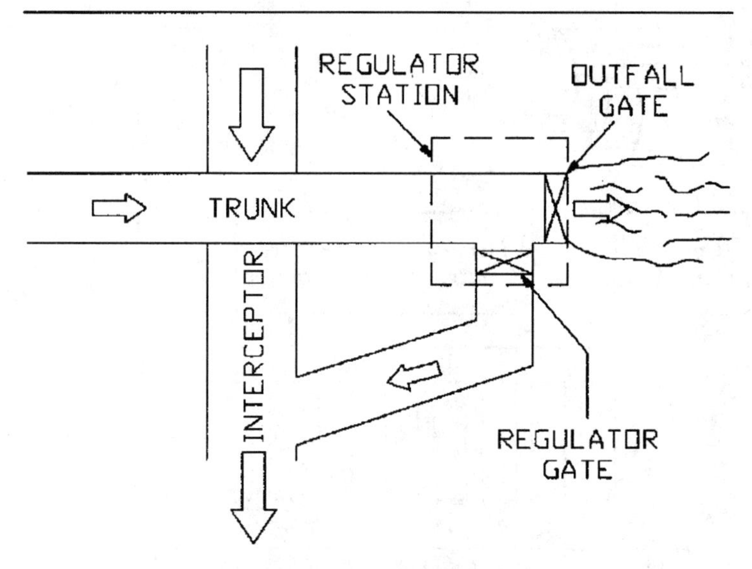


**Figure 6.8** Schematic of a typical regulator station.

in the trunk is exhausted. All gates can be modulated to different positions remotely, and are equipped with electronic position indicators. At most stations, an overflow weir is also used to release the flows to the receiving water. Most pump stations in the West Point system utilize variable speed pumps. The largest pump station is the Interbay pump station, which has a firm pumping capacity of 5.7 $m^3/s$. This station is particularly critical to the conveyance system, and has special controllers to ''pump down'' the interceptor upstream of the station when a storm approaches. Under current operations, the ''pump down'' mode is initiated when rain is detected in the basins upstream of the station.

Each pump and regulator station (referred to collectively as ''off-site'' stations) is equipped with a programmable logic controller (PLC), which stores readings from analog and digital instrumentation, as well as control logic for operating the station. Field sensors connected to the PLC include pump tachometers, level sensors, rain gauges, gate position indicators, and limit switches. Each PLC is connected by phone line and modem to a control computer located at the treatment plant.

The Supervisory Control and Data Acquisition System (SCADA) system at the treatment plant was obtained through contract with Forney International of Dallas, Texas. The general architecture of the West Point SCADA system is shown in Figure 6.9.

Each PLC is programmed internally to operate the station based on a pre-determined setpoint, in what is referred to as ''local'' control. In

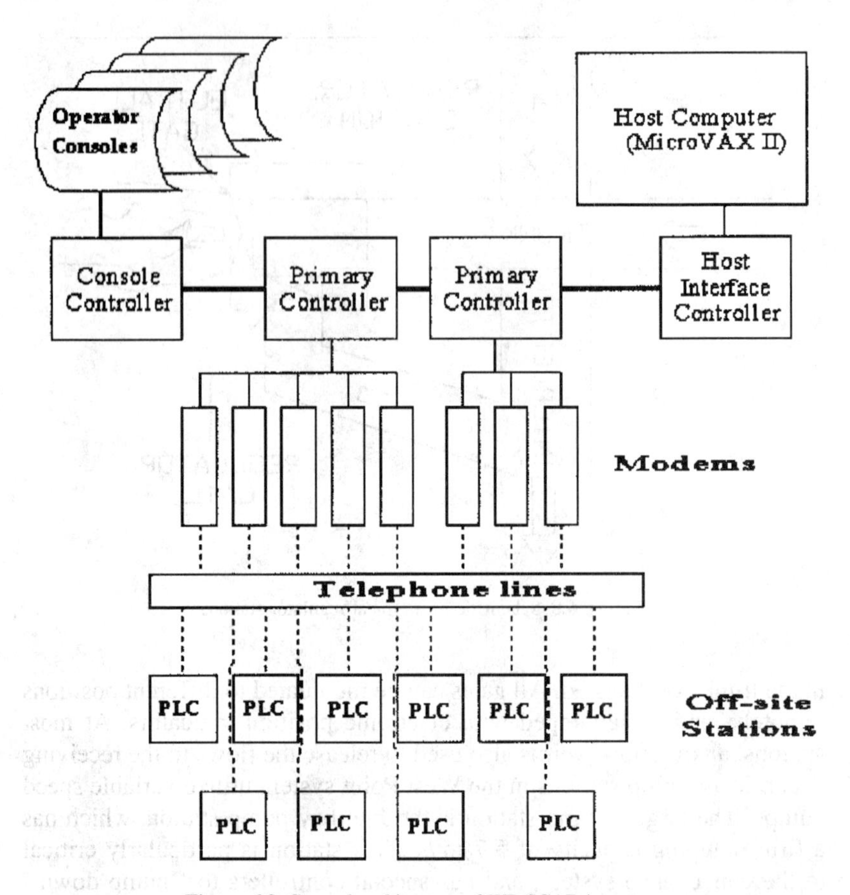


**Figure 6.9** General architecture of WP SCADA system.

addition, the PLC can adjust the setpoint by remote commands from the central control system (located at the treatment plant).

The SCADA vendor supplied the software for the primary controllers, which process all inputs and outputs. Boolean error checking, range checking, and status checking are done at this level. Alarm priorities are kept in an on-line database so that alarms can be sorted for display to the operator.

The primary controllers have capacity only for limited trending. No historical data is maintained on the primary controllers. For this reason, the SCADA system is equipped with a host computer and a controller interface. The host computer stores all historical data, processes reports, and performs system-wide calculations.

The host computer (Micro VAX II) was delivered with software to perform several real-time functions:

(1) Data acquisition from the controllers
(2) Statistical processing
(3) Data download to controllers
(4) Historical archiving and reporting
(5) Redundant system checking

This software was written in FORTRAN and designed to run in concert with the VAX VMS operating system from Digital Equipment Corporation.

Ever since Metro's first SCADA system was developed in the early 1970's, software has calculated flow rate information for each off-site station on a real-time basis, utilizing measurements such as levels, gate positions, and pump speeds. This software was adapted to run with the new control system during the 1988–1990 time frame. Algorithms that were originally coded in a mixture of FORTRAN and assembly language were re-written to be FORTRAN-77 compliant and interfaced with the real-time system software provided by the SCADA vendor. Although the primary focus of this task was to provide the operators with current and historical flow records, it also allowed the project team to become familiar with the application programming environment on the host computer.

In addition to the flow calculations, real-time control algorithms developed with the original SCADA system were translated. These control strategies (described in Beck, 1974) were based on heuristic algorithms, and provided a reduction in CSO volumes compared to using local control (see Leiser, 1971). This application was adapted for the new SCADA system so that automatic control could be utilized until an improved control system based on optimization (described below) was completed.

A software application (database) was developed to monitor analog sensor values on a real-time basis. This database was located in the host computer and it contained all analog and digital values scanned from field devices. The program included quality checks; for example, it would be able to check if a level sensor is reading too high for "non-storm conditions." Some error checking was performed in the controllers, but couldn't easily accommodate complex logic, as the programming environment (ladder logic) was more primitive. The host computer maintains a statistical database for most analog points which tallies minimum, maximum, and average values for daily, weekly, monthly, and yearly bases. The sensor monitoring program uses this information to detect abnormalities in signals. Operations staff use daily printouts from this program to check for sensor irregularities. This program has also been modified to notify staff via electronic mail for certain conditions (e.g. a designated level rises to certain height under non-storm conditions).

### Control System Overview

The on-line control program (''Predictive control'') was developed to improve the performance of the existing automatic control program. Instead of using heuristic algorithms, Model Predictive Control theory (MPC) was applied (Ricker, 1990). In this method, an on-line optimization problem is updated and solved periodically, resulting in a control strategy which is continually changing to reflect current conditions. One of the main advantages of MPC is its ability to use forecasts of future inflows, i.e. the control strategy is predictive or adaptive in nature and not merely reactive (Trotta et al., 1977). Figure 6.10 shows a schematic diagram for these processes.

### Real-Time Operation

Automatic control of the off-site stations is available on a 24-hour basis. The control software, resident on the host computer, calculates optimal flow

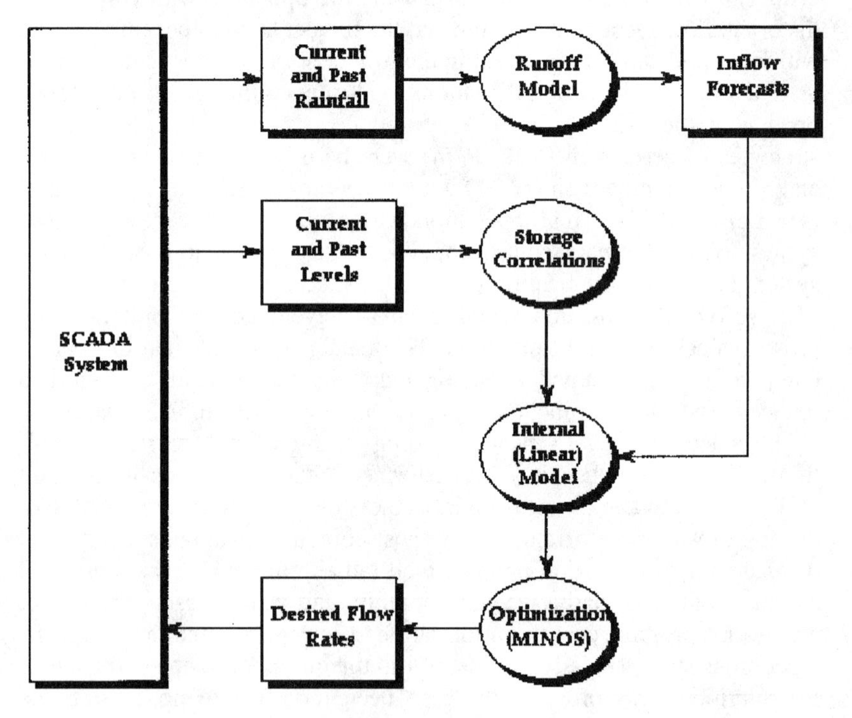


Figure 6.10

rates for the off-site stations every ten minutes. If the operator at the treatment plant has enabled automatic control at a particular station, flow calculation software (also resident in the host computer) calculates the correct gate position or pump output signal to produce the desired flow rate. The following section will briefly describe the software elements required to calculate and implement these optimal flow rates. For a more detailed description, refer to Ricker (1991) and Gelormino and Ricker (1991).

### Inflow Forecasts

In order to provide a simulation environment for Metro engineers to test different control strategies, mathematical models were developed that describe how the entire West Point collection system reacts to different input conditions (Vitasovic et al., 1990). A portion of this model was adapted for real-time use to provide the control software with "forecasts" of inflows at 46 upstream locations. These algorithms require rainfall information (available from the real-time database). Base flows are a function of the time of day. Both a basin model and a pipe model are utilized. These models use the kinematic wave assumptions to solve the equations of flow.

### Internal Model

A set of linear difference equations is used on-line as an "internal model" for the control program. These predict how the regulated flows (independent variables) will affect the flows and storage volumes in the system. There are twenty-four equations, each of which represents a portion of the collection system upstream of a regulator or pump station.

Any model-based control strategy relies on the accuracy of the internal model. However, the complexity of the model directly affects the calculation effort required by the optimizer (see below). The large-scale nature of this problem and the limited computing resources dictated the use of first-order (linear) difference equations.

The current value for storage volume in each section of the interceptor (modeled as a reservoir) is obtained by using level measurements and a non-linear polynomial function. Each "storage correlation" was developed using a fully dynamic hydraulic routing model (Vitasovic et al., 1990). These calculations are updated at every solution interval so that feedback from measurements are incorporated into the control strategy.

### Mathematical Solution

The purpose of the optimization program is to select regulated flows so that predicted storage volumes will achieve some optimal goal. The measure of optimality must be defined mathematically, in an ''objective function.'' It formulates the following goals in mathematical terms:

(1) Minimize overflow volumes.
(2) Empty the collection system quickly after a storm.
(3) Balance storage levels throughout the system.
(4) Maintain smooth rates of change of regulated flows.

The objective function is a scalar quantity, a weighted sum of terms that measure each of the above goals. Also, the solution must satisfy a set of constraints, which define a ''feasible'' operating region. They include the reservoir capacities, limits on regulated flow rates, and their rate of change. Constraints refer to both maximum and minimum. In this application, the operational constraints vary in time (e.g. possible flow through a regulator gate depends on the upstream head). The objective function and the internal model are fixed, however.

A commercial optimization package (MINOS version 5.1) is utilized to minimize the objective function while obeying the operational constraints and the internal model. This program is appropriate for large-scale linear and non-linear problems.

Flow calculation software in the host computer converts desired flow rates to an appropriate output signal. For regulator stations, a desired gate position is calculated. For pump stations, an output signal (0–100%) is calculated. The SCADA vendor provided software to download values from the VAX host to the primary controllers, which are programmed to send the values to the appropriate PLCs using modems. The host software monitors the movement of the control elements, and generates an alarm if the control action is not carried through.

At the largest (Interbay) pump station, special controls have been implemented to make use of storage in the influent pipe. The MPC program has integrated this functionality into its logic. Three separate modes of automatic control are defined for the pump station:

(1) Automatic Idle—local PLC runs pumps using the local setpoint
(2) Automatic CSO—local PLC uses pump-down mode controller
(3) Automatic Active—local PLC uses remote output signal from MPC

The station mode changes between these three alternatives without operator intervention.

### Operator Interface

For proper operation of the control strategy, the operator must place the stations into Automatic control, and monitor any field errors that may prevent the station from switching to remote control. The SCADA system was delivered with an operator interface consisting of four color monitors, a line printer, and a color printer. The monitors are PC based, and two are equipped with touch screen capability. A color printer allows for "snapshots" of operator screens. Metro engineers provided training to each shift of operators to familiarize them with the new control strategy. While the control screens remained much the same as for the previous automatic control, operators needed to know what to expect from the new system. For example, regulator gates will close during storm conditions at different times under MPC than under previous heuristically based Automatic control. Also, it was important to review alarm conditions and how to detect instrument or equipment failure at the station.

### System Testing and Startup

Initial testing of the control programs was done using a dynamic hydraulic routing model. This model solves the St. Venant equations of flow using an implicit technique that allows for time-steps in the range of 2 to 5 minutes. While Metro developed this model specifically for this project, it has also been used extensively to assess other impacts to the conveyance system, such as construction projects and sewer rehabilitation (see Swarner and Vitasovic, 1989). The model uses inflows generated from the runoff model and simulates control behavior by utilizing the same control programs used in the real-time system.

Investigations using this simulation tool led to decisions concerning the appropriate horizon interval and the detail of the rain forecasts. In late 1991, the programs were loaded on the host computer at the treatment plant, but were limited to solving for the desired flow rates only. These values were kept in the real-time database for analysis.

### Results of MPC in Seattle

The new Predictive control algorithms were tested on-line during storms in early 1992, and were introduced into normal operation in the fall of 1992. In order to test its effectiveness in the long term, simulations have been made to determine overflow volumes resulting from several storms. Figure 6.11 provides a summary view of the improvements provided by Predictive Control.

This figure shows the simulated overflow volume from each of seven "design storms", which have varying degrees of frequency. These storms are historically based (rainfall events from 1982–1983) and have been used for several years at Metro as a basis for CSO planning. The graph shows that Predictive Control reduces the system overflow for each storm.

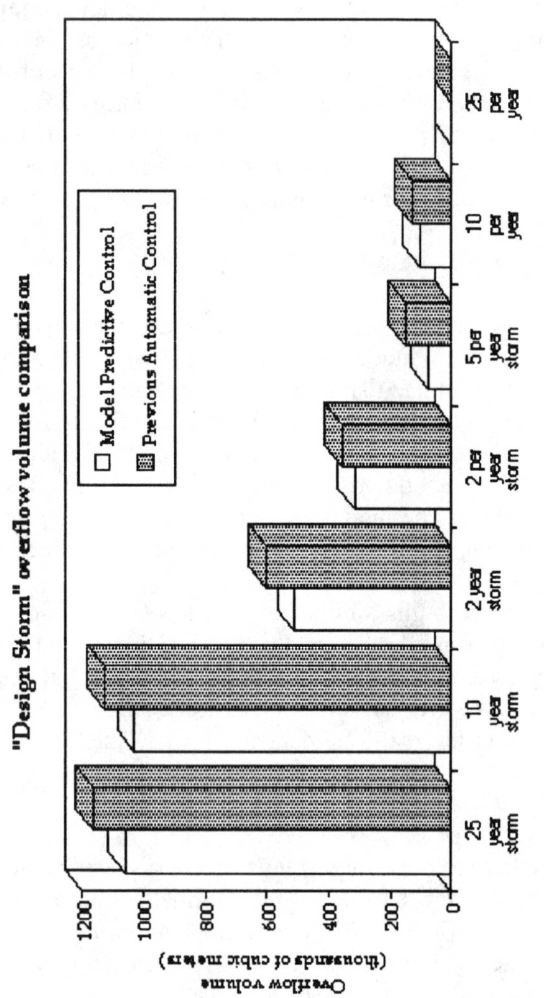


**Figure 6.11** Comparison: heuristic RTC vs. modeled-based predictive RTC.

The amount of reduction ranges from around 9% (for the two largest storms) to a 49% reduction for the 5 per year storm. The actual reduction realized will depend on the rainfall volume and patterns during each individual year. Metro expects to achieve a much greater reduction from Predictive control than was originally planned at the outset. The original project budget was $4.3 million with an expected reduction of 150 million gallons per year. The final project expenditures were $2.9 million, including level sensor and rain gauge installations and a software interface for planners (Speer and Vitasovic, 1991). The annual CSO reduction is expected to exceed 750,000 $m^3$, or roughly ten percent of the 1988 total CSO baseline figure.

Figure 6.12 illustrates the increased use of in-line storage using MPC during the 2 year design storm. This storm is characterized by two distinct periods of rainfall, where the latter period is the more intense.

The graph indicates that more in-line storage is used by MPC during the highest loading periods, and that it emptied the system more during the period between loadings due to the inclusion of this goal in the objective function.

### Summary of the Seattle Case Study

Metro has successfully implemented a real-time control system for a combined sewer system. Software utilizing Model Predictive Control and inflow forecasts directs control elements throughout the tributary area to utilize in-line storage during periods of high flow. The reduction in combined sewer overflow volumes is expected to exceed 750,000 cubic meters during an average year. More can be found about this work on the Internet at the following site: http://www.yahoo.com/Science/Engineering/Environmental Engineering/

## CASE STUDY 2: RTC IN HAMILTON, ONTARIO

The Regional Municipality of Hamilton-Wentworth (hereafter referred to as the Region) operates and maintains a combined sewer network which services about 250,000 people residing in an area of 54 $km^2$. The combined sewer system employs over 150 diversion structures to regulate the flow of combined sewage into the Western and Red Hill Creek Sanitary Interceptor sewers for conveyance to the Woodward Avenue Wastewater Treatment Plant (WWTP). Most of these diversions are static overflow structures located upstream of the interceptors. These structures cannot be quickly or easily manipulated, if at all, to control the volume of flow entering the interceptors. However, the combined sewer system also employs a number of motorized gates (13 in all) along the two interceptor sewers which can

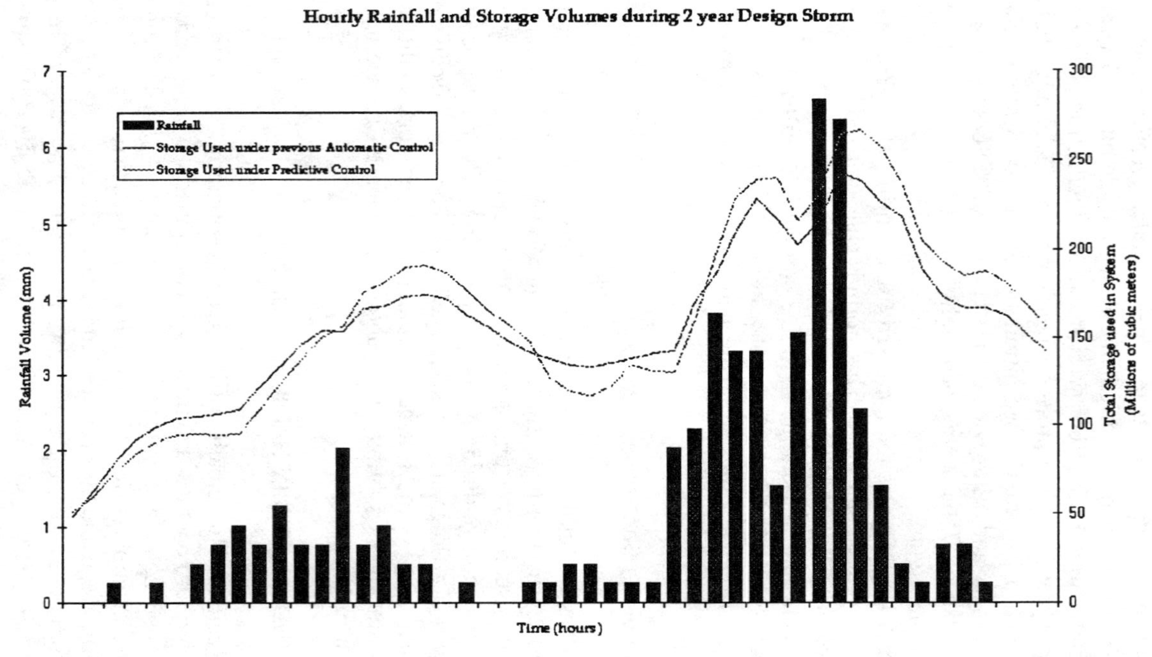


**Figure 6.12** Use of in-line storage using MPC.

be opened or closed by operators at the Woodward Avenue WWTP to dynamically regulate the amount of flow entering the interceptors and ensure the interceptors and the WWTP are not overloaded.

During dry weather and small storm events, these gates are left open and all sanitary and storm flow enters the interceptors which convey the flows to the WWTP where it receives full treatment being discharged into Hamilton Harbor. During large storm events, the inflows to the combined sewer system can exceed the capacity of the interceptors and/ or WWTP, causing combined sewer overflows to local receiving waters. The Region's combined sewer system discharges overflows to Hamilton Harbor, Cootes Paradise, Chedoke Creek and Red Hill Creek at up to 22 locations. These diversions are necessary in order to minimize basement flooding and overloading of the Woodward Avenue WWTP. On average, there are 23 overflow events per year for every outfall, diverting a total volume of approximately 4.33 million $m^3$ of untreated combined sewage to local receiving waters each year.

According to its Pollution Control Plan (PCP) (Stirrup, 1996) the Region undertook a number of activities in order to reduce the pollution caused by the overflows. The PCP includes the construction of 10–12 CSO storage facilities to detain overflows during periods of wet weather. During dry weather, the stored wastewater is pumped and/or drained back into the sanitary interceptors and conveyed to the Woodward Avenue WWTP for treatment. To date, the Region has completed construction of five underground CSO storage tanks (4 off-line, 1 in-line), as shown in Figure 6.13. Together, these facilities provide about 193,000 $m^3$ of additional CSO storage volume within Hamilton's combined sewer system.

As mentioned above, larger storm events generate inflows to the combined sewer system which can exceed the capacity of the interceptors and/ or the WWTP. This often necessitates closure of some or all of the 13 motorized control gates along the interceptors. Historically, when operators have deemed closure of the gates to be necessary, they have closed all the gates at once. Similarly, when sufficient conveyance and treatment capacity returned at the end of the storm, the standard practice was for the operators to reopen all the gates at the same time. In the past, the operators have had little or no information as to the state of the collection system and the varying conditions in different portions of the system, and could do little else. As part of its PCP program, the Region is implementing a computer-based real time control system to optimize the operation of these automatic regulator gates and the existing CSO storage facilities. The real time control system will continually adjust the level of existing regulator gates and storage facilities according to changing rainfall and flow conditions, to maximize the use of the storage capacity available (in-

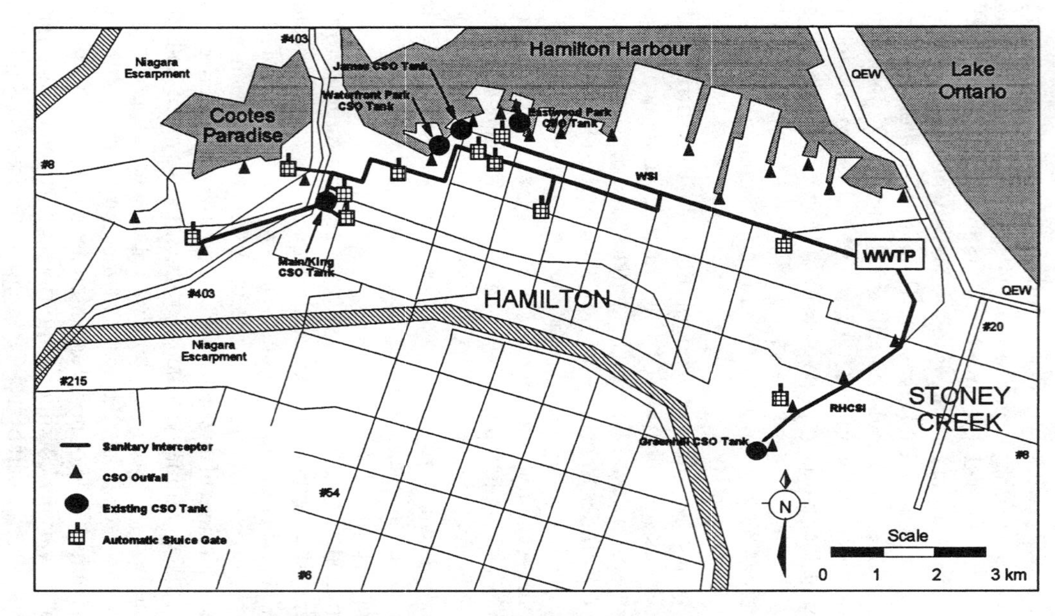


**Figure 6.13** Combined sewer network in Hamilton-Wentworth region.

line and off-line) within the combined sewer system and minimize overflow volumes during each storm event.

The major objectives of this real time control project are

- to reduce overall CSO from 22 discharges
- to control flooding of interceptors and the waste water treatment plant without over-discharging CSO
- to optimize the discharge locations according to different pollution protection priorities

To accomplish these objectives a computer modeling-based CSO control package is being developed and applied. By using the package, the system in-line and off-line storage can be optimally used and the subsequent release of combined sewage detained by the storage facilities can be properly regulated to avoid upsets at the waste water treatment plant.

In order to optimize the operation of the system, the entire duration of the storm event needs to be considered when the set points for gate openings at each control structure are established. Therefore, the control strategy includes a prediction of the system loads through the duration of event. The time frame for prediction is called a ''prediction horizon'' or ''simulation horizon.'' The control strategy defines the gate operation strategy for the entire prediction horizon, so that the diverted flows are maintained as close as possible to the desired flows at each station. The strategy is established at the beginning of the storm period, and an internal CSO control model is used to simulate the response of the system. In order to be applicable in practice, the CSO control model needs to have the ability to consider the following:

- several interconnected reservoirs in a variety of configurations
- realistic flow routing components that properly allow for time lag between control points
- backwater effects in flow routing

The CSO control model interacts with the optimization package in order to ensure the feasibility of the selected control solutions.

### Software Components for Hamilton's RTC System

The RTC package developed for the H–R Region includes four primary components. They are:

- rainfall forecasting model
- inflow prediction model
- CSO control optimization model
- dynamic sewer hydraulic model for unsteady flow

The relationship among the four model components is shown in Figure 6.14.

### Rainfall Forecasting Model

Since the CSO control optimization is performed using predictions for the duration of the storm event (prediction horizon), the hyetographs of rainfall intensities versus time for the period of the event are needed.

The two most common general forecasting approaches are: (1) The extended Kalman filter; and (2) the autoregressive moving-average transfer function models (Box and Jenkins, 1976). Both of these models are based on stochastic, or so-called ''black box'' techniques. An alternative to these models has been developed using neural network techniques (Amorocho and Wu, 1975).

The rainfall forecasting model developed for this project is a back-propagation algorithm based on neural network theory. This algorithm includes rules for iteratively changing the weights in any feed-forward network to better describe the relationship between inputs and outputs, thus allowing the model to ''learn'' about the nature of the input-output relationships.

The model training process is divided into two phases:

(1) First, several typical historical rainfall events in Hamilton were used for training the hydrology model. After sufficient model training, the model can remember different rainfall patterns and recall them according to a specified input and time step.

(2) Second, the model input may again be updated in simulated real-time

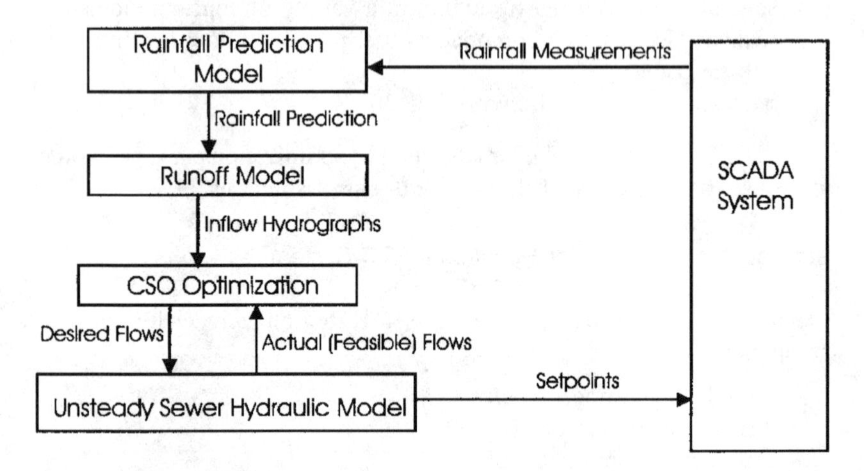


**Figure 6.14** Components of Hamilton-Wentworth RTC software.

as new data describing the current real-time event becomes available. Also, as this is a generally applicable method, any new rainfall data can be used for retraining the neural network rainfall forecasting model.

The rainfall forecasting model is designed to be a learning model that can efficiently incorporate new information into its structure as it becomes available. Succeeding forecasts ideally become better as the model is used to predict more storm events.

### Runoff Model

The overall simulation catchment in Hamilton is divided into 49 subcatchments and each subcatchment has a channel/pipe inlet. For a predicted rainfall hyetograph, the RUNOFF model (adopted from SWMM model) makes the calculation of 49 inlet hydrographs by accounting for infiltration losses in pervious areas, surface detention, overland flow, channel flow, and the constituents washed into the inlets. For the purpose of this project, only quantity is taken into account and water quality or pollutant buildup/washoff are not considered.

### CSO Control Optimization Model

The CSO control optimization model consists of 22 storage elements and 42 delay elements. Each storage element represents a small subnetwork with an overflow structure. Additionally some of these elements have storage capacity and discharge manipulation structures. The diverging or converging elements in the network are represented as storage elements with their storage parameters set to zero. As the network is modified and/or expanded, this model needs to be revised.

The state equation set and related constraint conditions function are solved using the software package MINOS 5.1. The solution can be obtained for either a linear or a nonlinear objective function. The MINOS software package of matrix solver contains four basic solving algorithms, i.e. the simplex method, a quasi-Newton method, the reduced-gradient method and a projected Lagrangian method.

The optimization program can determine the optimum regulator flows for each control point during the storm event so as to minimize the occurrence of overflows. If overflows are unavoidable, the pollution impact on receiving waters can be minimized by appropriately selecting the weights for the objective function in the system optimization model. The major output of the optimization model is a set of desired flow rates during the storm period at 13 regulator stations (automated sluice gates) and one pump station throughout the entire sewer network.

### Hydraulic Model for Unsteady Flow in Sewers

The flow routing (or direction) and flow lag time are not obtained from state equations of the optimization model, as this model only satisfies continuity and not momentum. The flow direction for every conveying element in the entire network is determined by the hydraulics model, and this information is used to construct the optimization model from the basic system elements (such as storage elements, delay elements, etc.).

The optimization model produces a set of desired flow rates for the control points in the system. To convert the desired flows to gate opening-based setpoints, an unsteady hydraulics model needs to be coupled with the CSO control optimization model. The hydraulic model in this control package is a dynamic routing model which employs an implicit finite difference scheme to solve St. Venant equations, i.e. one dimensional momentum and continuity equations with a static hydraulic pressure assumption. The governing equations are similar to that used in the EXTRAN block in SWMM. However, the numerical scheme used in the model solution reduces computational instabilities and permits use of much larger simulation time steps. The model can accurately simulate free surface and surcharge flows, backwater effects, gradually varied flow in pipes, rapidly changing flow in control structures, and hydraulics of looped networks.

### Model Calibration and Verification

The hydraulics model included within the CSO control package normally does not need extensive calibration for each different sewer network since the model is based on mechanistic principles (full momentum and continuity equations) and very few empirical relationships were introduced into the model. The primary calibration parameter in unsteady flow modelling of sewer systems is the pipe (or channel) roughness coefficient which is considered very reliable for most commonly used pipe materials (Labadie et al., 1980). The performance of the hydraulic model was compared with field data and SWMM Extran model results in many large urban sewer networks such as Winnipeg, New York, Liverpool, Vancouver, and London, England. Fairly good agreement between model and field flow data was documented by Ji et al. (1995).

The optimization model includes 22 state equations and an objective function which could be both linear or nonlinear depending on the memory of the computer used to run the model. Some modifications to the model will be made to consider the new Catharine/Ferguson CSO tank which will be added. As mentioned above, the results of the optimization model cannot be directly used for sewer system control. One major reason for this is that the optimized gate flows required by the optimization algorithm only consider continuity and not full hydrodynamics, and thus may not be realistic for the real physical system due to the insufficient calibration of delay elements and storage volumes in the system.

In order to generate desired flows which are realistic for the changing hydraulic conditions in the network, four major parameters in the optimization model need to be calibrated based on the information from both the physical system and the hydraulic model. These parameters include: the delay elements in the system, storage capacity of 22 reservoirs, the upper limit of each diverted flow (through the gate to be conveyed to the WWTP), and the upper limit of overflows (CSOs). The timing, or ''phase'' of the flow pattern generated by the optimization model is very sensitive to the values of storage capacity for reservoirs and the duration of the delay elements in the CSO optimization model.

This concept can be demonstrated on a simple example including a single transport/delay element and one gate. The delay element represents the time which the sewage needs to travel through a conduit between point A and Point B. A hydrograph is introduced into the upstream node A, and a desired flow is computed at gate B. As seen from the Figure 6.15, the

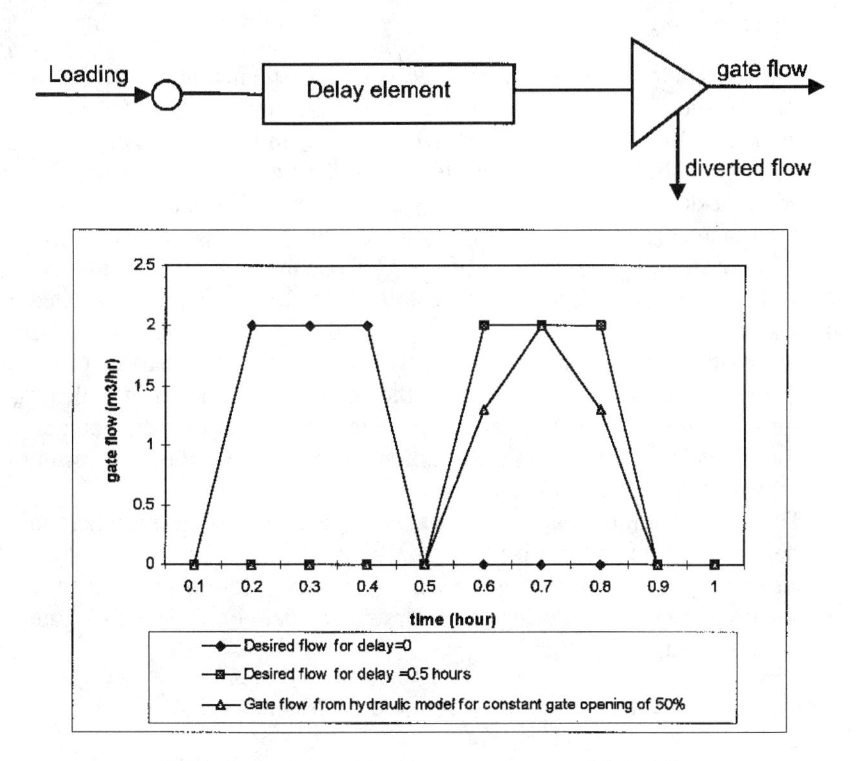


**Figure 6.15** Effect of delay elements on pattern of desired flow.

timing, (or "phase") of the flow pattern at Point B depends on the size (duration) of the delay element: the longer the delay time, the larger the "phase shift" in time between hydrograph at point A and the hydrograph at point B. For a delay of 0.4 hours, the two hydrographs will appear as shifted in phase by 0.4 hours as shown in Figure 6.15; if the delay is set to zero, the hydrograph appears instantly at point B.

If the optimization algorithm generates a desired flow based on the assumption that the hydrograph will travel with no delay, the desired flow cannot be met at gate at node B because the real system does include a transport delay, and the water simply has not reached node B yet. If the optimization model includes the delay, than the difference between the desired and actual flow can be managed by gate actions.

Figure 6.15 shows the following:

(1) A desired flow profile at gate B assuming no delay between node A and node B
(2) A desired flow profile at gate B assuming delay of 0.4 hours between nodes A and B
(3) The gate flow predicted by a hydraulic model, assuming a constant gate opening

Although both desired flows can theoretically result in minimized CSO's at tank B, the desired flow predicted by optimization model with delay element equal to zero cannot be achieved either in the hydraulics model simulation or the real physical system regardless of the gate opening.

With the delay element set to 0.4 hours, the desired flow can be achieved with an increased gate opening. In both the hydraulics model and the real physical system, the adjustment of the gate opening can only change the flow rate at the gate while the phase shift of the flow profile is controlled by the storage and lag elements (in the model) or by the layout of the network (in the actual physical system). In order to ensure that the phase of desired flow profile generated by the optimization model is realistic with respect to the physical system, the delay and the storage elements in the optimization model need to be calibrated using the information from the physical system and the hydraulic model.

The flow direction in each of the connecting elements of the optimization model must be pre-determined by the hydraulic model and cannot be changed during the simulation process. If reverse flow happens at any gate within the system during the hydraulic model simulation, this gate must be pre-set as closed.

### Coupling the Optimization Model with the Hydraulic Model

In order to obtain optimized gate settings based on realistic and verified desired flows, the optimization model must be coupled with the hydraulic model. At the beginning of the hydraulic model simulation, 22 gate opening profiles are assumed for the 22 modeling control gates for the entire simulation period. At the next step the hydraulic model predicts the gate flows during the entire simulation period (20 hours), based on the assumed gate openings. The results are then compared with the desired flow profiles generated by the optimization model. If any phase error is detected between the hydraulics modeling results and the desired flow profiles, the delay elements and the local system storage elements should be adjusted in the optimization model. The model is then rerun to generate a new set of desired flow profiles.

Once the phase error is eliminated, the assumed gate opening matrix needs to be modified by imposing small opening increments (DH), which are proportional to the flow differences between the desired gate flow and the flow predicted by the hydraulic model (DQ).

### Modeling Results

Figure 6.16 shows the predicted gate flow rates and gate opening profiles at the Greenhill CSO Tank. This underground tank can store approximatly 70,000 $m^3$ of combined sewage before overflowing to Redhill Creek. The results of the hydraulic model simulation show that the desired gate flow generated by the optimization model can be accomplished by the optimized gate opening strategy.

Figure 6.17 presents the desired gate flow, the gate flow predicted by the hydraulic model, and the optimized gate settings at the Main/King CSO tank. The storage provided by the Main/King CSO tank is about 75,000 $m^3$. Ten iterations of comparisons between the desired flow and flow generated by the hydraulic model were required.

For an identical rainfall event, the profiles in Figure 6.18 present three different WWTP flows due to different sluice gate control strategies in Hamilton sewer system.

The profile 1 is the result of the hydraulic model simulation for the case when all the gates are closed. In this case there is no risk of flooding either the treatment plant or the interceptors. However, the treatment plant experienced full loading for only about 4 hours since all 13 gates were closed at the beginning of the storm. Obviously, the CSOs in this case were substantially larger than necessary. The profile 2 gives the treatment plant inflow for a case with all the gates fully open. The extra peak loading could be bypassed at treatment plant through the plant spillway. The period during which the treatment plant experienced full loading was about 8 hours. This strategy gives a high risk of flooding the treatment plant and the interceptor.

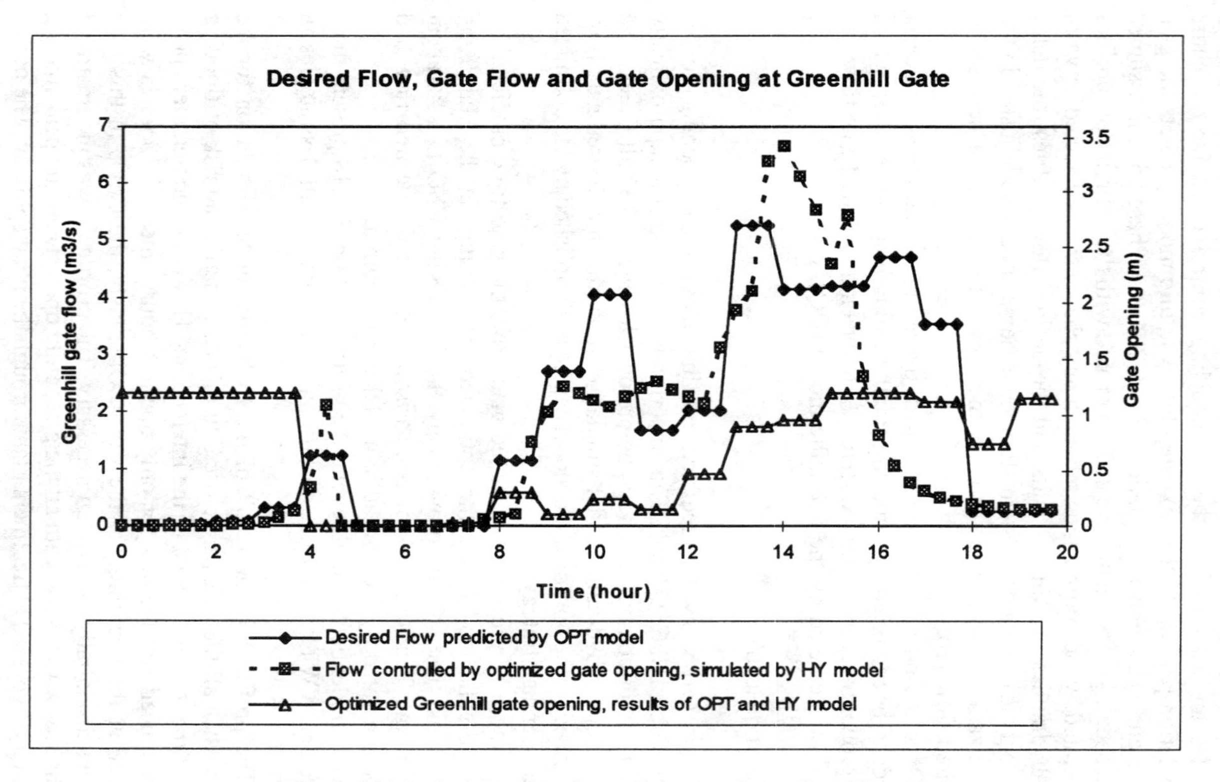


**Figure 6.16** Desired flow, gate flow and gate opening at Greenhill gate.

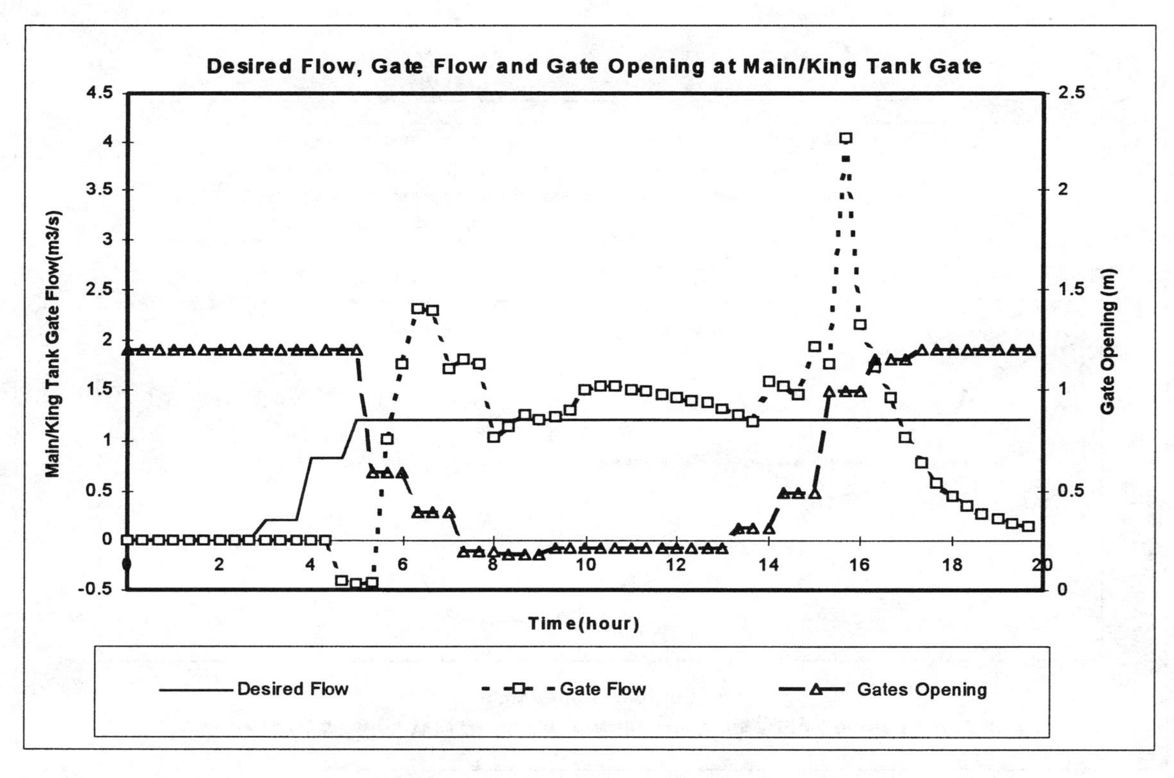


Figure 6.17 Desired flow, gate flow and gate opening at Main/King gate.

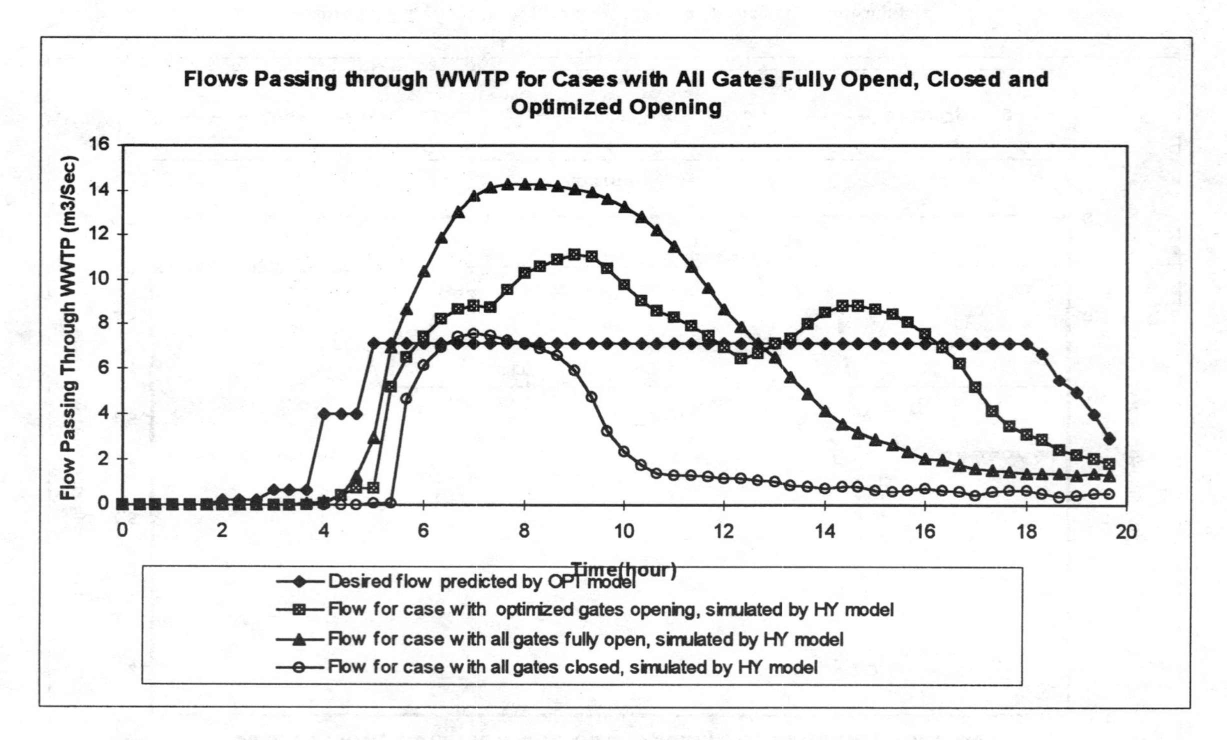


**Figure 6.18** Flows to WWTP for three gate control strategies.

The profile 3 presents the plant influent flow predicted by the hydraulic model coupled with an optimization model. The results indicate that for this rainfall event, the existing storage volume of the Hamilton sewer network should provide the treatment plant a full influent loading ($Q_{plant}$ = 7.1 $m^3/s$) for more than 16 hours by optimizing the operation of all gates.

### System Implementation

At the time of writing this section, the software development for the Hamilton-Wentworth RTC has been largely completed. The modeling and simulation system was in place, and the programs for RTC were tested off-line using simulations. The instrumentation and communication issues were close to completion, and the on-line implementation was scheduled for the second half of 1997.

The problem of rainfall forecasts is common to all applications of optimization to sewer systems. The simulation results show that the optimization algorithm developed for the Hamilton-Wentworth Region works well when loads to the system can be predicted into the future, but in reality these forecasts will not be easily obtainable. The neural network rainfall forecasting algorithm developed for this project shows improvement over other stochastic methods previously used, and it produces reasonable results when the size of the event is roughly estimated (small, medium, or large storm) ahead of time.

The forecasting model is trained on historical records. Implementation will include experiments with iterative methods using updates based on on-line measurements and the use of remote rainfall gauges. While inaccurate forecasts will have some impact on the efficiency of the control strategy and thus the actual control strategy will not be truly optimal, it will still be a significant improvement over the existing operating practice.

## REFERENCES

Amorocho, J., and Wu, B., ''Mathematical Models for the Simulation of Cyclonic Storm Sequences and Precipitation Fields,'' *Proceedings of the National Symposium on Precipitation Analysis for Hydraulogic Modeling,* American Geophysical Union, June, 1975, pp. 210–255.

Beck, R. W. (1974) ''Computer Management of a Combined Sewer System.'' Metro report.

Box, G. E. P., Jenkins, G. M. (1976) *Time Series Analysis Forecasting and Control.* Holder Day.

Brueck, T. M., Knudsen, D. I., Peterson, D. F., (1981) Automatic Computer-Based Control of a Combined Sewer System, *Wat. Sci. Techn.,* Vol. 13, No. 8, 103–109.

Buczek, T. S., Chantrill, C. S., (1984) A Computer-Based System for Reduction of Combined Sewer Overflow in a Metropolitan Wastewater Collection System, *57th Annual WPCF Conf.,* New Orleans, LA, Oct 1–4, 1984.

Callery, R. L., (1971) Dispatching System for Control of Combined Sewer Overflows, USEPA, Report No. WQO-11020-FAQ-03/71, NTIS-PB 203678.

Cantrell, C. J., Godsey, A. H., Schnipke, D., Frutchey, R. W., (1996) Analysis of the City of Lima Real Time Control System Utilizing a State of the Art Hydraulics Computer Model, *Proceedings of Urban Wet Weather Pollution: Controlling Sewer Overflows and Stormwater Runoff,* WEF Speciality Conference, Quebec City

CH2M Hill (1986). "CATAD Improvement Study," report for Metro Seattle.

Colorado State University (1970). Unsteady Free-Surface Flow in a Circular Long Drain, Colorado State University Hydrology Papers, Nos. 43–46, Vol. II. 1643–1646.

Gelormino, M. S., and N. L. Ricker (1991) "Model Predictive Control of Large-Scale Systems," Canadian Chemical Engineering Conference.

Ji, Zhong, Strand, Eric, Vitasovic, Zdenko, Zhou, Siping, (1995) "The Development and Application of a Fast Hydrodynamic Model for Large Looped Sewer/Channel Systems," *WEFTEC' 95, 68th Annual Conference and Exposition,* October, Miami, Florida.

Leiser, Curtis P. (1974) *Computer Management of a Combined Sewer System.* EPA report No. 670/2-74-022, NTIS PB 235717.

Liggett, J. A. (1975) Basic Equations of Unsteady Flow, from *Unsteady Flow in Open Channels,* Yevjevic, V., and Mahmood, K., editors, Water Resources Publications, Fort Collins, CO, Volume 1.

Ricker, N. L. (1990) "Model-Predictive Control of Processes with Many Inputs and Outputs," *Advances in Control and Dynamic Systems,* C. T. Leondes, (ed), vol. 37.

Ricker, N. L. (1990) "Model predictive control of processes with many inputs and outputs," Control and Dynamic Systems, Academic Press, C. T. Leondes, ed., Vol 37, pp. 217–267.

Ricker, N. L. (1991) "Model-Predictive Control: State of the Art." *Proceedings of the Fourth International Conference on Chemical Process Control,* Padre Island, Texas.

Ricker, N. L. (1991) "Model-predictive control: State of the art." Chemical Process Control—CPC IV, Y. Arkun and H. Ray, eds., AIChE/CACHE, pp. 271–296.

Roesner, L. A., Aldrich, J. A., and Dickinson, R. E. (1988) Storm Water Management Model User's Manual Version 4: EXTRAN Addendum I, EXTRAN. Cooperative Agreement CR-811607, U.S. EPA, Cincinnati, OH.

Schilling, Wolfgang (1984) Application of Real Time Control in Combined Sewer Systems, Interim Report, Colorado State University, Fort Collins, CO, USA

Speer, E. S.; Vitasovic, Z. (1991) "AutoCAD based GIS Tool." *Proceedings of the Water Environment Federation Speciality Conference on GIS in Public Utilities,* Orlando, Florida.

Stirrup, M., 1996, "Implementation of Hamilton-Wentworth Region's Pollution Control Plan," *Water Qual. Res. J. Canada,* Volume 31, No. 3.

Swarner, R.; Vitasovic, Z. (1989). "Simulation of Large Urban Wastewater Conveyance System." *Proceedings of 16th Annual Conference of Water Resources and Planning Management,* Sacramento, CA, pp. 222–225.

Trotta, P. D.; Labadie, J. W.; Grigg, N. S. (1977) "Automatic Control Strategies for Urban Stormwater," *Journal of the Hyd. Div., ASCE,* Volume 103, pp. 1443–1459.

Vitasovic, Zdenko; Swarner, Robert; Speer, Edward (1990) "Real Time Control Systems for CSO Reduction," *Water, Environment, and Technology,* v. 2–3, pp. 58–65.

Watt, T. R., Skrentner, R. G., Davanzo, A. C., (1975) *Sewerage System Monitoring and Remote Control,* USEPA Report No. 670/2-75-020.

Winn, C. B. and Morrre, J. B., 1973, ''The Application of Optimal Linear Regulator Theory to a Problem in Water Pollution,'' *IEEE Transactions on Systems, MAN, and Cybernetics,* Vol. SMC-3, No. 5, Sept., pp. 450–455.

Wixcey, J. R., Lewy, M., Price, R. K. (1985) Computational Modeling of Highly Looped Networks of Storm Sewers, Wallingford Software, Howbery Park, Wallingford, Oxfordshire, OX10 8BA, U.K.

CHAPTER 7

# The Integrated Computer Control System: A Comprehensive, Model-Based Control Technology

## INTRODUCTION

### BACKGROUND

AN advanced model-based control system for wastewater treatment facilities has been developed. The Integrated Computer Control System ($IC^2S$) is based on the premise that an integrated computer-based approach to wastewater treatment plant operation and control can have a significant impact on the performance of a plant, including:

- reduction in the duration and frequency of water quality effluent excursions
- reduction in energy costs
- deferred capital expenditures
- optimal use of existing facilities
- ability to cope with unusual plant operation conditions
- reduction in the overall pollutant loadings to receiving water bodies

These impacts can be realized through the use of modern instrumentation and automatic control technology, and by using knowledge of the dynamic behavior of wastewater treatment processes. While treatment plants are usually designed based on steady-state conditions, their performance is sensitive to time-varying loads and environmental factors, which are both beyond the control of operators.

Imre Takács, Gilles Patry, Bruce Gall, and Jasmin Patry, Hydromantis, Inc., 1685 Main St. West, Suite 302, Hamilton, Ontario, L8S 1G5 Canada.

Existing control systems are often based on isolated control loops that do not operate in an integrated manner. By using a dynamic model as the centerpiece of a comprehensive control system, integrated control becomes possible.

$IC^2S$ is connected to the plant Supervisory Control And Data Acquisition (SCADA) system by means of a software link ("driver" or "bridge"). The relationship between the main structural modules and the treatment plant is shown in Figure 7.1.

## $IC^2S$ LAYERS

In Figure 7.1, there are four main *layers:*

(1) Plant process (WWTP)—the physical plant along with control actuators and sensors. Broadly defined, sensors include both on-line and off-line sources of data.

(2) SCADA layer—the existing plant SCADA system, which typically includes a database of historical data, as well as linkages to the sensors and actuators. The SCADA layer is capable of providing automatic, on-line, distributed control of selected plant operating variables.

(3) $IC^2S$/SCADA bridge/(driver)—this component is customized to the type of SCADA system in use at the plant. The bridge links $IC^2S$ to the SCADA database, and allows $IC^2S$ to send operator-approved control actions to the plant via the SCADA system.

(4) $IC^2S$ executive—contains the building blocks and functional modules

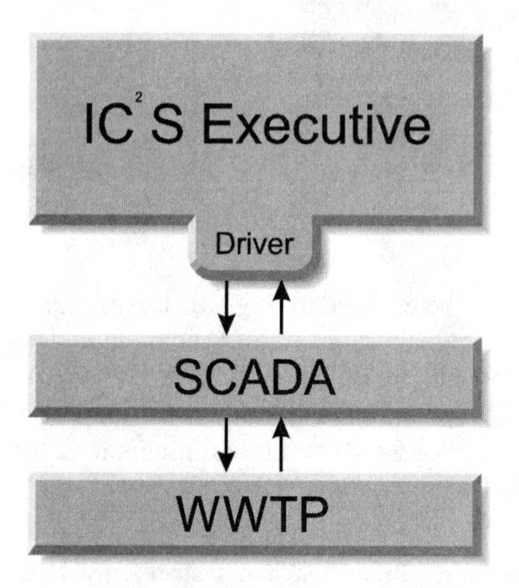


**Figure 7.1** $IC^2S$ and the plant.

that will be defined and described in detail throughout the rest of this chapter.

## THE BUILDING BLOCKS OF IC$^2$S

There are several "low level" components within the IC$^2$S Executive. These "building blocks" are used to create the functional modules of the control system; it is the synchronized operation of these modules that constitutes IC$^2$S. The IC$^2$S building blocks are:

(1) Adaptive data filter (ADF)—treatment plant data can be noisy and/or biased. These problems can sometimes mask important trends. The ADF is used to identify important trends contained within a noisy time series. Smooth data trends are necessary to ensure stable operation of modules such as the Dynamic Parameter Estimator.
(2) Signal tracking and alarm (SigTrack)—provides low level signal tracking to identify and report outliers, noisy or missing data and excessive rate of change.
(3) Respirometric parameter evaluator (RespEval)—uses advanced pattern recognition to determine the biological activity of activated sludge and the completeness of waste component oxidation (i.e., wastewater and sludge characteristics). The measurement of these characteristics improves the accuracy of the dynamic plant model.
(4) Dynamic parameter estimator (DPE)—automatically estimates the time-varying nature of parameters that are usually regarded as constants in off-line modelling studies. Along with respirometric parameter identification, the DPE will help to ensure a high degree of simulation model accuracy. In addition, changes in parameter values can provide a basis for problem detection and diagnosis.
(5) Advanced control design (GMI)—allows the creation of linearized dynamic models based on the non-linear calibrated simulation model. The resulting linear model can be imported into GMI (running within Matlab) to design process controllers; these controllers can then be tested using the non-linear simulation model.
(6) Simulation model—the model of the plant being controlled. The simulation model is created with the GPS-X simulator program.
(7) Operator interface (scenario manager)—the Operator Interface (or On-line Scenario Manager) manages all of the IC$^2$S functionality, schedules the various modules and provides a graphical user interface for the operator.

In addition to the blocks listed above, the IC$^2$S Kernel also has intrinsic bridges that link the various functional modules.

## FUNCTIONAL MODULES INSIDE IC²S

The configuration of the IC²S is flexible: building blocks (e.g., ADF, DPE, SigTrack, etc.) are implemented only if needed for a particular plant. A potential functional implementation of the IC²S is shown in Figure 7.2 containing the following elements:

(1) Auto-calibration module—IC²S uses an on-line calibrated dynamic model of the plant being controlled. The current calibration status has to be verified and calibration constants updated if necessary. IC²S, using the Dynamic Parameter Estimator (DPE) block, can dynamically optimize effluent suspended solids, sludge blanket height, or other relevant variables by varying the model parameters that have the greatest effect on the selected monitored variables. The Autocalibration Module consists of several on-line, continuously running optimizations, each maintaining calibration of respective parameters and communicating the best values to each other and to a calibration database.

(2) Advanced fault detection module—In addition to the low level signal tracking and alarm generation blocks, more advanced, intelligent sensor and process fault detection is offered in this module. The Advanced Sensor Fault Detection works by continuously comparing the output

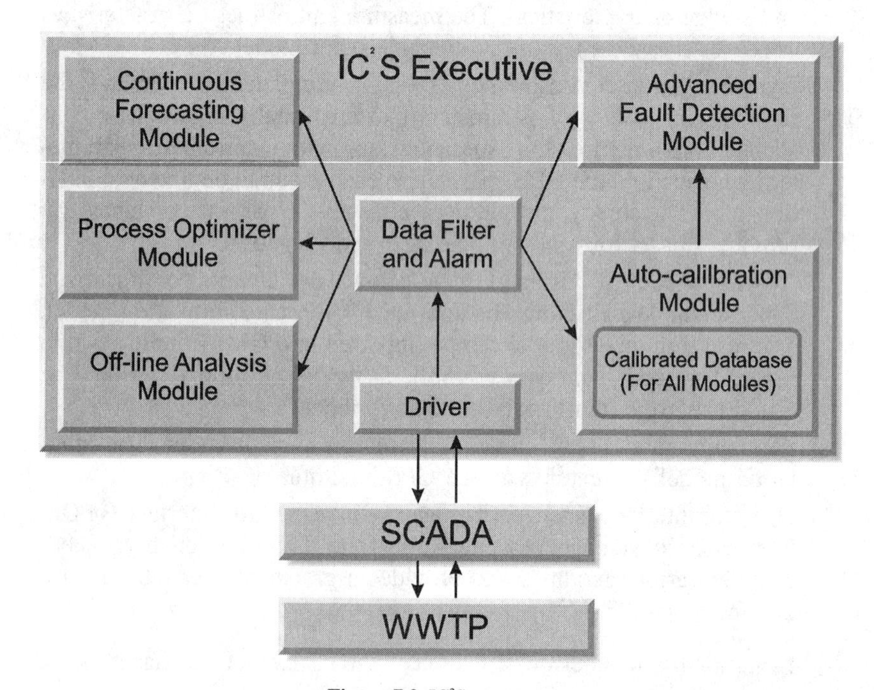


**Figure 7.2** IC²S structure.

of the calibrated model with data provided from the SCADA system. Discrepancies between these two signals either indicate a sensor fault, or, in some cases, a fault in maintaining self-calibration. The cause can be decided based on mass-balance calculations, typical parameter and sensor ranges, etc. The Process Fault Detection tracks auto-calibrated parameters and generates alarms for sudden changes or slow drift.

(3) Process optimizer module—This module (conceptually shown in Figure 7.3) uses several instances of the calibrated dynamic model of the plant to perform process optimizations in real-time. The Process Optimization Module uses the same numerical technique (the Dynamic Parameter Estimator) as the Auto-Calibration Module, but it optimizes process variables as opposed to model parameters.

(4) Continuous forecasting module—This module is a separate simulation module which automatically runs using predicted or historical forcing functions (influent patterns). The output from the Continuous Forecast Module is used to evaluate the long-term effect of desired process changes on effluent quality and other key operational parameters.

(5) Off-line analysis module—This module consists of a plant model which has access to the calibration database. This module can be used by plant staff for tasks such as specific forecasting, plant analysis, or design.

## THE BUILDING BLOCKS OF IC²S

### ADAPTIVE DATA FILTER (ADF)

Signals originating from sensors typically contain noise and spikes that are not representative of the variable being monitored. On-line data can be filtered in a number of ways, but simple filtering techniques (e.g., low-pass) are not always adequate since they do not guarantee extraction of the maximum amount of information from the signal. The adaptive data filter was developed to address this problem.

The adaptive data filter (ADF) constructs a cubic spline for a given set of data. The construction adjusts the number of *knots* (connection point between two cubic segments) and the location of these knots to maximize the information content of the resulting spline, as defined by the Akaike Information Criterion (AIC) (Brannigan, 1990). The splines that result can therefore be assumed to be the best estimate of the real behavior of the measured variable.

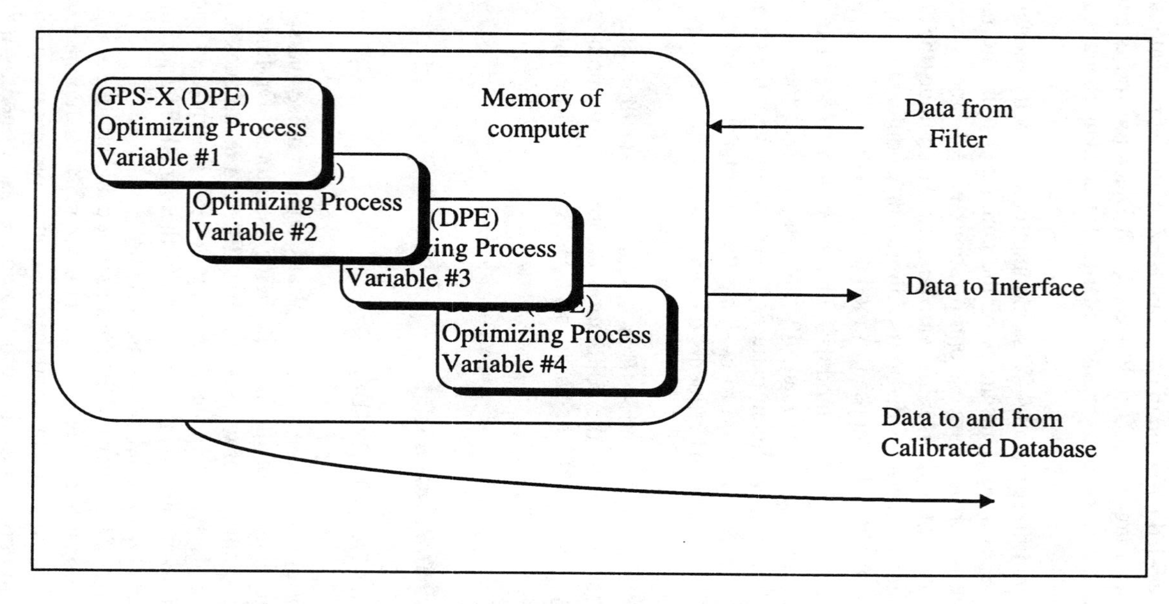


**Figure 7.3** Structure of the optimization module.

The role of the ADF within the overall IC²S framework is shown in Figure 7.2. The ADF serves two important purposes in the IC²S. The spline calculated by the ADF, rather than the actual (raw) data, can be used by the dynamic parameter estimation (DPE) routine. This is an effective way of addressing the problems that result when optimization tools are used together with noisy data. Since the ADF spline is a continuous function, problems associated with DPE implementation are also resolved; the user can select any size of window for the optimization process. The Respirogram Evaluator (RespEval) module also benefits from the ADF routine. RespEval requires information about the first and second derivatives of the oxygen utilization rate to indicate transitions between components. Since the ADF supplies a (twice-differentiable) continuous signal, RespEval can use this well-conditioned signal to get the required information.

The following screen-shots from GPS-X illustrate the functionality of the adaptive data filter (ADF). Figure 7.4 shows the ADF being defined for influent flow and influent BOD. The ADF has been added into the GPS-X *DEFINE* function, so that a data set for any parameter or variable can be filtered.

Figure 7.5 shows a typical set of noisy data generated at a wastewater treatment facility. In this case the data shown is influent carbonaceous $BOD_5$. While this data set is somewhat noisy, it is also apparent that it contains trends.

The results of ADF filtering are shown in Figures 7.6 and 7.7. Figure 7.6 shows the ADF fit when the user specifies a tight fit between the original data and the fitted curve. Figure 7.7 shows the ADF fit when the user specifies a looser fit between the original data and the fitted curve, which results in a smoother filtered curve. The degree of fit is controlled

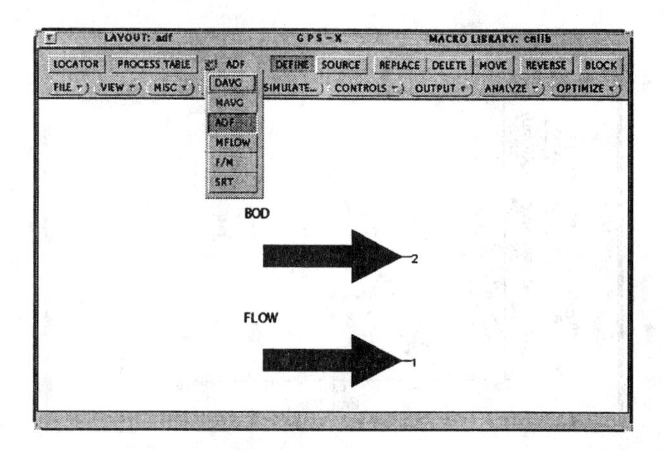


**Figure 7.4** Defining the ADF for BOD and flow.

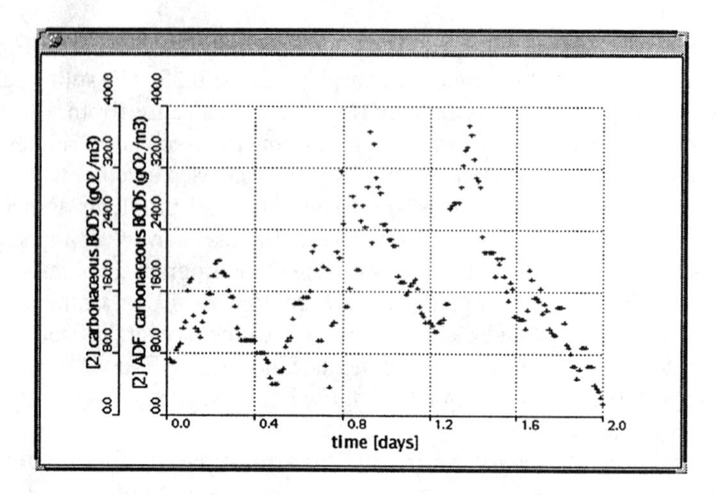


**Figure 7.5** Raw data.

by a tightness factor which can be varied from 0 (very smooth curve) to 1 (very tight-fitting curve). A tight-fitting curve will have a larger number of knots than a smooth curve. From these graphs it becomes apparent that if the fit is too smooth, we lose important trends and information from the data. If the fit is too tight, we essentially duplicate the original data and the filtering effect of the ADF is minimized.

Finally, Figure 7.8 shows a comparison between raw data, ADF filtered data, and a moving average. In this figure, the moving average data displays a lag due to the effect of the type of moving average performed (the value at a given point is based on an average of previous data points in a

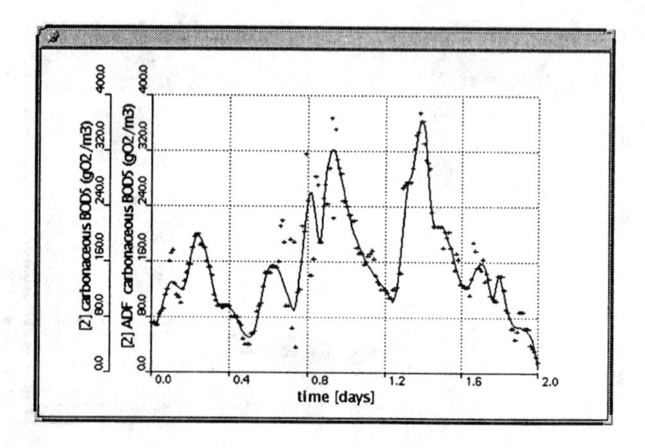


**Figure 7.6** Noisy data with filtered signal (tight fit).

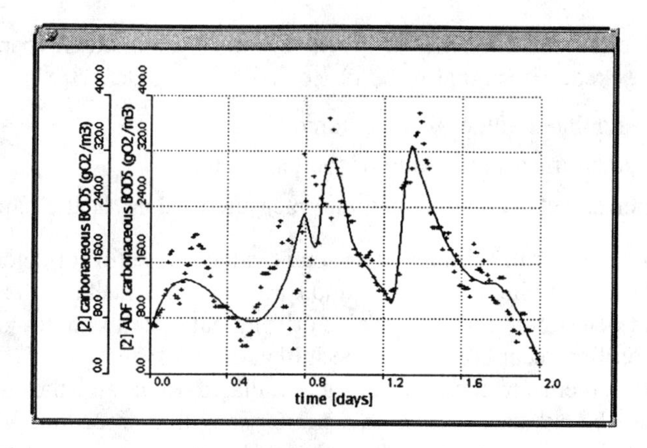


**Figure 7.7** Noisy data with filtered signal (smooth fit).

user defined window). The ADF also avoids the over-fitting problems commonly associated with standard filtering techniques.

## SIGNAL TRACKING AND ALARM (SIGTRACK)

This component allows the user to assign a "tracking function" to any parameter of interest; this tracking function has the ability to detect several

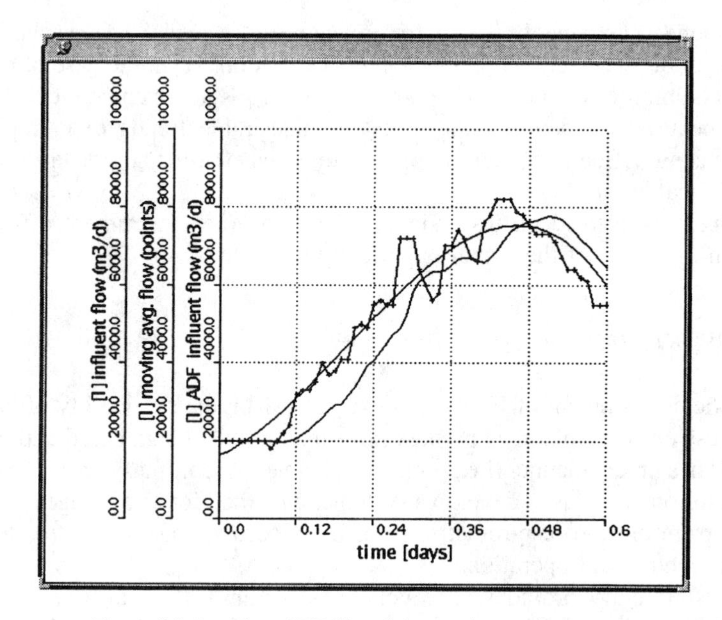


**Figure 7.8** Raw data, ADF filtered data, and a moving average.

types of error conditions. In a given layout, any number of parameters can be tracked. The signal to be processed can include:

(1) The simulated value of the parameter
(2) The actual measured value of the parameter
(3) The error between simulated and measured values of the parameter

Tracking either the simulated or actual values is useful for determining if the plant is moving towards an undesirable state. Checking the error is useful for determining if the model is drifting out of calibration—in which case an on-line recalibration can be initiated.

To implement error checking, the simulated value and the measured value must be evaluated based on a similar sampling procedure. Since the simulated signal is continuous, while sampled signals are often based on a form of averaging procedure, it was necessary to simulate this sampling procedure by creating a function which ''samples'' the simulated signal. This function can be used in other aspects of $IC^2S$, including calibration.

Three types of signal tracking can be enabled simultaneously for the chosen signal:

(1) Value bounds checking (upper and lower)
(2) Rate of change checking
(3) Noise level checking

Bounds checking is useful for indicating if a signal has reached an undesirable state (e.g., high effluent concentration, or, in the case of error signals, high error indicating poor calibration). Rate of change checking can be used to identify dangerous or undesirable trends (e.g., rapidly increasing effluent concentration, on-line sensor drift, or deteriorating model calibration).

SigTrack also identifies a signal as missing (i.e., instrument off-line) when the value of the signal is perfectly constant.

## RESPIROGRAM EVALUATOR (RespEval)

The development of $IC^2S$ relies extensively on the availability of basic process information. This process information must be gathered and analysed in a timely manner (i.e., pseudo-real-time). Accordingly, the approach to respirometry was to develop a systematic method for knowledge extraction from on-line respirometric data in support of wastewater treatment plant control and operation.

Respirometry provides an excellent opportunity for real-time control of biological wastewater treatment plants (see Chapter 4). Respirometry can be defined as the measurement of the oxygen consumption rate (mg/L · h) of microorganisms during the oxidation of organic matter and ammonia. Respirometry is a well-known tool for extracting information from the activated sludge processes. By monitoring the respiration rate of the microorganisms in the process, it is possible to make inferences regarding the activity of the biomass and the state of the system.

RespEval uses a model-based approach for the analysis of respirometric data and for the design of respirometric experiments that maximize the information content of respirograms. For example, on-site experiments involving the addition of settled sewage and/or other substrates (sodium acetate, ammonium chloride, etc.) can provide more insight into the characteristics of the mixed liquor and allow for a more efficient operation of the plant.

Although the usefulness of respirometry is recognised and has been demonstrated, experimental methods that extract maximum information—i.e., as many parameters (wastewater and sludge characteristics) as possible—still need to be developed and improved. The amount of information generated by respirometry likely depends on the specific apparatus and the experimental set-up (see Chapter 4).

RespEval can be operated in two modes: (1) calibrate and (2) evaluate.

Calibrate mode is used to determine the "oxygen to concentration conversion factor" when a component concentration is known. Organism yield can be directly computed from this factor. The word "component" refers to substrates such as carbonaceous substrate, ammonia, or volatile fatty acids. This is generally performed on a sample of mixed liquor that is brought to an endogenous phase and then dosed with a known quantity of a known substrate. Typical output from the calibration phase is shown in Figure 7.9.

After this conversion factor has been determined, RespEval can be used to calculate the concentration of the component for any given respirogram, such as the one shown in Figure 7.10. The first step in evaluating the respirogram is to determine how many substrate components it represents. One method of doing this is to look at the second derivatives of the respiration rate as shown in Figure 7.11. Generally whenever the second derivative goes from a negative value to a positive value, this indicates that the oxidation of a substrate is nearly complete. Therefore the number of times that the second derivative goes from negative to positive is the number of substrate components in the dose.

The evaluator can then determine the half saturation coefficient, the maximum respiration rate, and the concentration of these substrates based on the conversion factor determined in calibration mode. The results of an evaluation for a single component are shown in Figure 7.12.

```
RESPEVAL CALIBRATION MODE

Input data files:
respeval_real_1995_03_24.dat

Calibration started at:            5.0 min
First endogenous detected at:      5.0 min
Dosage detected at:                15.1 min
Endogenous reached at:             62.3 min
Evaluation stopped at:             90.0 min


Respirogram evaluation results
                         Time       OUR
                         [min]      [mgO2/L/h]
First endogenous:        5.0        12.0
Sample dosage:           15.1       12.0
Peak:                    15.4       107.0
Sample oxidized:         62.3       10.8
Last endogenous:         90.0       10.6

Time to treatment:                 47.2 min
Short term BOD:                    481.8 mgO2/L
Measured conversion:    4.23
```

**Figure 7.9** Typical calibration output for one component.

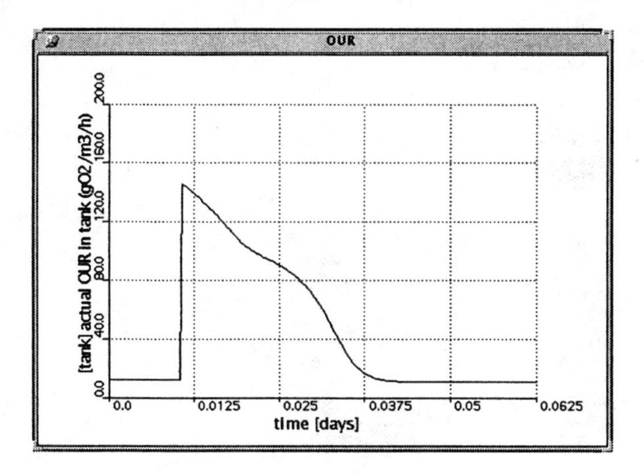


**Figure 7.10** Two component respirogram.

## DYNAMIC PARAMETER ESTIMATOR (DPE)

DPE allows automatic estimation of time-varying parameters, which have hitherto been considered to be relatively constant. On-line estimation of parameters is required to compensate for immeasurable and confounded state variables and to update the model calibration as new information is made available. The confounded state variables are the components of the system for which insufficient knowledge is available to produce a meaningful and/or useful model. As an example, consider the IAWQ Activated Sludge Model No. 1 which is used to model the fate of the

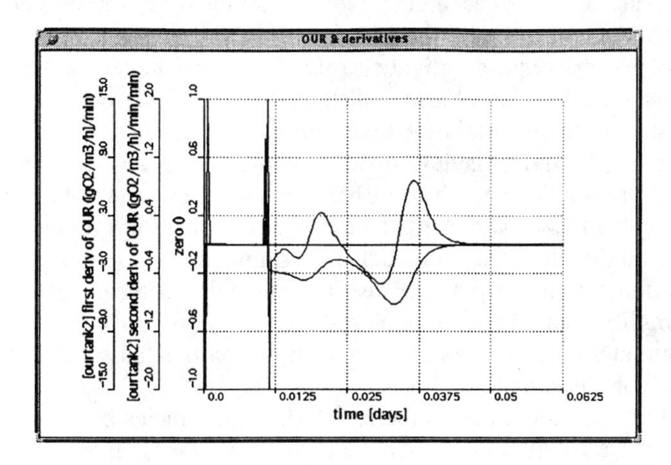


**Figure 7.11** Indentification of two components from the derivative of OUR.

```
RESPEVAL EVALUATION MODE

Evaluation started at:            5.0 min
First endogenous detected at:     5.0 min
Dosage detected at:               15.1 min
Endogenous reached at:            55.3 min
Evaluation stopped at:            90.0 min


Respirogram evaluation results
                           Time          OUR
                           [min]         [mgO2/L/h]
First endogenous:          5.0           12.0
Sample dosage:             15.1          12.0
Peak:                      15.5          103.4
Sample oxidized:           55.3          10.7
Last endogenous:           90.0          10.6

Time to treatment:              40.2 min
Short term BOD:                 345.6 mgO2/L
Concentration:                  82.3 mg/L

Maximum respiration rate:       106.2 mgO2/L/h
Half saturation:                1.2 mg/L
R2 (using 20 data points):      0.999

  Time from   Adjusted   Inferred    Real      R2
   Dosage       OUR        Conc      Conc     Conc
    0.9         92.1       8.8       7.2     *********
    1.9         91.2       8.4       6.8     *********
    2.9         90.3       8.0       6.5     *********
    3.9         89.3       7.6       6.1     *********
    4.9         88.3       7.2       5.8      0.9985
    5.9         87.4       6.8       5.4      0.9975
    6.9         86.3       6.4       5.1      0.9966
    7.9         85.2       6.1       4.8      0.9958
    8.9         84.0       5.7       4.4      0.9954
    9.9         82.7       5.3       4.1      0.9953
   10.9         81.2       4.9       3.8      0.9955
continued ...
```

**Figure 7.12** Typical evaluation output for one component.

carbonaceous and nitrogenous components. This model contains two types of bacteria categorized according to their function—heterotrophs and nitrifiers. This gross assumption was made since the real system has numerous types of bacteria whose individual roles are poorly understood.

If a model was developed which included all the various types of organisms it would certainly be very impractical for use. If, on the other hand, a simpler model is developed, it would not be very meaningful for the types of applications that are being pursued. For example, a simpler model probably could not predict the extent of nitrification very accurately. It is hoped that the types of models being implemented represent the best trade-off between the practical use of a model and meaning of results. The simplifications made to the working models to improve their practicality come at a cost of introducing hidden or confounded states which are usually not measurable on-line.

How, then, should these inherent model inaccuracies be dealt with? If the simplest solution is chosen (i.e., to do nothing), then it might be expected that the error in the model predictions may be very high. This is probably not acceptable, especially when the aim is to make forecasts using the model and develop operational strategies based on these forecasts, perhaps compensating for these errors in some manner. Another solution would be to make forecasts based on sensitivity analysis of the unknown model inputs. For example, if the operator was considering a change to the plant solids retention time (SRT), forecasts could be developed for different values of parameters such as autotrophic growth rate or yield and perhaps choose the prediction based on the most conservative results.

The ammonia forecasts produced are a function of the autotrophic growth rate and SRT. If the goal was to determine the shortest SRT to maintain an ammonia level below 5 mg/L, then the operator might choose an eight day SRT based on a lower growth rate ($0.25\ d^{-1}$) as a conservative estimate. The difficulty with this approach is that the ranges of uncertainty for the parameters are also unknown. Furthermore, this technique quickly becomes impractical for multiple parameters: three parameters each varied over three different levels implies a $3 \times 3 \times 3$ factorial design or 27 simulations followed by an analysis to decide which forecast to use.

A better approach is to use a technique of parameter estimation or updating. In this case, the value of the parameters used for a forecast represents the current values dictated by measured variables at the plant. With this methodology, the unknown parameters are continuously updated based on new process information. The Kalman filter is one method that updates unknown parameters. Details of this approach and its application to models are provided elsewhere (Beck and Chen, 1994). Another method to provide parameter estimation is the use of an optimization routine (Côté et al., 1995). The DPE uses an optimization approach.

When calibrating a model off-line to one set of observations (plant data), a rigorous approach is to use mathematical optimization to estimate immeasurable states or parameters to provide the best fit between model and data. One selects an appropriate optimization routine and defines the objective function and parameters to be optimized. The routine will then estimate the best parameter values.

This method has been extended for use with a dynamic data set; that is, one that is continuously updated. The approach is identical to the case of the static data set with the following exceptions. First a moving time window is specified. The objective function is specified (i.e., the error between model predictions and observed data). The DPE will then optimize the identified model parameters over this time window to minimize the objective function. As an example, if the nitrification growth rate has been identified as the parameter for optimization and the time window is 12 hours, then the DPE will optimize the fit between the model and data target variables based on the preceding twelve hours of data.

The next application of the DPE (twelve hours later) will again be based on the preceeding twelve hours data. The time window thus serves as the weighting factor, so that if it is longer, more weight is placed on the current data and less on the previous history (which is stored in the initial conditions for that period). If the time window is made shorter then more weight is placed on the history and less on the most recent data. Shorter time windows require a smoothing method like the ADF to prevent fitting for noise in individual data points. The mathematical optimization routine used is the Nelder-Mead Simplex method (Press et al., 1986) modified to include bounds on the parameter values and a method to prevent parameters from sticking to those bounds.

In addition to the length of time window mentioned above, a number of factors can also affect the performance of the DPE. These factors include: (1) noise in data; (2) formulation of objective function; (3) number and identifiability of parameters to be estimated; (4) initial step size of optimization routine; (5) termination criteria for optimizer routine; and (6) starting point of simulation. Each of these factors and their effects on the DPE are described below.

The quality and frequency of the data used to drive the model are critical factors. Some noise is expected with all data that is measured at the plant. There are numerous types and sources of error which are described elsewhere (Olsson and Piani, 1992). The effect of the noise is a function of the length of time window. If the time window is too short, the noise may dominate, so that the DPE will try to fit the noise rather than the underlying data trend. Another solution, besides increasing the duration of the DPE window, is to filter the data with the ADF prior to invoking the DPE.

The formulation of the objective function also affects the performance of the parameter estimator. The objective function is defined before the optimization begins. Forms of the objective function are shown as follows:

Absolute difference:

$$F = \sum_{j=1}^{S} \sum_{i=1}^{N} |\Phi_{i,j}^{m} - \Phi_{i,j}^{d}|$$

Normalized difference:

$$F = \sum_{j=1}^{S} \sum_{i=1}^{N} \frac{|\Phi_{ij}^{m} - \Phi_{ij}^{d}|}{\Phi_{ij}^{d}}$$

Sum of Squares

$$F = \sum_{j=1}^{S} \sum_{i=1}^{N} (\Phi_{ij}^{m} - \Phi_{ij}^{d})^2$$

where:

$F$ = the objective function,
$\Phi_{ij}^{m}$ = the $i$th model data point in the $j$th series,
$\Phi_{ij}^{d}$ = the $i$th data point in the $j$th series,
$N$ = the number of observed data points
$S$ = the number of observed data series

The number of parameters identified for optimization is important. They must be identifiable (i.e., have some measurable effect of the objective function) and be selectively chosen. It is not practical to select all of the parameters of the model as optimization variables since this will slow the update process, and will likely result in meaningless values for the model parameters. It is preferable to choose only those parameters that have the greatest effect on the model mismatch. For example, if the mismatch is in the effluent ammonia, then only the growth rate of the nitrifiers should be included and not the settling parameters.

Since the optimization routine will begin its search with the current parameter value, the initial step size will determine how quickly the optimum is found as well as affecting the probability of locating the global optimum (i.e., a large step size may break away from a local optimum). If the initial step size taken is too small, then the optimizer may require an excessive number of steps to update the parameter, especially if the new parameter value is considerably different from the current parameter. However, if the parameter has not significantly changed, then the small step size will be beneficial.

The termination criteria of the optimizer are:

(1) Reaching the specified objective (some acceptably small amount of model mismatch)
(2) Small change in parameter values
(3) Small change in objective function (meeting this criteria without satisfying criteria 1 or 2 may indicate a poor parameter selection)
(4) Maximum number of iterations

The values of these criteria provide a trade-off between the length of the optimizer run and the proximity to the optimum.

The following results were produced using the DPE. In the first case, the data to which the DPE fit the model parameters was noise-free. The data was generated with a model which avoids the problem of poor model identification. The case represents a change in biosolids clarification ability caused by the introduction of some ionic substances affecting the zeta potential of the individual flocs and their ability to flocculate. This could be the change in the addition rate of a coagulant. Although other micro-scale models examining the interactive energies between particles (Stern-Gouy-Chapman model) should be able to predict the change in clarification, little work has been carried out in this area; therefore, in this case, it is more desirable to adjust the clarification parameters using the DPE.

The data was generated for a four-day period using a biological excess phosphorus removal activated sludge model coupled with the one-dimensional reactive final settler model with diurnal influent. The model used for simulating the plant during parameter optimization was a modified ASM#1 activated sludge model with a non-reactive one-dimensional settler model. A layout of the model is shown in Figure 7.13.

The input flow rate is shown in Figure 7.14. A sinewave was used to introduce some dynamics to the simulation. Figure 7.15 shows the other inputs to the simulation model supplied from the plant model. These include influent COD, TSS, TKN, influent flow rate, primary sludge pumped, air flow rate, recycled biosolids flow rate, and wasted biosolids flow rate.

Initially, the clarification parameter was 0.4 (on a qualitative scale of 0.0 to 1.0 where 0.0 represents poor clarification and 1.0 represents good

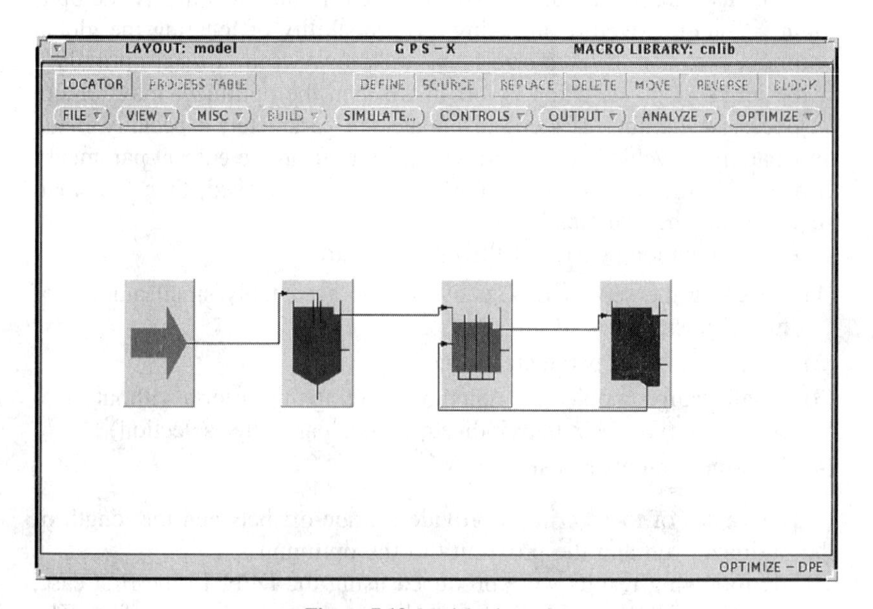


**Figure 7.13** Model schematic.

Infl
[PI] flow (x10+3) (m3/d)
0.0
200.0
400.0
600.0
800.0
1000.0
0.0
0.8
1.6
2.4
3.2
4.0
time [days]

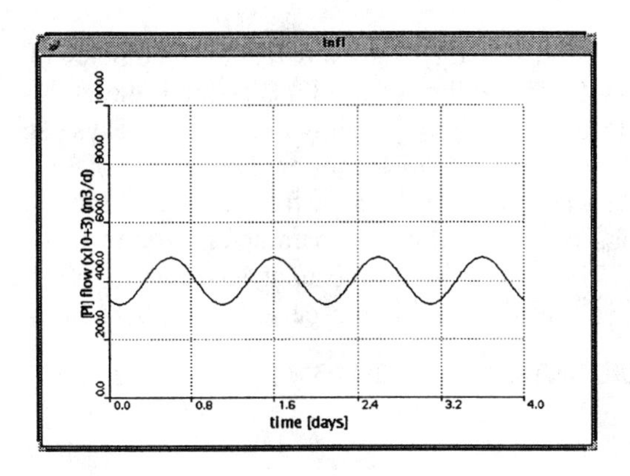

**Figure 7.14** Diurnal influent flow rate.

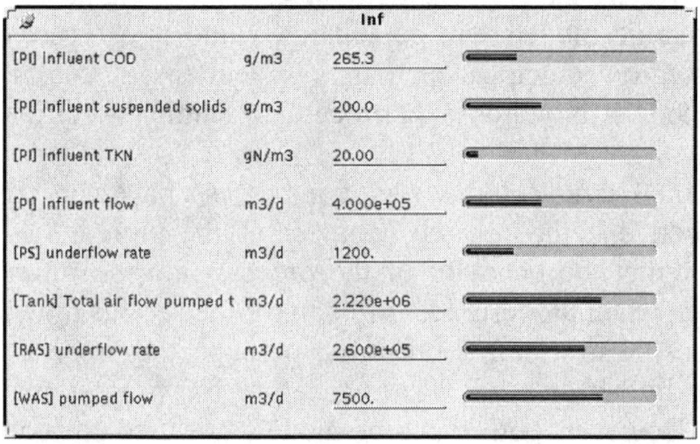

Inf
[PI] influent COD g/m3 265.3
[PI] influent suspended solids g/m3 200.0
[PI] influent TKN gN/m3 20.00
[PI] influent flow m3/d 4.000e+05
[PS] underflow rate m3/d 1200.
[Tank] Total air flow pumped t m3/d 2.220e+06
[RAS] underflow rate m3/d 2.600e+05
[WAS] pumped flow m3/d 7500.

**Figure 7.15** Inputs to the model.

Clarification Parameter
[SE] Clarification (0 – bad, 1 (-)
0.0
0.2
0.4
0.6
0.8
1.0
0.0
0.8
1.6
2.4
3.2
4.0
time [days]

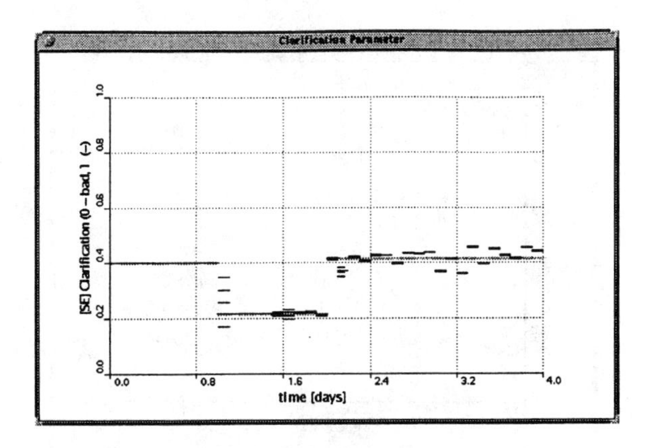

**Figure 7.16** Clarification parameters tracked by DPE.

clarification). After one day, the clarification parameter in the plant was decreased to 0.2. The ability of the DPE to track these changes is shown in Figure 7.16 using a moving time window of 0.1 days. The clarification factor was increased back to 0.4 after 2.0 days. However, at that time random noise was added to the data from the plant model. This resulted in fluctuations in the clarification parameter as seen in the later half of Figure 7.16. The values of the effluent suspended solids concentration over the entire four-day simulation period are shown in Figure 7.17.

## ADVANCED CONTROL DESIGN (GMI)

### GMI Purpose

The design of advanced control systems requires dynamic mathematical models that reflect the important characteristics of the process to be controlled. Dynamic models can be separated into two types: fundamental models (based on principles such as mass and energy conservation), and empirical models (based on experimental dynamic data) (See Chapters 1 and 2).

Fundamental dynamic models of a process have the advantages of providing both insight into the behavior of the process, as well as the ability to predict the behavior of the process into the future. Empirical models, even when properly identified, have the serious drawback of only being applicable to a single operating point in the process. Changes in process conditions or other disturbances to the process may invalidate the model. However, both types of models are used for control design; fundamental models because of their power, and empirical models because of their simplicity.

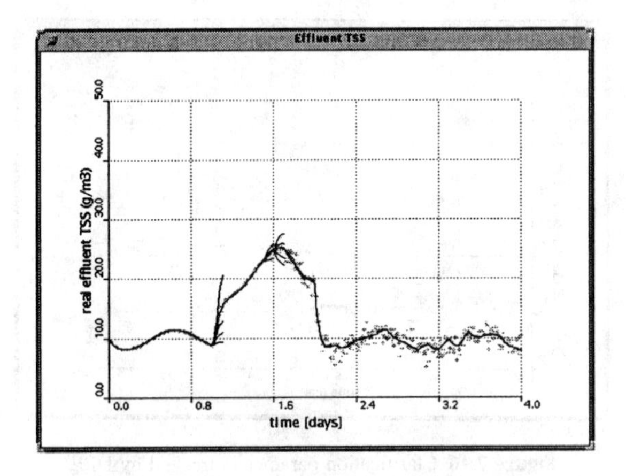


**Figure 7.17** Effluent TSS concentration.

For wastewater treatment processes, identifying dynamic models is not a simple task. Fundamental models of the activated sludge process, for example, can contain dozens, or even hundreds, of state variables. This makes "pen-and-paper" modeling of the process prohibitive. Empirical modeling presents its own set of problems: since treatment processes are constantly subject to influent flow and load disturbances, isolating the dynamic relationship between any two process variables is difficult. This problem is aggravated by the long time constants that characterize much of the dynamic behavior of wastewater treatment processes.

The introduction of computer-based modeling tools for wastewater treatment processes has given the process engineer the opportunity to manipulate a full non-linear model of the wastewater treatment process to aid in tasks such as process operation and optimization. However, nonlinear models are infrequently used by engineers for control design, since much of control theory is based on the use of linear time-invariant (LTI) models.

GMI[1] was developed to provide a "bridge" between process simulation and modeling tools (e.g., GPS-X) and control design software (e.g., Matlab[2] and the Control System Toolbox). The synergy between these two tools is substantial: GPS-X can be used to develop a calibrated fundamental dynamic model of a treatment process, GMI and Matlab can be used to design controllers, and GPS-X can then be used to evaluate the controller design. This design process is shown in Figure 7.18.

### GMI Description

GMI makes use of *state-space models* to represent both the process model and the process controllers. A state-space model of the process is generated in GPS-X and imported into GMI for controller design. Users can use the state-space process model to design controllers in one of two ways:

- to quickly and easily design MIMO[3] feedforward/feedback PI controllers with a customized graphical user interface built-in to GMI
- to design more complex controllers in the Matlab environment

#### *State-Space Models*

Processes and controllers are represented by state-space models in GMI. State-space models are LTI models with the form:

[1] GMI is an acronym for GPS-X/Matlab Interface.

[2] Matlab is a registered trademark of The Mathworks, Inc.

[3] Muli-Input Multi-Output. Variations include SISO (single-input single-output), SIMO (single-input multiple output), and MISO (multiple-input single-output).

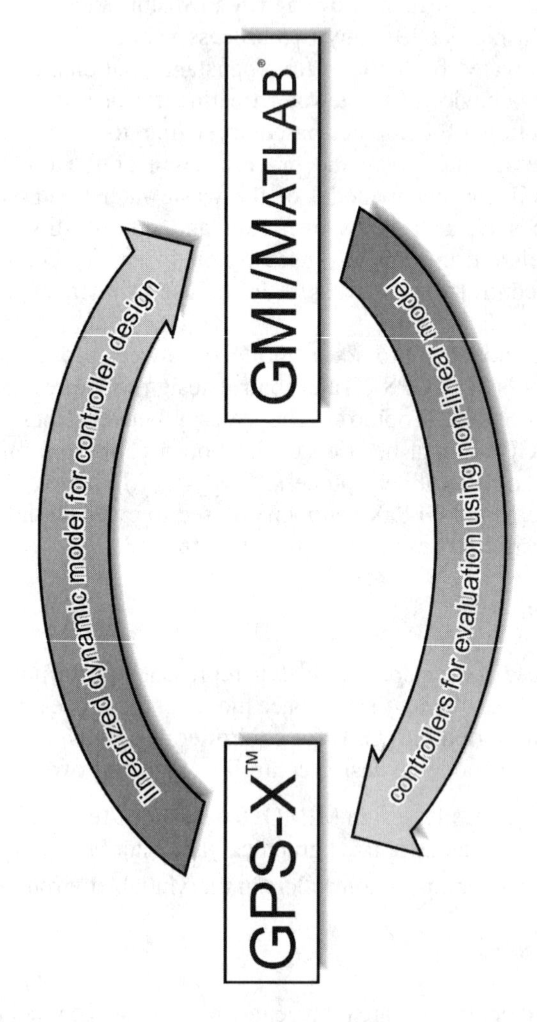


**Figure 7.18** The interrelationships between GPS-X and GMI/Matlab.

$$\frac{dx(t)}{dt} = Ax(t) + Bu(t)$$

$$y(t) = Cx(t) + Bu(t)$$

where:

$x(t)$ = a vector of the states of the system $[x_1(t)\ x_2(t)\ \ldots\ x_p(t)]^T$
$u(t)$ = a vector of the inputs to the system $[u_1(t)\ u_2(t)\ \ldots\ u_q(t)]^T$
$y(t)$ = a vector of the outputs of the system $[y_1(t)\ y_2(t)\ \ldots\ y_r(t)]^T$
A,B,C,D = constant matrices

State-space models are particularly useful for dealing with large MIMO process models, which frequently occur when modeling wastewater treatment processes. For more information on the use state-space models in controller design, the reader is invited to consult Franklin et al. (1990).

### *Designing Controllers with GMI*

The steps involved in designing controllers with GMI are shown in Figure 7.19. As shown in this figure, the design process consists of six steps, summarized below:

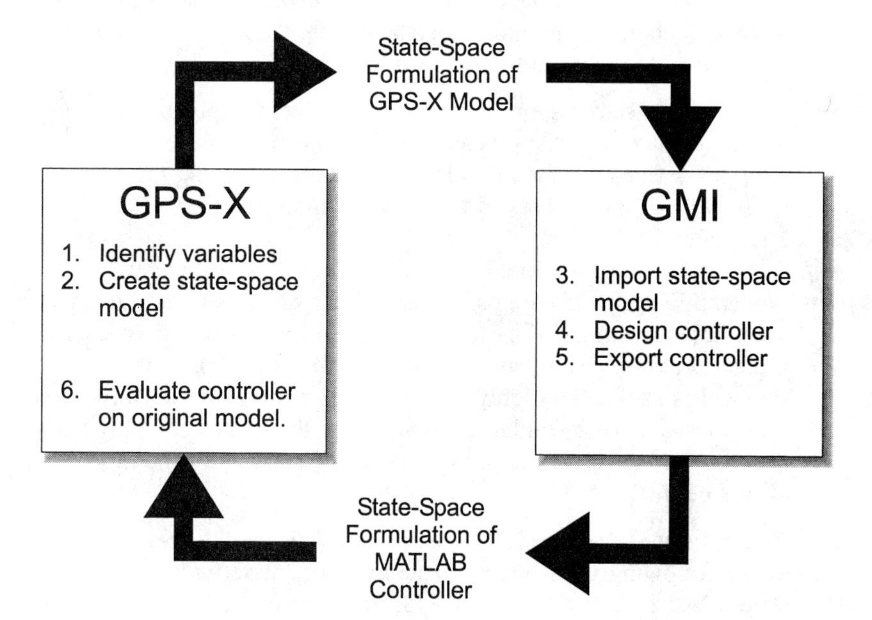


**Figure 7.19** The controller design process.

(1) Identify Variables and Perform Step Tests. GMI distinguishes between three different types of variables that are used by process controllers: *plant output variables, manipulated variables,* and *disturbance variables.*

Plant output variables (also known as *controlled variables*) are outputs of the process model, and cannot be directly manipulated. Some examples are dissolved oxygen concentration, MLSS, and effluent ammonia-nitrogen concentration.

Manipulated variables are model inputs that are used (manipulated) by the controller to keep selected plant output variables as close as possible to their respective setpoints. In order for control to be effective, it is vital that *causal relationships* exist between the manipulated variables and plant output variables that are to be controlled.

Disturbance variables are model inputs that affect the plant output variables but cannot be directly manipulated. These variables can be used as inputs to a controller; the controller uses this measurement in an attempt to *cancel out* the effect of changes in the disturbance variables by adjusting the manipulated variables. This type of control arrangement is commonly known as *feedforward control.*

Once the variables have been identified, step tests should be conducted. Step tests help to establish the dynamic nature of the relationship between the identified variables. The step tests involve perturbing the manipulated and disturbance variables in both the positive and negative directions about the nominal operating point.

The step tests yield the following information:

- a matrix of step responses of each of the plant inputs (manipulated and disturbance variables) to each of the plant output variables. This provides information regarding the behavior of the process in terms of its dominant rates of response.
- an indication of the linearity of the process about the chosen operating point. If the positive and negative steps in any given variable are similar (i.e., nearly identical in all characteristics other than the "sign" or direction of the response), then the model is said to be highly linear. If the positive and negative step responses are not similar, however, then the model is highly non-linear, and achieving a good linear representation of the model may be difficult.

(2) Create a state-space model of the process. This includes the execution of the following GPS-X commands: (1) the model is run to steady-state; (2) the eigenvalues of the Jacobian are checked to ensure that they are all negative (i.e., the model is stable); and (3) the state-space model is created.

(3) Import the state-space model into GMI. The GMI command, "Import from GPS-X", loads the state-space model into GMI.
(4) Design the controller. GMI has built-in tools that allow for the design of feedforward and feedback controllers. Advanced users familiar with the Matlab Control Systems Toolbox can design controllers using the wide array of tools included in the Toolbox.
(5) Export the controller from GMI to GPS-X. The GMI command, "Export to GPS-X", creates the files that are required to implement the designed controller in GPS-X.
(6) Evaluate the controller in GPS-X. Since the controller design tools in GMI (or Matlab) use a linear approximation of the full non-linear GPS-X model, it cannot be guaranteed that the controller will perform as expected when used in conjunction with a non-linear model. Therefore, this step is crucial in order to ensure that suitable controllers have been designed. If the controller fails to perform in a satisfactory manner, it is necessary to repeat steps 4 to 6 (or 1 to 6 if a change in the variable selection is required) until the design is deemed acceptable.

### GMI Example: Sludge Blanket Height Control

As an example of how GMI might be used to design a process controller using a GPS-X model, consider the following scenario. A fictional activated sludge plant with MLSS of slightly below-average settleability experiences solids "washout" during storm events (see Table 7.1 for a description of the process, and Figure 7.20 for the layout of this plant). The high flow experienced during the storm events transfers solids from the aeration tank to the clarifier, which cannot handle the increased solids load;

TABLE 7.1. **Process Description.**

| Process Type: | Activated Sludge |
|---|---|
| Average influent flow rate | 18,500 $m^3$/d |
| Aeration tank volume | 9200 $m^3$ |
| Number of cells in aeration tank | 4 |
| Dissolved oxygen | 2.0 mg/L |
| Average MLSS | 2750 mg/L |
| SVI | 190 mL/g |
| Clarifier surface area | 920 $m^2$ |
| Clarifer depth | 3 m |
| Average sludge blanket height | 0.8 m |
| Average RAS flow | 9000 $m^3$/d |
| Average WAS flow | 250 $m^3$/d |
| Average effluent SS | 12 mg/L |
| Existing control strategies | D.O. Control (using air flow); MLSS Control (using WAS flow) |

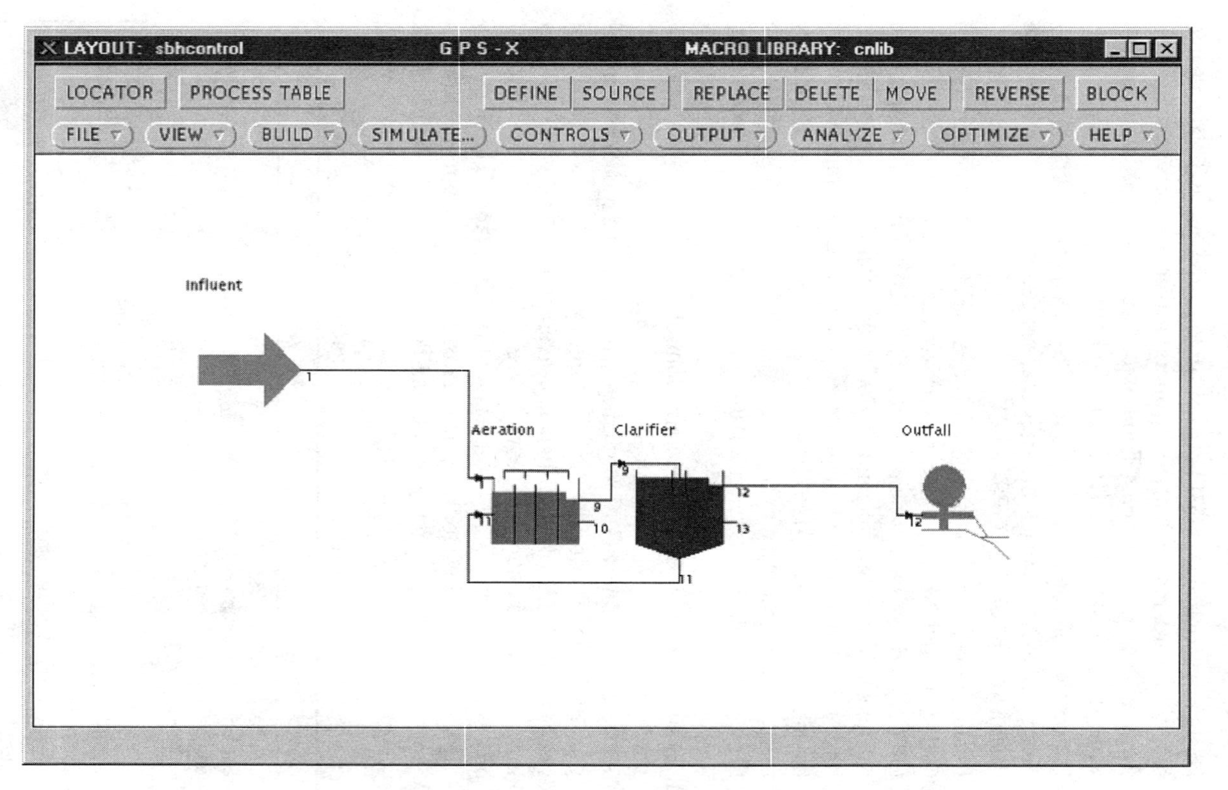


**Figure 7.20** GPS-X layout of an activated sludge plant.

the effluent suspended solids concentration increases as a result. Furthermore, solids are lost, reducing the MLSS and the ability of the process to oxidize substrate. It is desired to minimize this solids loss in order to meet effluent quality requirements as well as to preserve the solids inventory of the process.

The increased effluent suspended solids concentration during storm events is due to an elevated sludge blanket height (SBH). It is obvious that if the SBH could be maintained at a constant low level during storm events, then solids washout would be minimized. The manipulated variable which has the greatest *immediate* effect on the sludge blanket height is the return activated sludge (RAS) flow rate; therefore, it is proposed to use the RAS flow rate to control the SBH in the clarifier.

Without SBH control, the process behaves as shown in Figure 7.21.[4] It can be seen that the clarifier "fills up" during both storm events (as indicated by the high SBH); this is accompanied by elevated effluent suspended solids concentrations (peaking at approximately 1600 mg/L).

For future comparison, the integral of the absolute error[5] (IAE) without SBH control is 4.1 m·d. The total mass of solids leaving in the effluent is 14,670 kg. The performance of the controllers designed below will be compared against these values.

It should be noted that some form of WAS flow rate adjustment must be present in order to make the control strategies discussed below behave in a stable manner. For this example, the WAS flow rate is adjusted by a PI controller (see below) which attempts to maintain the MLSS equal to its setpoint on a long-term basis. If this is not present, the process will slowly become unstable (tending to very high or very low MLSS concentrations).

### *Feedback Controller Design*

Feedback controllers work by adjusting inputs to a process based on measurements of the outputs, i.e., the outputs are *fed back* to the inputs of the plant. This forms what is known as a *closed-loop system.* Feedback controllers are discussed in more detail by Olsson in Chapter 3 of this book.

For this process, the plant output that we wish to control is the sludge blanket height in the clarifier. The manipulated variable that the controller will use to try to keep the SBH equal to its setpoint is the RAS flow rate. A schematic of this closed-loop system is shown in Figure 7.22.

The dynamic relationship between the manipulated and plant output

[4] The influent data for this example is taken from Spanjers et al. (1997).
[5] The integral of the absolute error is defined as $\int|\text{Setpoint} - \text{Plant Output}|dt$. It is a measure of how much the plant output variable deviates from its setpoint.

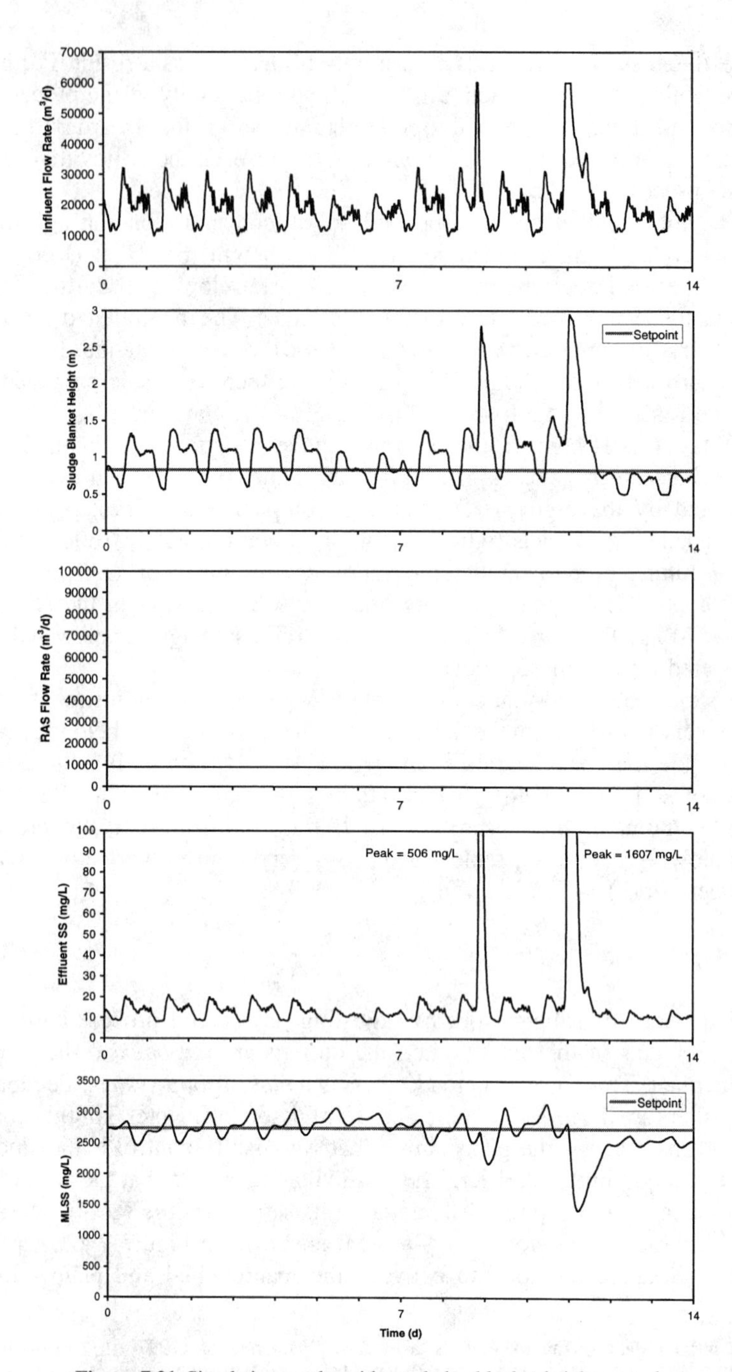


**Figure 7.21** Simulation results without sludge blanket height control.

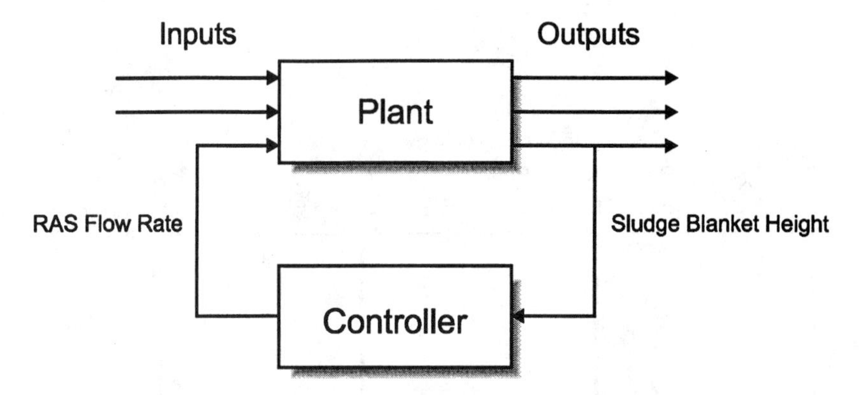


**Figure 7.22** Schematic of the closed-loop feedback control system for sludge blanket height control.

variables limits the achievable control performance for the closed-loop system. (In practice, the dynamics of the sensors and final elements (valves) also influence the achievable control performance.) As discussed above, step tests can be used to determine the nature of this dynamic relationship.

The results of a 10% step in the RAS flow is shown in Figure 7.23. The SBH responds quickly to the step in the RAS flow, reaching its steady-state value in slightly more than 0.1 d. Therefore, it is reasonable to expect relatively good control performance by using the RAS flow rate to control the SBH.

Now that the dynamic relationship between the manipulated and plant output variables has been confirmed, the state-space model of the process can be generated. The state-space model of the process is used to design the feedback controller. For this example, we will use GMI's built-in tools to design a PI (proportional-integral) feedback controller. The PI controller is a special case of the PID controller discussed in Chapters 1 and 3 of this text. In continuous form,[6] the PI controller algorithm is given by (Marlin, 1995 and Chapter 1 of this volume):

$$MV(t) = K_c\left(E(t) + \frac{1}{T_I}\int_0^t E(\tau)d\tau\right) + I$$

where:

$MV$ = the manipulated variable

[6] In practice, this algorithm is usually implemented in *discrete* form. However, if the time between controller executions is small enough compared to the dynamics of the process, then the discrete form closely approximates the continuous form shown here.

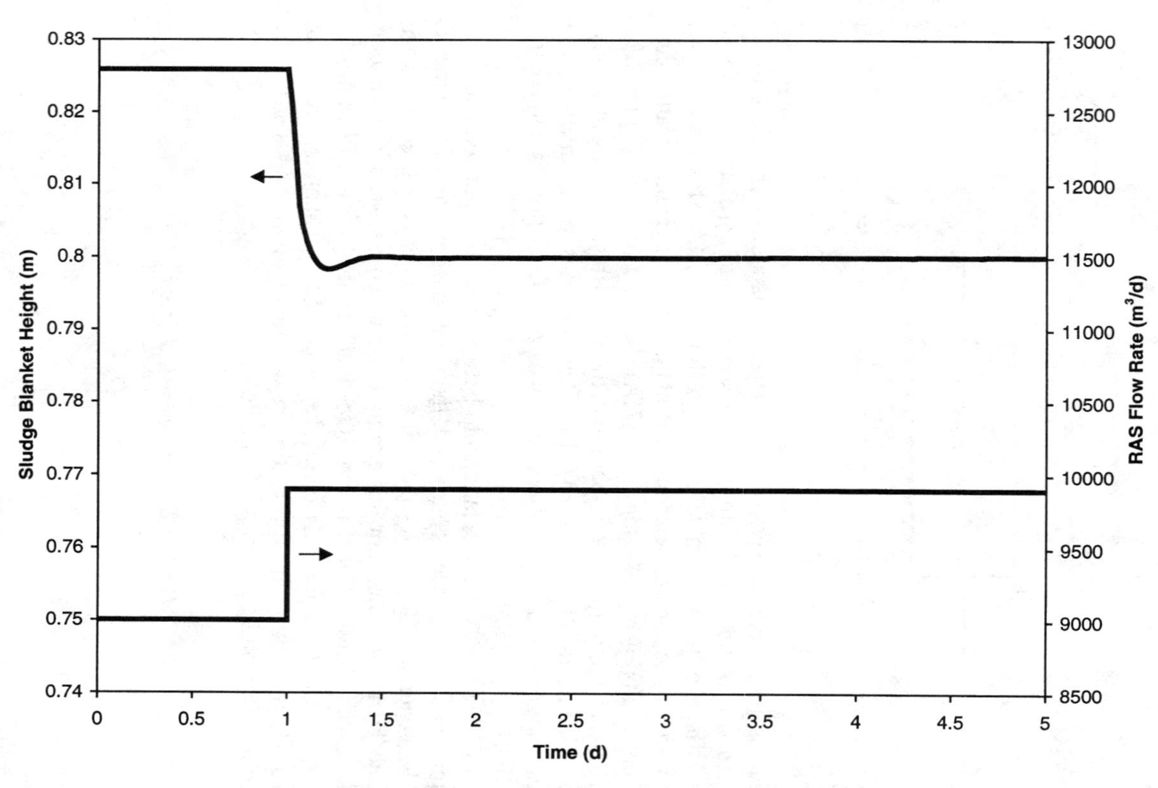


**Figure 7.23** Results of a 10% step in the RAS flow rate.

$E$ = the error on the plant output, and is equal to (setpoint-plant output)
$K_c$ = the proportional gain, a controller tuning constant
$T_I$ = the integral time, a controller tuning constant
$I$ = the initialization constant

The PI controller has two different *modes:* a proportional mode, given by

$$MV_P(t) = K_c E(t) + I_P$$

and an integral mode, given by

$$MV_I(t) = \frac{K_c}{T_I} \int_0^{\tau} E(\tau) d\tau + I_I$$

The overall controller output is the sum of these two modes. Conceptually, the proportional mode is a ''fast'' mode (i.e., it responds directly, or (proportionally) to changes in the plant output variable), while the integral mode is a ''persistent'' mode (i.e., it continues to adjust the manipulated variable until the difference between the plant output variable and the setpoint is zero).

Figure 7.24 shows GMI's feedback controller design window. GMI uses a design algorithm that selects the controller tuning parameters such that the closed-loop response of the system to a setpoint change most closely approximates a target first-order response. Therefore, instead of adjusting two tuning constants (three in the case of PID controllers) to find the desired closed-loop response, the user can simply adjust the target response time constant until the closed-loop response is satisfactory.

When tuning a feedback controller, it is important to consider not only the response of the plant output variable, but also the response of the manipulated variable. For the final tuning, the time required for the plant output to reach its setpoint has been sacrificed somewhat in order to achieve reduced variability in the manipulated variable. The performance is still good, however; the SBH (nearly) achieves its setpoint after only 0.05 d (70 minutes). The overshoot of the manipulated variable is limited to approximately 50% of its final value. A more aggressive control design would lead to increased overshoot of the manipulated variable and would result in greater variability in this variable when implemented in the plant.

With the feedback controller design completed, the controller can now be exported to GPS-X to evaluate its performance on the non-linear model using actual plant data. Figure 7.25 shows the results of the simulation with the feedback controller active. It is clear that the variability in the sludge blanket height is reduced; furthermore, the effluent suspended solids

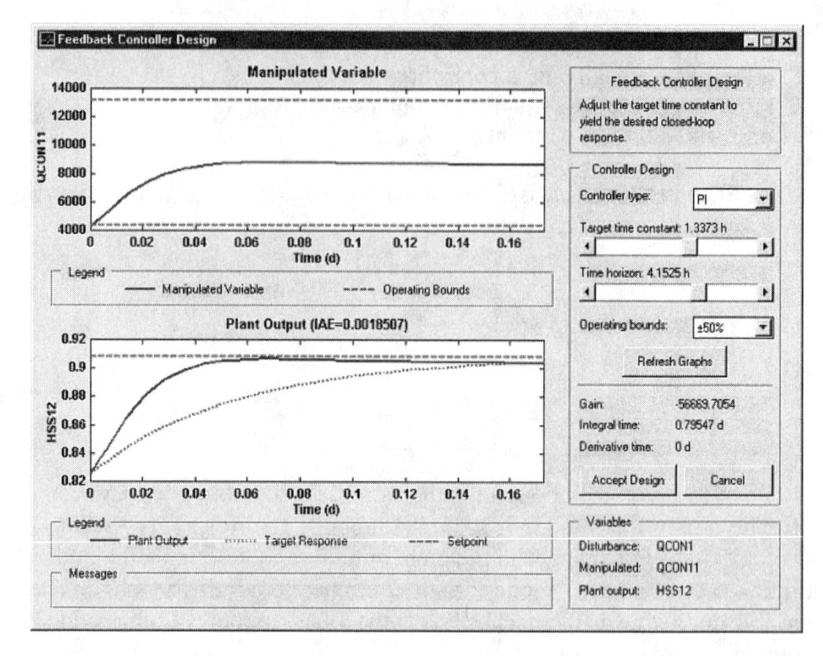


**Figure 7.24** GMI's feedback controller design window.

concentrations during the two storm events have been significantly reduced (the peak effluent SS concentration has been reduced from 1600 mg/L to approximately 60 mg/L.

The IAE for the two-week simulation with the feedback controller active is 1.04, approximately one-quarter of the value without control. This indicates that the controller is successful in reducing the deviation of the plant output from its setpoint, as expected. Furthermore, the total mass of solids in the effluent has been reduced by a factor of more than one-third, to 4330 kg. Therefore, the feedback controller has been successful in reducing the washout of solids during the simulated storm events.

### *Enhancing the Feedback Controller: Feedforward Control*

With a feedforward controller successfully designed, it is now possible to examine ways to improve the performance of the controller. As mentioned above, the influent flow rate variability is the most important disturbance affecting SBH. Since nearly all plants measure their influent flow rates, why not make use of this measurement to improve the performance of the controller?

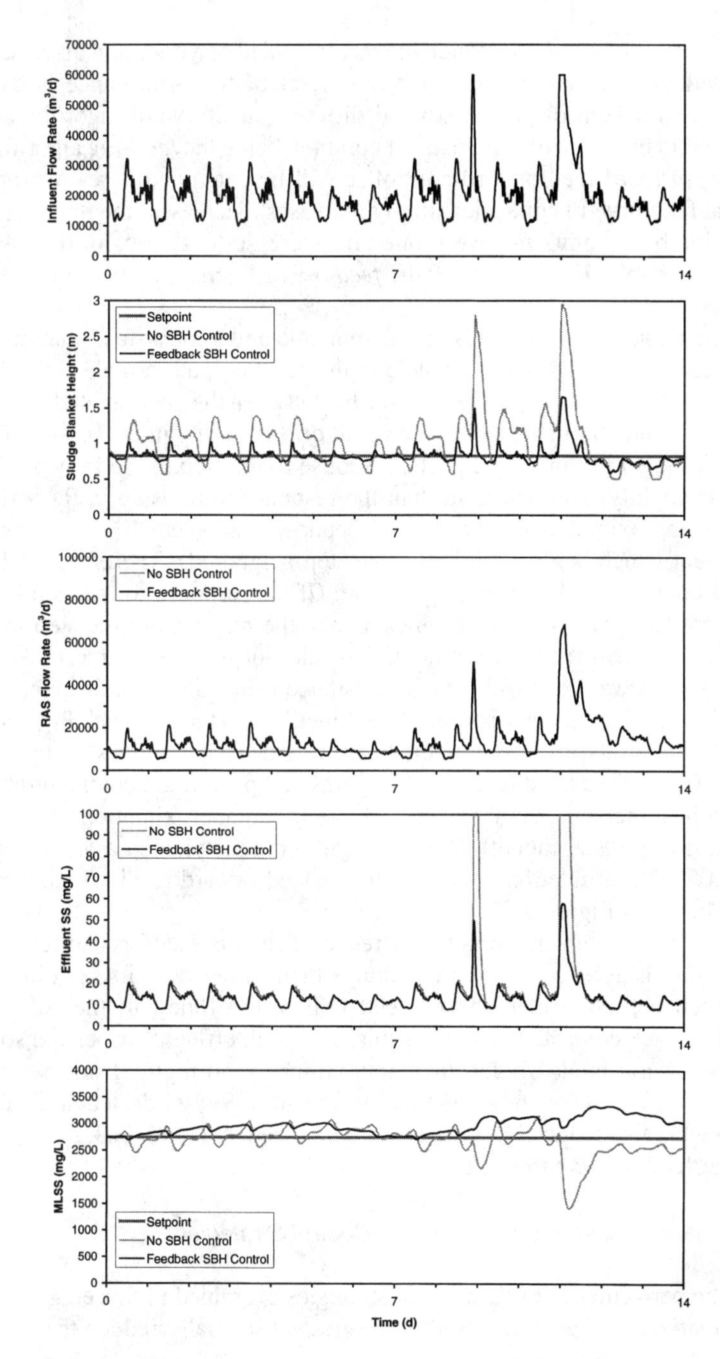


**Figure 7.25** Simulation results with feedback sludge blanket height control.

This type of control in which the measurement of a common disturbance is used to proactively counteract the effect of the disturbance is called *feedforward* control. While several different feedforward algorithms are utilized in practice (one of the most common being the *lead-lag* algorithm), a *proportional* feedforward controller will be examined here. A proportional feedforward controller simply increases (decreases) the manipulated variable by $K_{ff}$ units for every one-unit increase (decrease) in the disturbance variable. $K_{ff}$ is known as the *feedforward gain,* and may be positive or negative.

The choice of $K_{ff}$ depends on the dynamic and steady-state relationship between the disturbance variable and the plant output variable, as well as the dynamic and steady-state relationship between the manipulated variable and the plant output variable. The result of a 10% step in the influent flow rate is shown in Figure 7.26. The response to the step in the influent flow rate is slightly more sluggish than the response to the step in RAS flow, taking approximately 0.3 d to reach (pseudo-) steady-state (if the ''slower'' transient which begins 0.5 d after the step is ignored). Therefore feedforward control should be fairly effective. (If the response to the step in the influent flow rate was much quicker than the response to the step in the RAS flow, then feedforward control would not be as effective.)

A feedforward controller can be designed using GMI's built-in feedforward controller design tools and ''combined'' with the feedback PI controller designed above. The default feedforward design suggested by GMI ($K_{ff}$ = 0.35[7]) is based on steady-state gains and proved to be not aggressive enough (remember, the linear model is only an approximation of the full non-linear GPS-X model); a few design iterations yielded a value for $K_{ff}$ of 1.05. The simulation results with this feedforward/feedback controller are shown in Figure 7.27.

It is clear from the simulation results that this feedforward/feedback controller is able to maintain the sludge blanket closer to its setpoint than the feedback controller; the reduced IAE of 0.5 (one-half the value for the feedback controller) confirms this. The peak effluent suspended solids concentration is also reduced to approximately 50 mg/L (from 60 mg/L for the feedback controller). Finally, the mass of solids leaving in the effluent is also reduced slightly to 4230 kg (a reduction of 100 kg compared to the feedback controller).

### *Summary of Sludge Blanket Height Control Strategies*

The performance of the control strategies examined in this example are summarized in Table 7.2. Both strategies substantially reduce the loss of

[7] $K_{ff}$ is unitless in this case because it is the ratio of two flows with units of $m^3/d$.

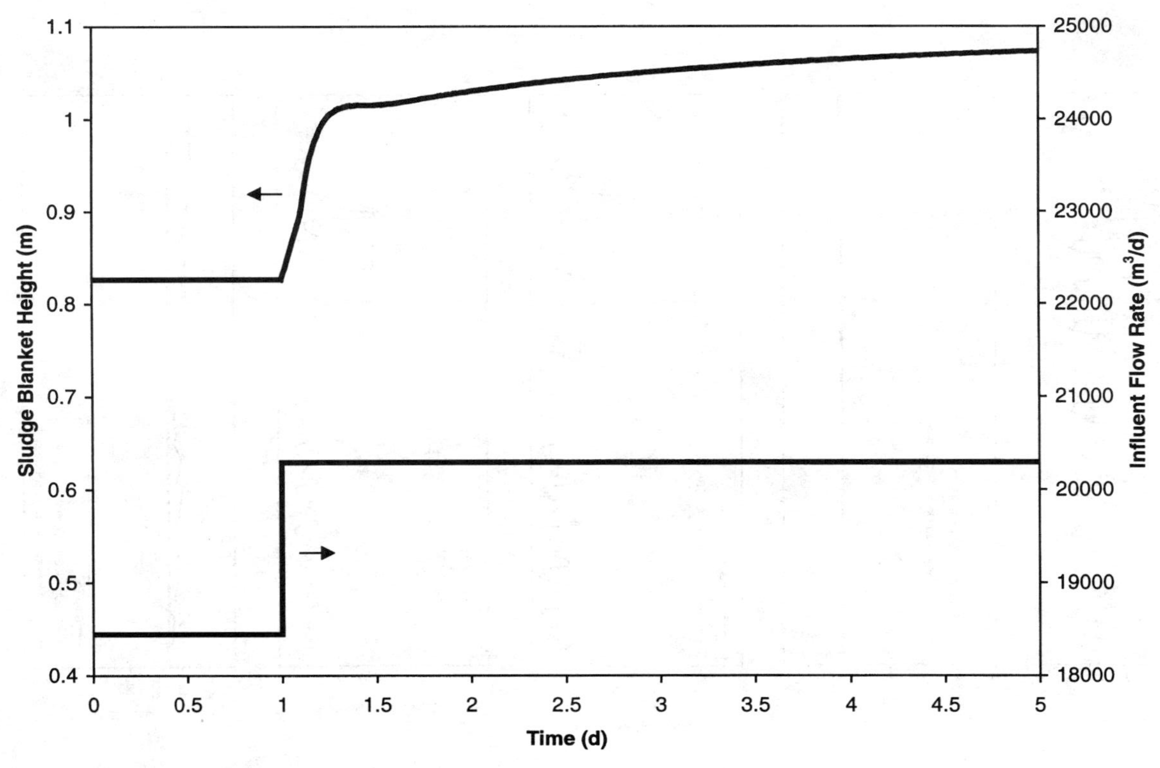


**Figure 7.26** Results of a 10% step in the influent flow rate.

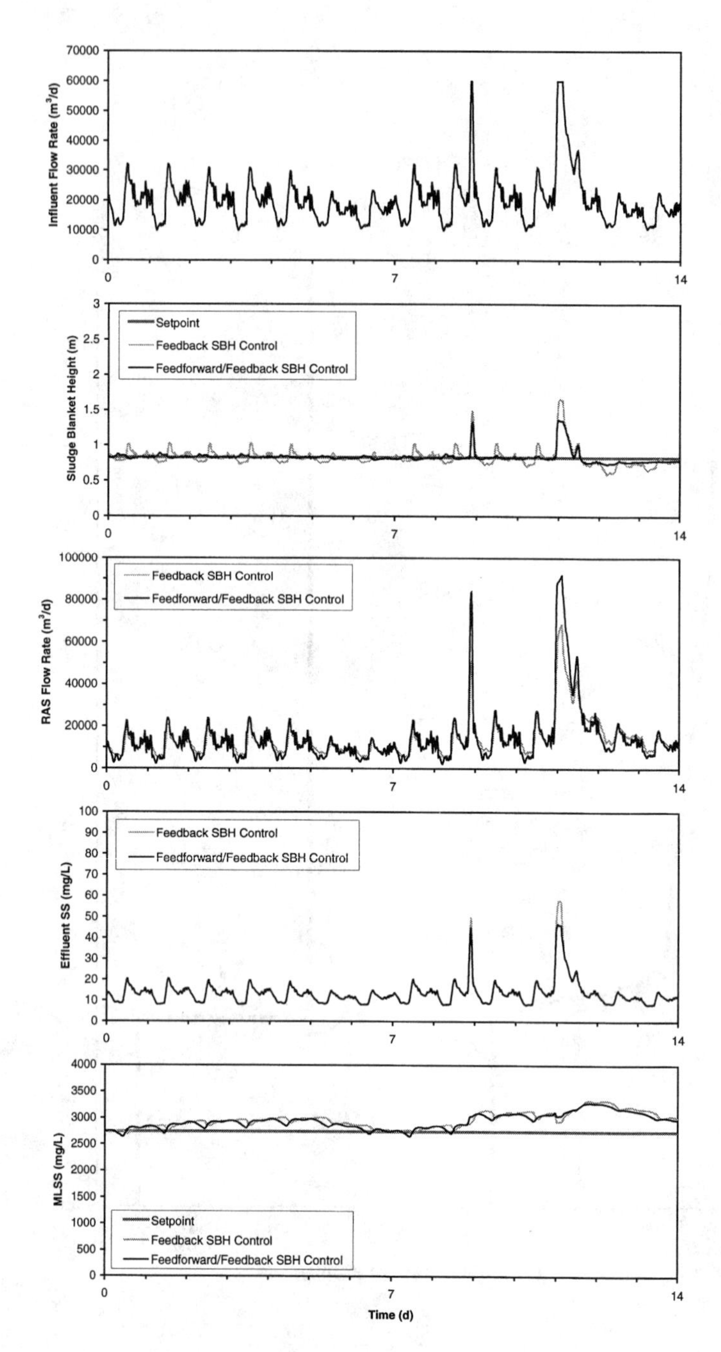


**Figure 7.27** Simulation results with feedforward/feedback controller.

TABLE 7.2. Summary of Control Performance.

| | Integral of Absolute Error (IAE) (m·d) | Mass of Solids Leaving with Effluent (kg) |
|---|---|---|
| No Control | 4.1 | 14,670 |
| Feedback Control | 1.04 | 4330 |
| Feedforward/Feedback Control | 0.49 | 4227 |

solids in the effluent when compared to the case without control; the feedforward/feedback controller slightly outperformed the feedback controller in this respect. Therefore, if the influent flow measurement is already available (and expensive re-configuration is not required in order to make it available to the control system, for example), then it would be useful to make use of the feedforward portion of the controller design. Otherwise, good performance could still be obtained by using feedback control only.

## SIMULATION MODEL (GPS-X)

GPS-X is a model development, calibration, and simulation tool which includes an extensive, modifiable library of models for most of the treatment units used in the wastewater engineering field. GPS-X, shown in Figure 7.20, runs either on personal computers running the Windows NT operating system, or on UNIX workstations running Solaris, HP UX, AIX, or OSF.

The simulator model is graphically developed with tools available in GPS-X. The user builds a flow-sheet of the facility by selecting unit process objects and control point icons from the Process Table (shown in Figure 7.28) and placing them in the drawing board of the main window. The objects which these icons represent include influents, splitters and combiners, several different types of biological reactors, sludge treatment processes, and tertiary treatment processes. Each of these objects contains one or more models, the majority of which are mechanistic type models. In addition, GPS-X contains non-mechanistic models including linear and non-linear regression and multi-layer, feed-forward neural network models.

With the desired objects placed on the drawing board, the user specifies the flow paths between objects and selects a model for each object. Pop-up menus are used to specify physical dimensions of process objects being modelled (e.g., clarifier surface area and depth), operational parameters (e.g., underflow rates), process kinetic rates, and stoichiometric relationships. Automatic controllers can also be specified for operational parameters to model the effects of on-line control systems (e.g., automatic chemical feed control).

After the model has been developed and populated with the appropriate

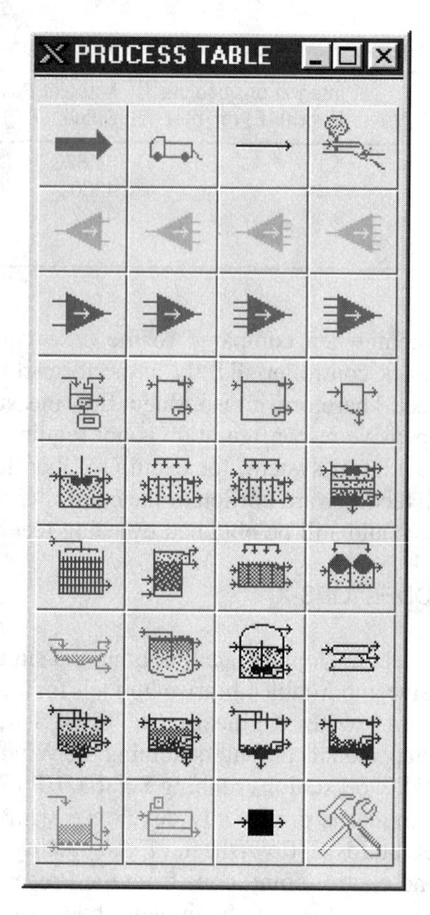


**Figure 7.28** GPS-X process table.

data, dynamic, interactive simulations can be conducted. The user can prepare time series and bar chart displays for visualization of simulation results and create, for any model independent variable, an interactive control that can be changed while the simulation proceeds. GPS-X contains additional tools for steady-state and dynamic sensitivity analysis, and nonlinear multi-parameter optimization.

## OPERATOR INTERFACE (SCENARIO MANAGER)

The Scenario Manager was developed to provide a user-friendly interface to GPS-X and the IC²S building blocks (Figure 7.29). Only those functions necessary to obtain simulation results are included in the interface. The objective was to simplify access and set-up of the IC²S building blocks without minimizing simulation and control functionality. Within the context of IC²S, the operator interface allows users to:

(1) Manage GPS-X layouts contained in the on-line system.
(2) Select the duration of the simulation (in simulated time).
(3) Save the status of the simulation (snap-shot of all GPS-X states at the time save is invoked) for later use.
(4) View a text window that describes the scenario.
(5) Change the speed of the simulation.
(6) Change pre-selected plant conditions as the simulation is running using GPS-X interactive controllers.
(7) View pre-selected outputs both during and at the end of the simulation using GPS-X output display graphics.

## BRIDGE

A bridge is a software link that provides communications between the IC²S modules, including the SCADA system, to ensure efficient data flow. When IC²S is installed at a plant, it is linked to the plant SCADA system. There are two ways to acquire data from a SCADA system:

**Figure 7.29** The scenario manager.

(1) Directly read data coming into the SCADA system from on-line sensors.
(2) Continuously check the database(s) used by the SCADA system to store data.

The main disadvantage of option 2 is that stored data is often filtered and smoothed, which may be incompatible with some IC²S data requirements. However, option 2 seems to allow a greater opportunity for standardization since the variety of data bases is relatively small and file manipulation or reformatting of data records can be tackled through software. This bridge will need to be customized for each installation.

## IC²S FUNCTIONAL MODULES

Different customized functional modules can be built using the basic building blocks described in the previous section. Influent characteristics and variability, process configurations and effluent criteria vary from plant to plant. The IC²S functional modules are custom made for the needs of the particular plant where it is implemented. The following plant characteristics will be considered when designing the on-line modules for the plant:

- Process configuration: Different wastewater treatment technologies and even different unit processes within the plant require special considerations. The list of available models, their complexity, reliability and track record must be evaluated.
- On-line data availability: For every process there is a list of critical inputs without which calibration becomes impossible. Flow, suspended solids concentrations, and ammonia concentrations are examples of critical variables which must be measured in a nitrifying plant at several locations.
- Effluent guidelines and permit levels: Depending on factors such as the local laws, the physical location of the plant, and the receiving water body, tolerated effluent discharge levels (and thus treatment performance objectives) can vary greatly.
- Overall objectives of the plant in terms of the three levels of control: Maintain operation, meet permit levels, and minimize operating costs.

The modules contain different basic IC²S building blocks implemented in GPS-X, and are operated and synchronized by the GPS-X Scenario Manager. Different modules are usually implemented in different instances of the Scenario Manager, though this is not necessary and with a small implementation a single central application may be easier to set up.

The following sections describe the function and setup of typical IC²S modules that can be built from the basic software blocks.

## AUTO-CALIBRATION MODULE

Since the IC²S engine is a calibrated dynamic model of the plant, the current calibration status has to be verified and, if necessary, calibration constants continuously modified. This module is critical in most implementations, as several other modules (and ultimately the successful operation of the control system) depend on it.

This module typically consist of several model layouts of the full plant as well as some partial layouts (e.g., settlers) where the calibration task can be sufficiently separated from the full plant layout. Each of these layouts is set up with one DPE block continuously optimizing one or more model parameters to fit one target variable. Typical target variables are MLSS and RAS concentration, effluent (or even better, mid-process) ammonia, primary and final effluent suspended solids, DO or air flow and many more. To fit these variables, the model parameters that affect the variables must be identified. This identification and setup step requires an experienced modeling engineer who will perform sensitivity analyses on the selected model parameters to verify that the dependency and sensitivity is sufficient to result in an effective auto-calibration.

As a simple example, consider varying clarification conditions due to varying influent and other process conditions on the plant. GPS-X, using the DPE building block, can dynamically optimize effluent suspended solids (and other relevant variables) by varying the model parameter that has an effect on the selected monitored variable. In the case of effluent suspended solids, the DPE run will generate a time-dependent flocculant settleability parameter.

The auto-calibration module consists of several on-line, continuously running DPE optimizations, maintaining calibration of respective parameters and communicating the best values to each other through a calibration database. The layouts are built in a hierarchical way; i.e., low level optimizations provide information for higher level layouts. The final layout, on the top of the hierarchy, will use all available signals and optimized parameters from the rest of the auto-calibration module layouts to perform its specific optimization task. An example is described in some detail later.

The above system requires large amounts of system resources to operate. Different configurations of this system were investigated, in which the separate parameter optimizations were combined into one large optimization problem. However, convergence problems were experienced—likely because of the greater complexity of the objective function and the increase in the number of degrees of freedom. Therefore, this approach was not pursued further at this time.

An important part and output of this module is a calibration database (containing a time-varying record of all optimized model parameters up to the current time). Other modules make use of this database when performing their respective tasks (the process optimization module, for example).

## ADVANCED FAULT DETECTION MODULE

Low level signal tracking and sensor fault detection, as well as advanced, intelligent sensor and process fault detection is performed in this module. The Advanced Fault Detection module consists of two $IC^2S$ building blocks:

(1) A model layout of the full plant (or section of the plant) for each signal on which fault detection is performed. This layout run in real-time (data-synchronized mode) using all available input and output signals as well as the calibrated database discussed in the next section.
(2) One or more signal tracking and alarm blocks in each of these layouts monitoring selected signals.

When the advanced fault detection module is applied to sensor fault detection, monitored data can be continuously compared with the output of the calibrated model. When a discrepancy develops between these two signals, this either points to a sensor fault or a fault in maintaining self-calibration. The cause can be decided based on mass-balance calculations, typical parameter and sensor values, etc. The detection algorithm has to be fine-tuned on a case-by-case basis, even though general guidelines are available. An example to detect inconsistent return flow and suspended solids signals will be discussed later.

When advanced fault detection is applied to signals generated by the auto-calibration module (i.e., calibrated parameter values), the module performs a process fault detection and early warning functionality. Consider for example the Vesilind hindered settling velocity, which can be extracted from on-line flow measurements around the clarifier, as well as MLSS, RAS and sludge blanket or solids profile signals. Before a bulking condition develops, the hindered settling velocity starts to increase (corresponding to decreasing settling velocities and increasing SVI values). In an initial stage the solids flux does not become limiting so there are no obvious signs of the impending failure. However, the signal tracking block is able to infer from the sustained positive slope of the hindered settling velocity parameter that bulking is imminent. This generates an alarm, potentially several days before the failure situation develops at the plant.

## PROCESS OPTIMIZATION MODULE

This module performs the required process optimizations in real-time. Plant control variables which can be optimized include:

- optimal feed loading to aeration (equalization)
- step feeding
- aeration—number of blowers, power uptake
- mixing requirements
- return activated sludge flow
- wastage flow
- nutrient (N, P) dosage
- polymer dosage for settleability
- operating costs

The Process Optimization Module uses the Dynamic Parameter Estimator, but it optimizes process variables as opposed to model parameters in the Self-Calibration Module. The advantage of this module over low level localized control loops is that it can optimize plant-wide objective functions like operating costs. This is due to the fact that the dynamic model is a global representation of the plant, and interactions between unit processes are truthfully simulated.

This module of $IC^2S$ has not been implemented yet on a full-scale plant and more experience is necessary to evaluate its reliability. In the current design the optimal value of the process variable is reported to the operators through a graphical display, digital output or printout. Implementation of the recommended value is subject to operator approval. Once experience on the reliability and robustness of the Optimizer Module is collected, further development is planned to investigate closed loop applicability.

## CONTINUOUS FORECASTING MODULE

This module consists of several instances of the full plant layout running simulations automatically. The number of simultaneous simulations (layout instances) depend on the number of typical tasks the module is set up to handle. The module is managed by the Scenario Manager. There are three major sources of input to the simulations.

(1) Driving functions: Predicted (if an influent model is available) or historical forcing functions (influent patterns for flow and concentrations). Layouts can be set up to select from a set of different forcing functions; e.g., if dry weather is expected over the duration of the simulation, the appropriate influent files or model is selected. The same model can also be initiated to use wet weather flows or a specific storm event.

(2) The calibrated database: The last known best values for parameters, or observed, extracted trends are loaded and used in the forecast simulations.

(3) The operators have flexibility in changing the desired runs. Length of the predictions can be adjusted. Typical runs are next day (1 d), next week (7 d) or next month (30 d). Long-term (steady-state) runs can also be selected. Different operational strategies can be tested by selecting the appropriate values for the input parameters. For example, if the objective is to establish the minimum necessary recycle, an input panel is used with a slider or buttons to continuously or discontinuously (step-wise) increment the flow.

The output from the Continuous Forecasting Module is used to verify the validity and long-term effect of desired process changes on effluent quality and other key operational parameters.

## IC²S DEMONSTRATION EXAMPLE

In this section a real-life example of the IC²S is demonstrated. In the previous sections of this chapter the basic building blocks and functional modules of IC²S were described. Of these elements, the process and control engineers together build the application, which is specific to the wastewater treatment plant. Just as all plants are different, all IC²S installations must be custom-built for the particular plant. It is not possible to build a generic model-based control system, which would apply to all plants, even when only the activated sludge process is considered. The configuration, loading, effluent criteria, instrumentation of the plant, as well as the potential final elements of control all vary from site to site.

Due to the large number of supporting models in GPS-X, which act as the engine of the IC²S implementation, and the variety of basic building blocks and functional modules, IC²S is unique in that it can be configured to suit practically any real-world situation.

In this section the following items are described.

- the wastewater treatment plant used for demonstration
- the model layouts used in the on-line system
- the SCADA link
- auto-calibration module
- fault detection module

In this particular installation the Adaptive Data Filter and Respirogram Evaluator building blocks were not utilized.

### THE WASTEWATER TREATMENT PLANT USED FOR DEMONSTRATION

The plant selected for demonstration is an activated sludge system consisting of two parallel trains. The process is designed for BOD removal, nitrification and denitrification. The primary facilities (screens, grit removal, primary settling) were not modeled, thus the process models use primary effluent as input.

The process starts with a common anoxic tank where nitrate from the return stream is denitrified. Two parallel aeration tanks follow. After the biological treatment there is a single settler for secondary solids/liquid separation.

Aeration is mechanical, with three rotors per aeration tanks.

The system is frequently subjected to substantial wet weather flows. The plant is loaded by a sewer and septage hauled in by tanker.

## THE MODEL LAYOUTS USED IN THE ON-LINE SYSTEM

The calibrated dynamic model of the WWTP as implemented in GPS-X is the engine of the $IC^2S$, and hence an integral component. The model includes the following three layouts:

(1) Full layout of the WWTP: The full layout, consisting of every single unit process on the demonstration plant, is not directly used in the on-line modules. This is due to limitations in on-line data availability. The layout is used for detailed off-line process analysis and optimization, as well as other studies, concerning for example plant expansion. The full layout for the demonstration WWTP is shown in Figure 7.30 for illustration purposes.

(2) Aggregated layout of the demonstration WWTP: This simplified layout contains all relevant processes in the plant. Wherever possible, parallel processes have been lumped together. The primary purpose of this layout is to serve as the final element of the auto-calibration module. At the same time use of this simplified model facilitates off-line process analysis by reducing the number of input parameters, while retaining full process modeling capabilities. The existing Aggregated Layout is shown in Figure 7.31.

(3) Settler layout: This layout contains the secondary settler only, with MLSS input from an influent object. The layout is used as the first step of the auto-calibration process to extract settling parameters and on-line flow measurement correction to be used in the more complex layouts. The Settler layout is shown in Figure 7.32.

The following mathematical models were used in the layouts.

(1) Influent: The BOD-based influent model contained in GPS-X. This model reads a select number of composite variables (influent suspended

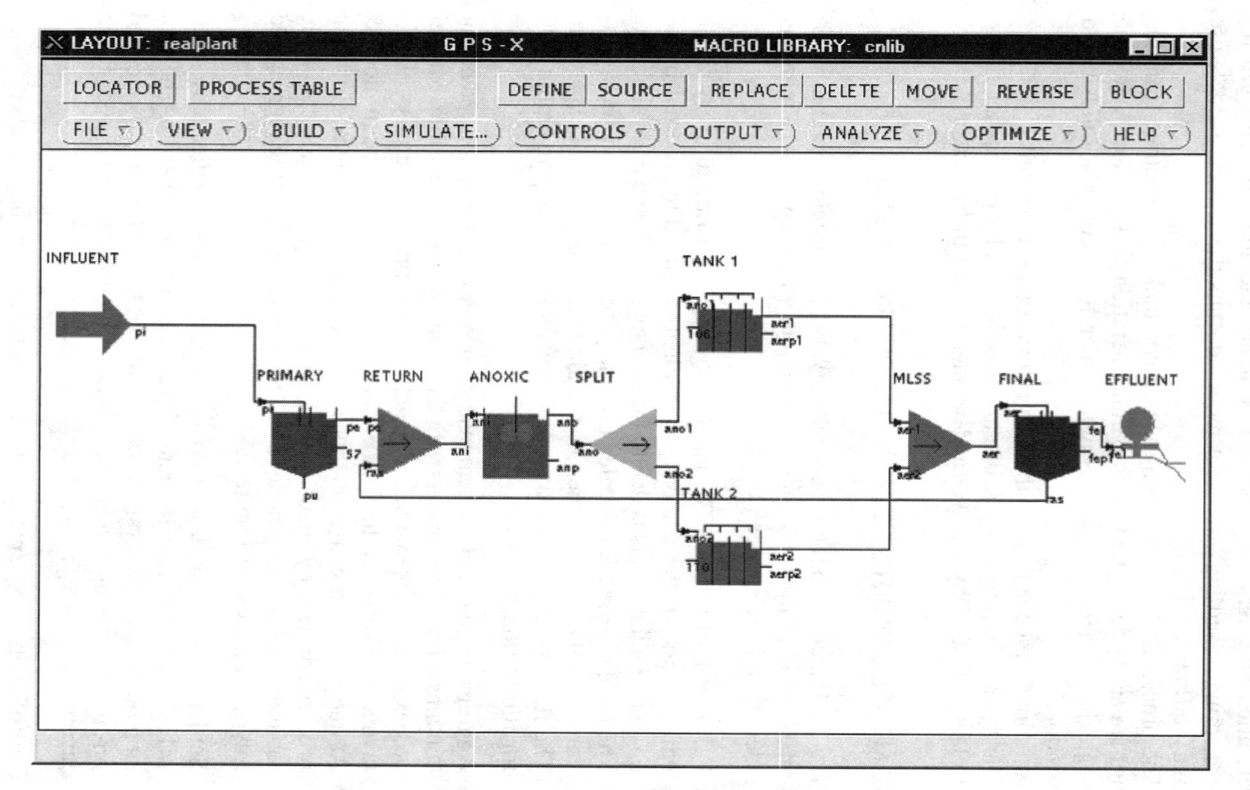


**Figure 7.30** Full layout of the demonstration wastewater treatment plant.

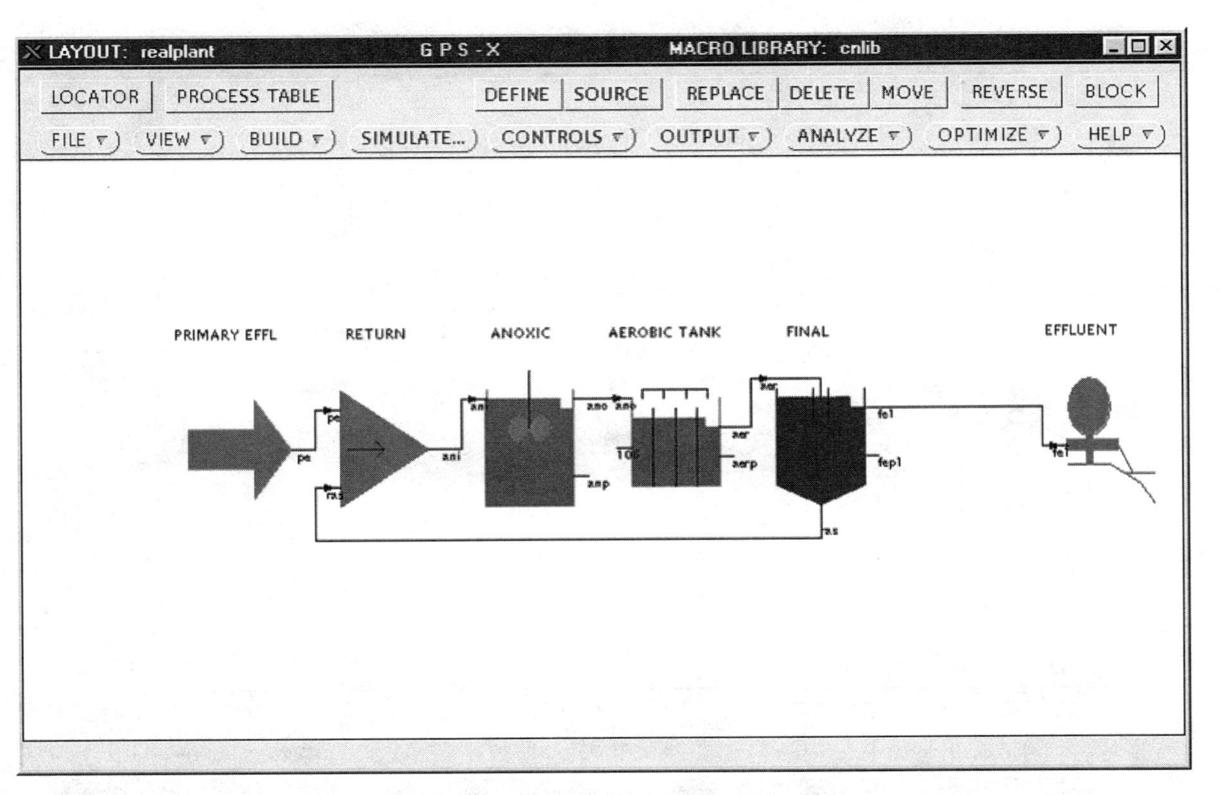


Figure 7.31 The aggregated layout of the demonstration wastewater treatment plant.

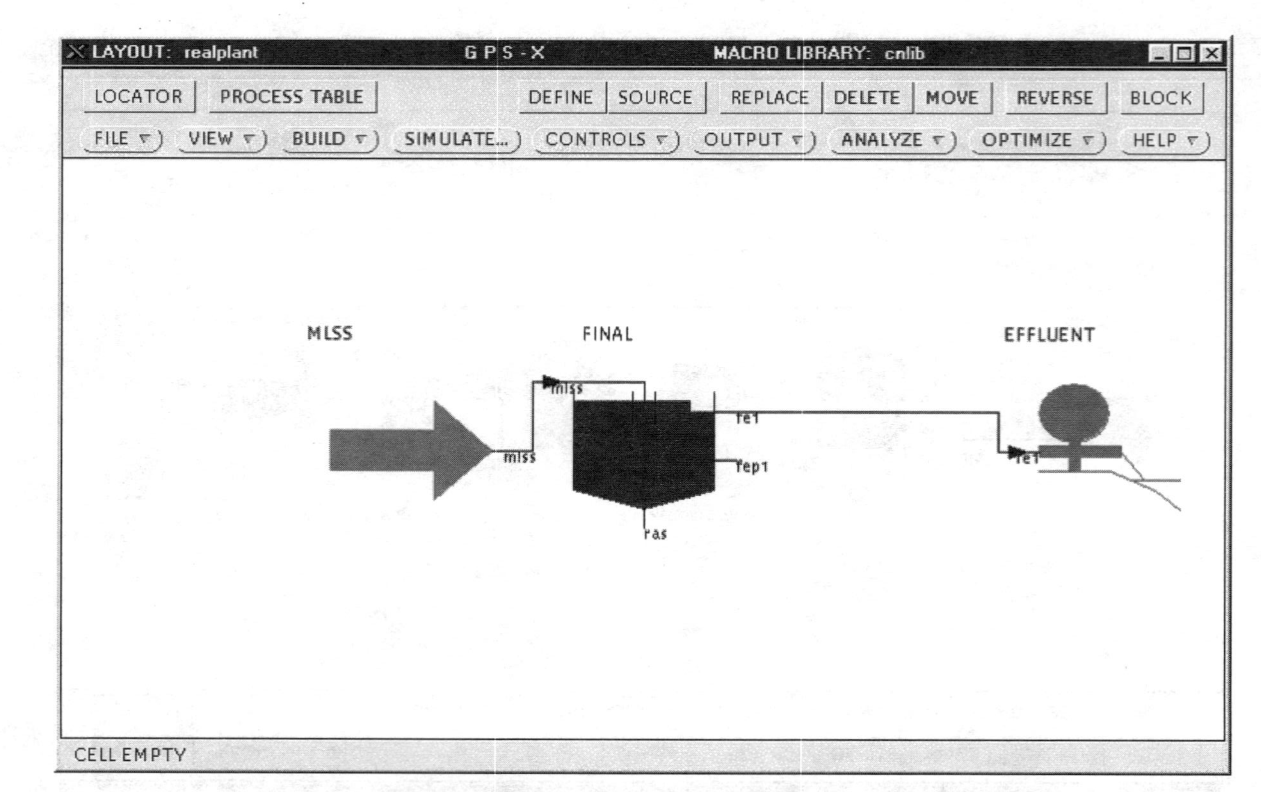


**Figure 7.32** Settler layout for the demonstration WWTP.

solids, carbonaceous $BOD_5$, TKN). The most important state variables in the models are then calculated from these variables based on the typical composition of the influent. The influent stoichiometry has to be established beforehand and periodically updated. In addition, several state variables which do not usually occur in influents or are expensive to measure (autotrophs, nitrate) are set to a small value or zero.

(2) Biological model. The popular IAWQ ASM1 model (Henze et al., 1987) was used to model the aeration tanks. The model equations were extended to include temperature dependency of major kinetic parameters, according to the Arrhenius equation.

(3) Final clarifier. The double exponential settler model implemented in a one-dimensional configuration (Takaćs et al., 1991) was used to simulate thickening and clarification. Twenty layers were used to correspond to the suspended solids data provided by the profiler instrument installed in the plant.

All layouts were customized to read in signals from the SCADA system in the units provided; units were converted to standard SI units used in GPS-X. Further customization included the implementation of a RAS flow correction factor, as well as DO, ammonia and SRT controllers for various investigations during the model building.

The plant model was initially calibrated based on steady-state and dynamic conditions. Three months of laboratory data was analyzed, outliers, upsets and other non-typical behavior discarded and the resulting data set averaged. A three-day intensive sampling campaign was also conducted. In addition to continuously monitored variables, influent and MLSS stoichiometry was also established during this period (Hydromantis, 1995).

The average values of the three month data set was considered to be equivalent to steady-state conditions; initial calibration was performed using these values. The focus was on establishing settling and nitrification parameters. Afterwards, the three-day dynamic data set was simulated and minor adjustments to the kinetic and settling constants were performed. Figures 7.33, 7.34 and 7.35 contain MLSS, RAS and DO simulation during this calibration period.

This pre-calibrated model was used to set up the on-line system.

## THE SCADA LINK

There are many different SCADA systems for Wastewater Treatment Plants. In this case a specific driver program was written to interrogate the database of the SCADA at the plant and send values of selected signals to the on-line system database every five minutes. The signals include influent flow, return flow, effluent flow, aeration power uptake, influent

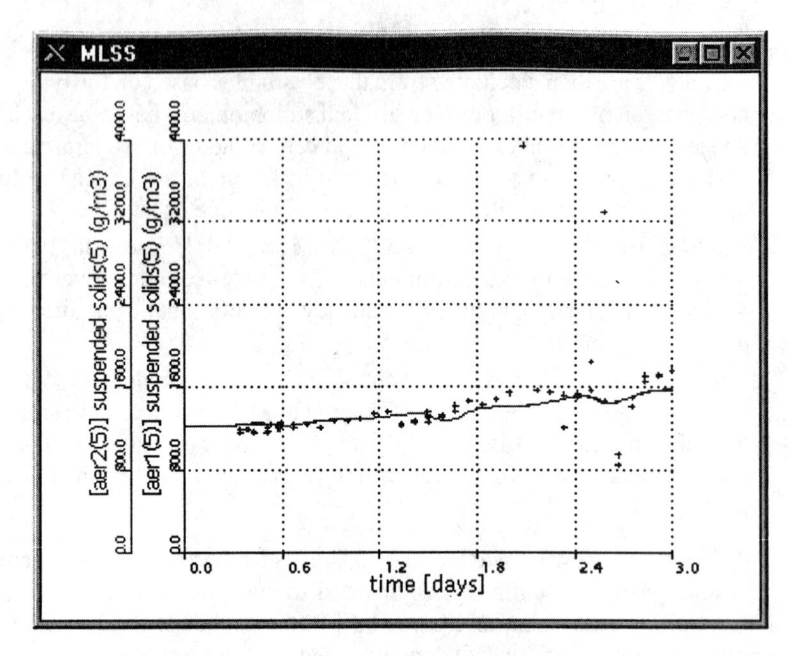


**Figure 7.33** MLSS fit with data during the 3-day calibration period.

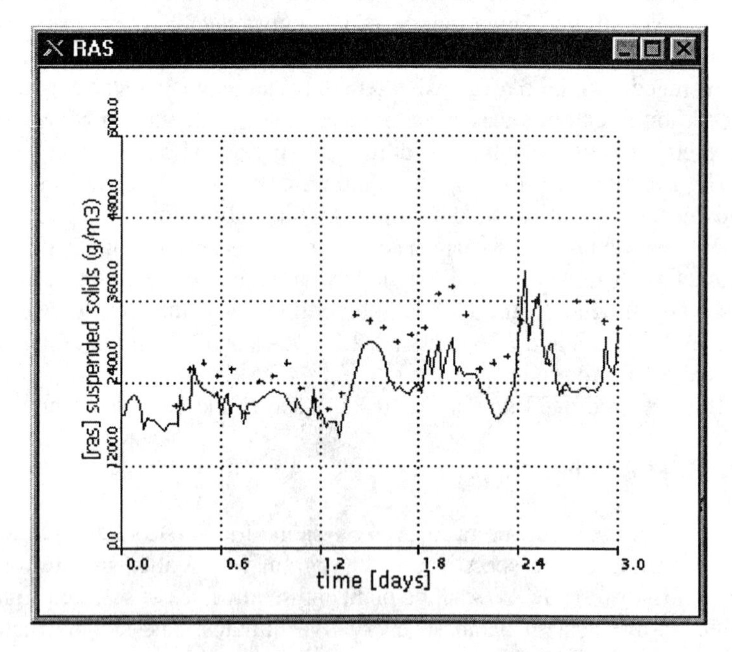


**Figure 7.34** RAS fit during the 3-day calibration period.

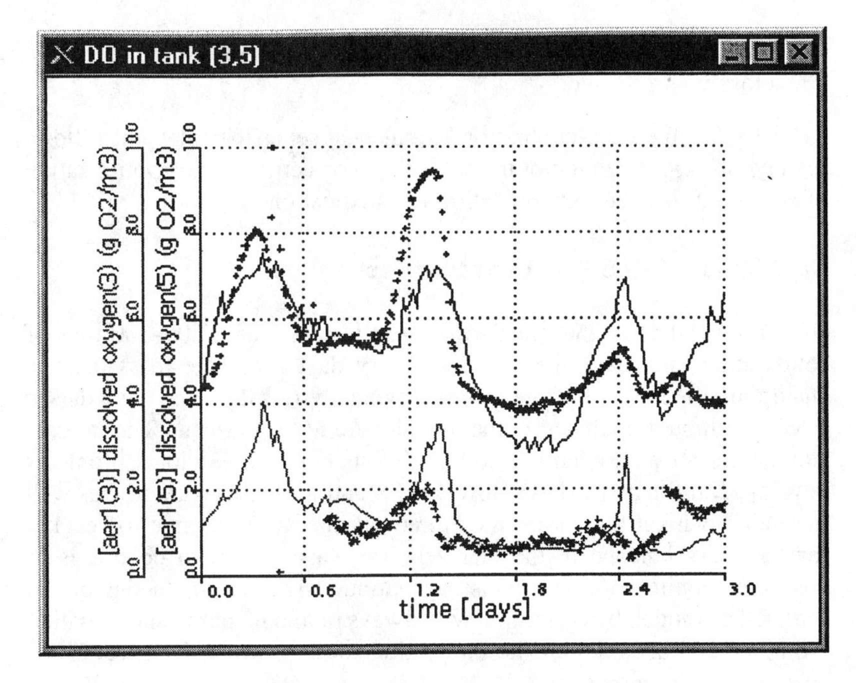


**Figure 7.35** DO fit during the 3-day calibration period.

and effluent ammonia, as well as ammonia midway down the aeration tank, DO's at two locations, sludge blanket (calculated from the profiler data), MLSS and RAS concentration. Data file format conforms to GPS-X requirements (Hydromantis, 1995), with new time-stamped files opened daily.

Special attention was given to missing or faulty signals. In an early implementation a marked reduction in speed performance was observed due to outliers and noise in input data. The solution of differential equations on noisy input data results in much shorter allowable timesteps even if the average signal value is sensible. Obvious outliers and missing data were replaced by a "typical" long term average value for the given parameter. This solution proved to be more stable than simply keeping values within predefined bounds.

## AUTO-CALIBRATION MODULE

The most important tasks of the on-line system, in this application, are to:

- simulate solids loading to predict and prevent clarifier failure on the plant

- predict the level of nitrification and required aeration energy as closely as possible

Therefore, the Auto-calibration Module was set up to contain the following layouts: Optimization of the RAS flow correction factor, Optimization of solids settling, and Optimization of nitrification.

### Optimization of RAS Flow Correction Factor

In a typical plant that treats 50,000 $m^3$/d of sewage, 200–300 tons of solids enter and leave the clarifiers every day. From the mass balance standpoint the mass of effluent solids (typically less than one ton a day or 0.5%)—although included in the model—are not significant. The internal storage capacity may amount to 20–30 tons before operational problems develop. If influent and RAS flow, MLSS and RAS concentration, as well as sludge blanket height are monitored on-line, it is possible to account for the mass balance in the clarifier. The easiest way to do this is to compare monitored data against a continuously running model of the settler. The model, by definition, will always maintain mass balance within the numerical accuracy of the integration routine, which far exceeds the accuracy of the monitors.

Typically we would not know which signal or signals cause the measured data to not close the mass balance. It is most likely an effect of several sensors being slightly off calibration. The practical approach for the model is to still close the mass balance by identifying a RAS-flow correction factor. By optimizing this factor to the real data the mass balance will be closed, the measured and modeled MLSS and RAS will fit much better and an estimate of overall mass balance error is gained around the clarifier.

Since the accepted measurement accuracy of flow metering is in the 10% range and suspended solids can be monitored even more accurately, a variation of around 10–15% in the mass balance is not critical and in fact can be observed on a continuous basis. When the correction factor changes to 20 or 30%, an alarm is generated in the advanced fault detection module.

An example of the fit between measured and corrected RAS concentration is shown in Figure 7.36 using the optimized RAS flow correction factors shown in Figure 7.37.

### Optimization of Sludge Settling

The task of this layout is to extract the hindered settling parameter from flow and solids signals around the settler. The solids settleability is a key

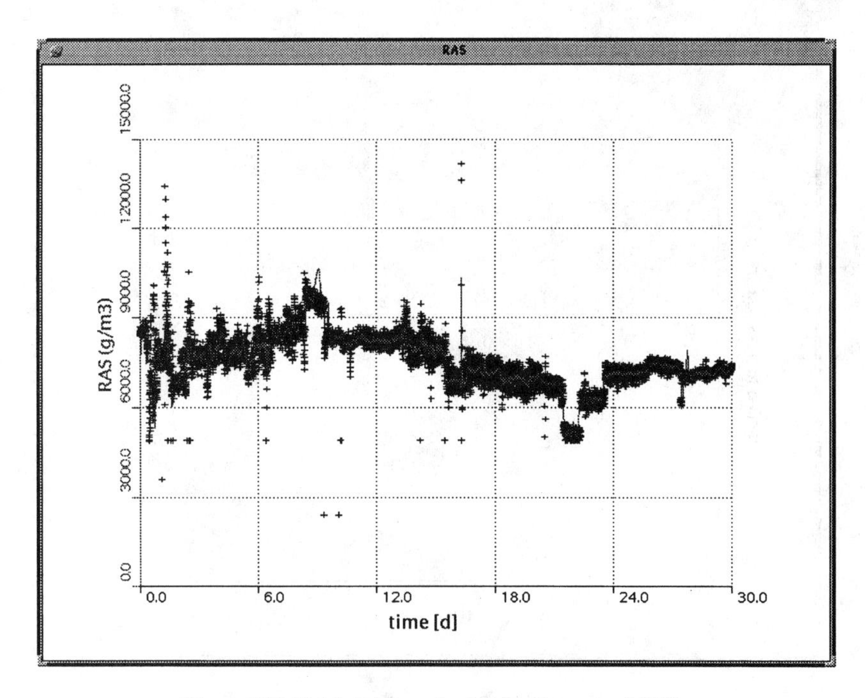


**Figure 7.36** Match between simulated and measured RAS.

parameter in the proper operation of the plant. This optimization makes use of the calculated RAS flow correction factor so does not have to account for mass balance errors in the measurements.

The layout is running in two different modes, depending on the level of sludge blanket. If the sludge blanket is negligible (the solids flux is enough to deliver all solids to the bottom of the clarifier), only an upper bound on the hindered settling velocity parameter can be identified. Under these conditions the plant is usually in a normal operational mode and settling is not a bottleneck.

If a sludge blanket height measurement is available and not negligible, the clarifier is close to the limiting flux conditions. This allows the optimizer to determine the exact value of the hindered settling parameter. It is under these conditions that settleability and proper predictions become critical.

## Optimization of Nitrification

The aggregated layout of the plant is used in this optimization task. The layout is reading the optimized RAS flow correction factor and the optimized hindered settling parameter as well.

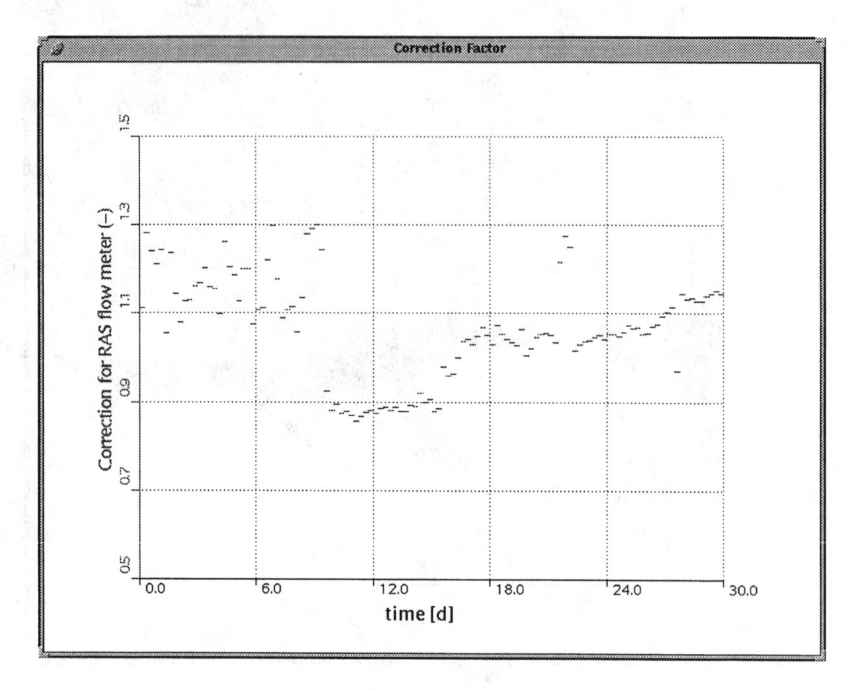


**Figure 7.37** RAS flow correction factor.

The layout was also reading in power uptake from rotors and calculating DO concentrations from the biochemical model and mass balances. While DO concentrations were not optimized, an alarm was placed on a difference between simulated and measured DO signals above 2 mg/L.

Similar to the sludge blanket optimization, it is impossible to optimize nitrifier growth rate based on only effluent ammonium levels. By the time the treated sewage travels most of the aeration tank it becomes fully nitrified and all the optimization can identify is a lower bound for the nitrifier growth rate.

Placing an ammonium probe into the aeration tank itself, approximately midway down the tank, alleviated this problem. At this position, the ammonium concentration typically varies between 1 to 4 mgN/L. The higher concentration periods (during the afternoon loading conditions) allow identification of the maximum specific nitrifier growth rate, while during the nights the half saturation concentration could be estimated. This latter method was not implemented on-line due to stability and accuracy problems.

Dynamic optimization of the nitrosomonas growth rate results in the simulated versus measured ammonia in Figure 7.38. The change in nitrobacter growth rate is shown in Figure 7.39. In this case the ammonia

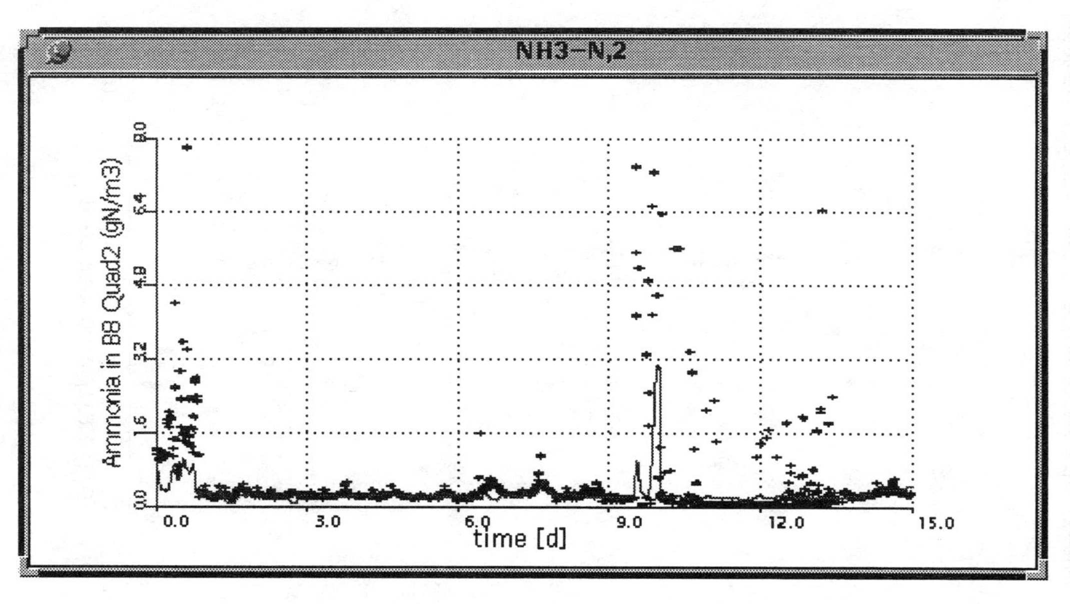


**Figure 7.38** Match between simulated and measured ammonia.

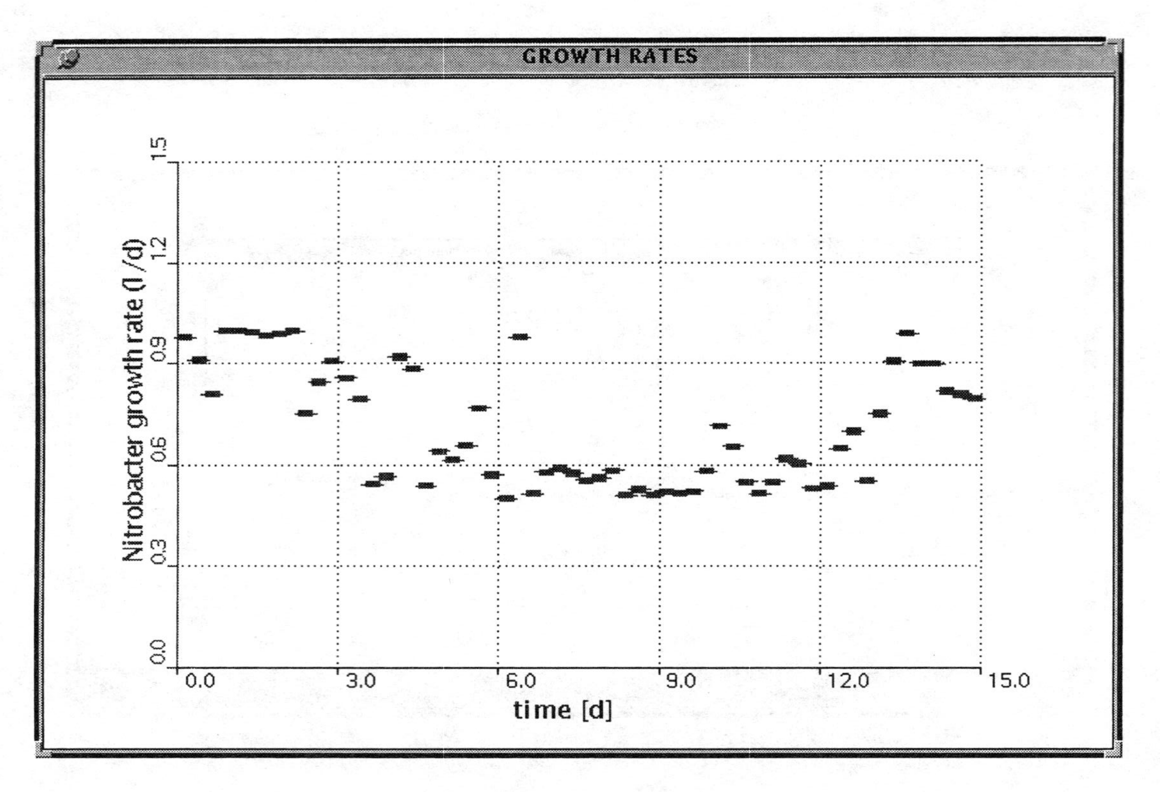


**Figure 7.39** Nitrobacter growth rate.

probe was placed at the end of the aeration basin, which results in very low ammonia variability.

## ADVANCED FAULT DETECTION MODULE

There are three elements of the fault detection system. Low level sensor fault detection is performed on monitored signals, advanced fault detection comparing simulated and measured outputs, and process fault detection based on dynamically changing model parameters generated in the Autocalibration module.

### Low Level Sensor Fault Detection

Low level sensor fault detection can be (and should be) implemented for all monitored signals which are fed to the on-line system. One GPS-X layout can hold any number of instances of the signal tracking tool, each of which monitors one signal. In this example an ammonia monitor installed to measure the influent ammonia concentration, is considered.

The low level fault detection is based on simple linear regression on a moving window. The number of data points or the width of the window can be specified. The method was described in more detail previously. From the operator's standpoint, the fault conditions can be directed either to the computer screen or a printer. Figure 7.40 shows a typical breakdown in influent ammonia signal and the resulting fault condition message.

Currently, fault conditions are manually processed. The operator has to acknowledge the alarm and remove the cause for the fault condition. There is an alarm grace period that can be set for different signals before a repetition of the alarm occurs. Once the signal has returned to an acceptable condition, the on-line system needs to be restarted if the alarm condition was serious. In some cases built-in protections (last known good value, lower and upper bounding) can help the on-line system to get through certain period with a missing or faulty signal.

### Advanced Sensor Fault Detection

While advanced sensor fault detection is numerically handled the same way as low-level sensor fault detection, using the signal-tracking building block, there are significant conceptual differences between the two levels.

In the low level signal analysis, the signal is considered from the numerical perspective. There is some possibility to qualify typical bounds, noisiness and rates of change that are specific to that particular signal.

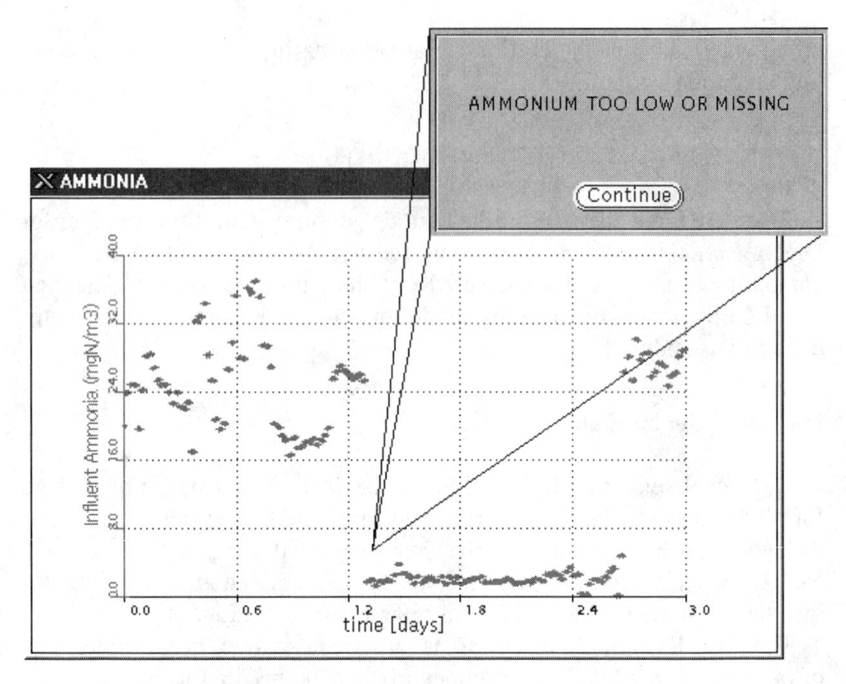


**Figure 7.40** Break in influent ammonium signal.

There is no other check to establish if the level and change in the signal is consistent with physical limitations that are placed on the real system.

The advanced sensor fault detection can use the intelligence built into the mechanistic mathematical models. A simple example will serve to highlight the advantages of this method. Assume the aeration tank biomass (MLSS) is simulated and measured in an on-line system. The advanced fault detection layout would track the difference (error) between simulated and measured suspended solids concentrations. As long as this difference is in a few hundred g/m$^3$ range and does not change rapidly, the condition is considered normal. Now consider a storm which suddenly transfers 30% of the MLSS from the aeration tanks to the final clarifiers. A low level sensor simply monitors the rate of change will signal a fault condition, even though the MLSS probe is in perfect working order. However, in the advanced fault detection unit, if the simulation can properly replicate the solids transfer due to increased hydraulic load, the error between the measured and simulated MLSS remains stable and no probe fault condition is reported. In fact a properly calibrated simulation model will handle the solids shifts with a large degree of accuracy and will be able to pinpoint failures in the probe reading, for example, when the mass drop is not consistent with mass balances around the aeration tank or clarifier.

A more advanced instance of the above method was implemented in the Auto-calibration Module. As described previously, one of the auto-calibrating layouts is continuously tracking (optimizing) the mass balance around the final clarifier. The result of the optimization is a RAS flow correction factor. This is the factor by which the RAS flow must be multiplied to close the mass balance around the clarifier. In an ideal situation this correction factor is expected to be 1.0 (no correction necessary). In reality the factor will move around the value of 1.0 due to two reasons:

- simplifications in the model structure, most notably the one-dimensional clarifier model used in this on-line system
- the combined measurement error of the flow, suspended solids and sludge blanket monitors: Calibration of these meters identifies the expected accuracy of the mass balance calculated from these measurements around the clarifier. Even if the faulty meter cannot be identified without taking further samples, the error in the mass balance identifies a faulty meter condition.

### Process Fault Detection

The Process Fault Detection element of the Fault Detection Module takes advantage of the information extracted by the Auto-calibration Module and stored in the calibration database. Sudden changes of these parameters may indicated process problems, (e.g., a sudden reduction in settleability may point to an impending settler failure). Another typical scenario is a slow drifting of the process towards an undesired status. The generated alarm is a useful early warning for the operators to prevent process failures resulting in violation of effluent criteria or temporary loss of treatment capacity.

The available signals for this purpose are the outputs from the Auto-calibration Module, i.e., calibration database. In this example two signals were available: the hindered settling parameter (rhindered), and the maximum specific nitrification growth rate.

The hindered settling parameter of the double exponential settling function (the Vesilind exponential settling parameter) has a close relationship to SVI—it determines the settling velocity of solids under higher solids concentrations (i.e., in the sludge blanket). A bulking event shows up as a characteristic increase in the hindered settling parameter. Note that part of the time the exact value of rhindered cannot be determined, and only an upper bound can be found. During these periods the settling flux is high enough to deliver all solids to the bottom of the clarifier and no appreciable sludge blanket develops. Therefore, these periods are not crucial for process fault detection.

Based on observations of hindered settling velocity dynamics and the way the auto-calibration layout is set up, the parameter is available daily for tracking. The settling process alarm consists of the following checks:

- Instantaneous (daily) values are tested against a critical upper bound.
- Five-day trends are analyzed by linear regression and any undesirable tendency (shift to lower settleability) is detected.
- A noisy settling parameter signal (low $R^2$ on the linear regression) is interpreted as unstable process or sensor condition.

The maximum specific nitrifier growth rate is also extracted and analyzed for lower and upper bound. A value of lower than 0.27 was found to signify substantially reduced nitrification capacity (though this condition did not develop during the testing period and thus the value is model based). Values higher than 0.55 most likely point to sensor or auto-calibration failure.

## SUMMARY

This chapter reviewed new computer-based modelling tools that can be used as part of a wastewater treatment plant control system. These tools were developed on the premise that an integrated computer-based approach to wastewater treatment plant operation and control can have a significant impact on the performance of a plant. Impacts include improved effluent quality, prevention of plant failures, reduced energy costs, and deferred capital expenditures.

These modelling tools include model parameter estimation and auto-calibration techniques, advanced controller design, on-line process optimization, process and sensor fault detection, and forecasting. The actual implementation at any given plant must be customized based on process configuration, data availability, effluent requirements, and other treatment and control objectives.

While testing is on-going, results prove that information which is inaccessible with traditional control methos can be deducted on-line using the described techniques. This deduced information can be used for determining and supporting control decisions.

## ACKNOWLEDGEMENTS

This project was funded by Hydromantis, Inc. and by the Ministry of Environment and Energy, Ontario, Canada, through the Environmental Technologies Program (ETP—project ET255WS).

## REFERENCES


Barnett, M. W. and Takács, I. (1994). Dynamic modelling of a large-scale wastewater treatment facility. In *Proceedings of the Chemical Engineering Chemputers Conference,* McGraw-Hill Science and Technology Group, Chemical Engineering Magazine, February 1–2, Houston, Texas.

Beck, M. B. and Chen, J. (1994). System identification, parameter estimation and the analysis of uncertainty: a review. *IAWQ Specialised Seminar: Modelling and Control of Activated Sludge Processes.* Copenhagen, Denmark, 22–24 August.

Brannigan, M. (1990). An adaptive data analyzer. *The C Users Journal.* May, pp. 113–118.

Côté, M., Grandjean, B. P. A., Lessard, P., and Thibault, J. (1995). Dynamic modelling of the activated sludge process: improving predictions using neural networks. *Water Research,* Vol. 29, No, 4, pp. 995–1004.

Dailly, C. (1994). *MATLAB/GPS-X User Manual.* Report No. H112/UD/001, v.1, Cambridge Control Hydromantis Ltd.

Dold, P. L. (1990). Incorporation of biological excess phosphorus removal in a general activated sludge model. *Proc. 13th Int. Symposium on Wastewater Treatment.* Montreal, Canada, pp. 83–113.

Franklin, G. F., Powell, J. D., and Workman, M. L. (1990). *Digital Control of Dynamic Systems.* Second Edition, Addison-Wesley, Reading, Massachusetts.

Henze, M., Grady Jr., C. P. L., Gujer, W., Marais, G.v.R., and Matsuo, T. (1987). *Activated Sludge Model No. 1,* IAWPRC Scientific and Technical Reports No. 1, IAWPRC, London, England.

Henze, M., Gujer, W., Takahashi, M., Matsuo, T., Wentzel, M., and Marais, G.v.R. (1994). Activated sludge model no. 2. *IAWQ Scientific and Technical Reports, IAWQ Specialised Seminar: Modelling and Control of Activated Sludge Processes.* August 22–24, Copenhagen, Denmark.

Hydromantis, Inc. (1995). *GPS-X User's Guide.* Hamilton, Ontario.

Hydromantis, Inc. (1995). *GPS-X Technical Reference.* Hamilton, Ontario.

Hydromantis, Inc. (1997). *IC²S: Integrated Computer Control System for Wastewater Treatment Plants.* Hamilton, Ontario

Marlin, T. E. (1995). *Designing Processes and Control Systems for Dynamic Performance.* McGraw Hill, New York.

Mitchell and Gauthier Associates Inc. (1986). *Advanced Continuous Simulation Language.* Mitchell and Gauthier Associates Inc., Concord, MA.

Olsson, G. and Piani, G. (1992). *Computer Systems for Automation and Control.* Prentice Hall, International Series in Systems Control and Engineering, New York, NY.

Patry, G. G., Spanjers, H., Giroux, É., Nguyen, K. (1997). Respirometry: A key component in the development of an integrated system for wastewater treatment plant control. *IEEE Instrumentation and Measurement Technology Conference,* Ottawa, Canada, May 19–21.

Press, W. H., Flannery, B. P., Teukolsky, S. A., and Vetterling, W. T. (1986). *Numerical Recipes: The Art of Scientific Computing.* Cambridge University Press. Cambridge.

Spanjers, H., Vanrolleghem, P., Nguyen, K., Vanhooren, H., and Patry, G. G. (1997). Towards a simulation-benchmark for evaluating respirometry-based control strategies. *Preprints 7th International Workshop: Instrumentation, Control and Automation of Water and Wastewater Treatment and Transport Systems.* Brighton, UK, 7–9 July 1997, pp. 101–109.

Takács, I., Patry, G. G., and Nolasco, D. (1991). A dynamic model of the clarification-thickening process. *Water Research,* Vol. 25, No. 10, pp. 1263–1271.

Watson, B., Takács, I., and Patry, G. (1994). Dynamic model parameter estimates: How constant are the ''constants''? *Proceedings of the Water Environment Federation 67th Annual Conference and Exposition.* Chicago, Illinois. October 15–19.

CHAPTER 8

# Application of Intelligent Control in Wastewater Treatment

## INTRODUCTION

**WASTEWATER** treatment processes, especially biological treatment processes, have very characteristic features from the standpoint of controller design. The following lists the major aspects we should take into account.

### INTRINSIC UNSTEADINESS

The inflow volume and concentration do not remain constant. Inflow volume to municipal wastewater treatment systems can be easily doubled or tripled under storm weather conditions. If we think about activated sludge processes, the adaptability of activated sludge microbiological populations to influent variation is the major strong point of these processes. To make things worse, occasionally, chemicals inhibitory to activated sludge may slip into the influent. This means that, in many practical situations, the often-used steady-state assumption for control system development is useless.

### NONLINEARITY

The reactions of the activated sludge process often reach pseudo-steady-state when substrates, nutrients or oxygen are limited under various environments. This suggests that the system often stays in the nonlinear area

Takayuki Ohtsuki, Tetsuya Kawazoe, and Takaaki Masui, Kurita Water Industries Ltd., 7-1, Wakamiya, Morinosato, Atsugi-city, Kanagawa, 243-01, Japan.

of the operation state space, which is usually described by a Monod type equation, and this makes the application of modern control methodologies more difficult.

## COMPLEXITY OF THE PROCESS

There are many elementary reactions involved in the activated sludge process and the rate limiting reactions change according to environmental condition transitions. This is the another aspect that annoys control engineers. A single controller cannot take care of every situation. Once the rate limiting equation changes, we need another different controller. For example the activated sludge model No. 1 proposed by the IAWQ task group, that is a somewhat simplified model of carbon and nitrogen biological removal, has 8 elementary reactions. Most of these reactions have multiple nonlinear terms that are expressed by Monod type equations, and each of these terms can limit the overall reaction. In reality, there are many other reactions that are not incorporated into such theoretical model, but many of them still have q large effect on effluent qualities.

## POOR PROCESS UNDERSTANDING AND DEPENDENCE ON EMPIRICAL KNOWLEDGE

We have a poor theoretical understanding of phenomena such as bulking and foaming, which vitally affect treatment efficiency. Therefore operators necessarily rely on their empirical know-how when these phenomena are critical to performance. They are always on the watch for signs of bulking and foaming, for example clearness of the sludge blanket interface, colors, foam size, and even smell intensity. Based on these observations, operators make situational judgements utilizing their know-how or they use qualitative criteria from textbooks to change operating conditions such as loading and aeration intensity. Utilization of ORP measurement provides a useful example. We know that the ORP index can tell us what kind of oxidation-reduction reaction is taking place because many of the treatment reactions are oxidation-reduction reactions. But the absolute ORP value varies according to the other oxidation-reduction reactions taking place. Usually plant operators interpret ORP values and variation patterns based on their empirical knowledge.

## LACK OF KINETIC INFORMATION OF THE PROCESS

Conventional management of the activated sludge process puts too much stress on the evaluation of effluent qualities from a management point of view. Little kinetic information that is important from control point of view, such as bacteriological activity, is recorded. If we can measure the maximum oxidation capacity for a specific contaminant, we can know the maximum load of that contaminant from the standpoint of bacteriological activity. If we measure the treatment time, that is how long it takes to oxidize wastewater to endogenous conditions, we can know how much hydraulic load can be applied to stabilize the effluent. Unfortunately this kind of information is rarely measured in the field. This situation impedes a deep understanding of the process conditions, and usually try-and-error methodology is the only operational alternative.

### CONTROL OBJECTIVE CHANGE

Usually, the control objective of a wastewater treatment system should be the treatment efficiency from the standpoint of water quality. However, under irregular conditions such as storm weather or activity deterioration, the control objective is often different. Operators may bypass the influent to the ocean to protect the activated sludge in their reactors. They may concentrate on the activity recovery, ignoring the effluent quality. This means that an operator may change the control objectives, at least for a short time, to deal with the current irregular circumstances.

All these characteristics show that a single conventional methodology for controller design cannot give adequate control in every possible situation. Here we want to discuss how to cope with these difficulties within the framework of intelligent control systems.

## MODEL REPRESENTATIONS FOR WASTEWATER TREATMENT SYSTEM

From the brief review in the previous section, it is clear that the overall modeling and controller design for wastewater treatment processes using a single representation method is almost impossible, especially when the target treatment system relies on biological processes. If we wish to develop a flexible control system that can manage various practical situations, we should think about utilizing various model representations that can complement each other to give a full description of the target system by making the most of each representation's strong points. The following lists some of the typical model representations we can use for this purpose.

### THEORETICAL DYNAMIC MODELS

Since the emergence of IAWQ model No. 1 and No. 2, many research projects in control systems of biological wastewater process have applied these dynamic models as their base theoretical models. Dynamic models can describe the unsteady, complex, and nonlinear nature of the biological treatment systems. Dynamic models can mimic the dynamics of the target system as long as the incorporated kinetic representations are valid. These models are generic in this sense and potentially very useful under various circumstances that arise in real situations. Some researcher uses such theoretical models as base models for the extraction of linealized models at specific operation point (Weijers 1997). These models can also be utilized as generic reference models for model based reasoning (Patry and Chapmann, 1989). These models are described by differential and algebraic equations of their kinetics and stoichiometry. Commercial numerical differential equation solvers like ACSL by MGA software (MGA Software, 1995) can be used for the evaluation of these models. If we use these numerical solvers, what we need to do is just to declare these equations. This declarative programming style is very assimilable with object-oriented software development methodologies and modern software user interface based on an object-oriented paradigm. Commercial model construction and analysis systems like GPS-X by Hydromantis Inc. (Hydromantis, 1997) can be used to build very complex models that precisely follow the flow sheets of real plants just by declaring process unit icons and relations between these units, that is flows, in a graphical user interface (Figure 8.20). In this system, model equations for each unit is tied up with its graphical unit icon and placing icons in the GUI environment results in the declaration of kinetics and stoichiometry equations. Connecting icons declares mass balance relationships. Rapid model development is very important from a practical point of view and recent software technology allows us to achieve this based on rigorous theoretical relationships.

## EMPIRICALLY EXTRACTED MODELS

Empirical model extraction methods, such as time series analysis (Box and Jenkins, 1976), neural nets (Hertz et al, 1993) and decision trees (Quinlan 1993), are useful when we don't have theoretical models. They can mimic the input and output relationships as far as actual examples are given. But we should note that application of these models to systems with a wide operation range is potentially dangerous, because we may not be able to obtain all possible empirical information beforehand. Unfortunately, this is exactly the case with wastewater treatment processes. In practical management of wastewater treatment plants, our main concern is how to return the process from abnormal conditions to normality. Abnormal conditions are usually rare cases in reality and these data are hardly available for empirical model extraction. If controllers are build based on models derived from insufficient empirical information, we cannot tell what happens when unexpected upset occur. Nevertheless, this approach is important for rapid model construction when we are certain about the operation range, and sometimes this is the only alternative. Adaptive model extraction methods like neural nets can be used for on-line extraction of empirical models. After long time application to real plants, they can learn from the abnormalities they have faced.

In another application, these empirical model representations can be used to express what-if scenarios that are obtained from theoretical models. We are free to do any experiment in the virtual model environment. We can set up every possible situation and get a comprehensive input-output relationship from theoretical models. Then we can use these relations to build empirical models such as neural nets. This approach may seem strange at first glance, but through this process we can check the state space of the theoretical models and if we wish, we can choose the control action and incorporate it explicitly in the control system beforehand. This approach can avoid potential problems relating real-time search in the nonlinear model space, and allows us to tune and prove the stability of controllers in the design phase. What-if scenarios extracted in this way can be viewed as an interpretation of generic theoretical models, and helps operators to utilize theoretical models in much more understandable ways. Low computational load and robust model evaluation after the installation are other merits of this approach. Real-time integration of differential equations may not be justified to fit in the required response time. If we want to build in numerical solvers in our control systems, we do need to prepare some measure to avoid potential problems that come from computational integration problems.

## LINGUISTIC MODELS

Linguistic model representation, like expert systems, is another way of empirical modeling based on heuristics and qualitative observations. Linguistic models are necessary to describe the phenomena such as bulking and foaming that are described mainly based on qualitative observations. Skilled operators utilize a wide range of qualitative information, for example colors and textures of foam. And they have their own judgement rules to interpret such qualitative information. Although this kind of modeling method has the well-known problems of required time for rule knowledge extraction and subjectivity of extracted rules, these representations are widely used in practical fields and utilized in many applications. Recently there is a lot of research works on the application of machine learning techniques to overcome the problems relating to rule extraction. Rule extraction through decision trees is a typical example (Quinlan 1993).

Linguistic model representations can also be used to describe how to judge circumstances and determine required control actions based on collective thinking. Issues, such as the combination of water quality and running cost judgements, and control objective changes, belong to all multi-dimensional evaluation problems. These collective thinking issues inevitably require heuristic judgements based on an engineer's sense of values. Rule representation can describe this kind of knowledge typically by weighing the value of each rule. The inference engine of an expert system usually utilizes these values for the selection of rules to be fired.

## FUZZY MODELS

Fuzzy methodologies allow us to bridge the gap between the qualitative and quantitative world. They are also another method for empirical model representation. Fuzzy control methods are very useful to incorporate the rules of thumb of operators. A typical fuzzy controller incorporates multiple fuzzy rules for control action, such as "If DO is low then increase aeration" and "If DO is high then decrease aeration." These rules are simple one step rules from a situation judgement to a control action. We know that many of the control rules used by skilled operators belong to this type. Firing conditions (if part) of each rule are described by fuzzy membership functions and are checked for a match with observed data. This matching level is used by the defuzzification process as a weighing factor for control actions. Fuzzy controllers can be useful for the smooth transition from manual operation to automatic control operation, taking into account the skills (rules) of the operators.

Fuzzy expert systems utilize methods of fuzzification to evaluate ambiguity and vagueness of judgement. Fuzzy methodologies can be a powerful way to describe the collective thinking that is mentioned in the last section. Some fuzzy expert systems like FLOPS do not select rules by weighing values, but make judgement by firing all possible rules in parallel and make an overall decision based on the certainty values (Siler and Tucker 1989). This is just like operators making final decisions considering every possibility. This method is also important to cope with the inconsistency problems between registered rules. For example, foaming can be ascribed to both too excessive aeration and too deficient aeration. The final decision depends on other judgements that comes from another rules, such as "If DO is very low, the aeration is deficient." Fuzzy expert systems permit description of such rules, and can place focus on consistent rules judgement using the fuzzy matching capability.

## INTELLIGENT CONTROL SYSTEM BASED ON BLACKBOARD ARCHITECTURE

To obtain a full description of the system, an organized strategy must be devised that combines the various modeling methods described in the last section and utilizes them efficiently. Recent active research on intelligent control systems provides one such approach (Medsker 1995) and its application to the control of wastewater treatment systems seems to provide a typical case study that can effectively show the advantages of this approach. One of the characteristic features of an intelligent system is its ability to utilize the wide range of modeling methods just like humans do. When a human operator starts up an activated sludge process he may rely on the theoretical knowledge of bacterial growth and activity values from respirometery to manage the organic loading. But once he finds out that the reactor is foaming he will rely on his empirical knowledge like "This seems to be organic overload to the current sludge." "From my experience I should lower the loading for a while." We want to realize this kind of flexibility in our automatic control system. In this example the operator utilized the common information of sludge loading such as 1.0 kg COD/kg VSS in both judgements. To manage such common data between various model representations, a blackboard architecture is adopted in this paper. The blackboard concept is a widely used control metaphor in the field of artificial intelligence, where multiple software agents must contribute collaboratively to achieve system objectives (Lesser and Erman 1988). This concept comes from an analogous situation where various experts, each with their own specialty, discuss a common problem by sharing data on a blackboard (Figure 8.1). Based on this concept all reusable data are stored in the abstracted database named blackboard (abbreviated as BB hereafter) in the developed system. The software

**Figure 8.1** Conceptual scheme of blackboard system. Multiple experts who have their own specialties collaborate with each other to solve common problems by sharing data on blackboard.

agents specialized in various modeling methods are named expert modules (abbreviated as EM hereafter).

This conceptual scheme offers the following advantages;

- Clear separation of data (BB) and modeling methods (EM) allows the construction of a data-centered system architecture, and ensures a longer service life time for the control system, because data has a longer life time than methods of data utilization (modeling methods). Plant operators may start building a control system for optimum aeration, but after succeeded he may wish to construct an automatic report generation system using the same data set used for aeration control such as influent load, effluent qualities, and required electricity for aeration. If these data were interwoven into the control system, the development of a report generation system would be a full project that needs to start with data collection programming from scratch.
- The limitation of data manipulation methods on BB allows the simplification and standardization of the interface to the database, ensures the exchangeability of EMs, and ensures the consistency of the database. In practical system development, much efforts is exhausted in the development of interfaces between various modules. Especially, when we think about utilizing various modeling methodologies together, this kind of attention is very important.
- The limitation of the communication method between EMs to data sharing on BB allows interdependency between EMs to be minimized, helps gradual system development, and ensures system robustness in avoiding the propagation of one module's system failure to other systems.
- Centralized database allows easy maintenance of common and shared data. One of the most important issues of practical control system development is how to protect the valuable data from accidental loss. In the blackboard system, all common and permanent data is stored in BB. Therefore, we can focus on the maintenance of BB for this purpose. BB can be easily implemented using a commercial relational database. Once BB is built on such a system, the commercial system offers every utility required for database management; including data backup, retrieval, duplication, and even remote maintenance facilities.

Many of these advantages coincide with the merits of an object-oriented software development strategy, which has been devised and widely used in the field of software engineering to promote reusability and extendibility (Meyer 1994) of computer programs. A typical automatic control system based on this system concept is shown in Figure 8.2. Please note that the user interface and sequencer interface modules are also assumed to be one kind of EM, which allows the simple and uniform system configuration.

Other characteristic features of the developed system are:

- Quantization of data: In a dynamic system all the data need to be stored with related time information. To enable rapid data retrieval, the BB database automatically converts the raw time information to a unique index in the time space with a fixed time interval (Figure 8.3). The time space lookup table works as a conversion table between specific time and the time space index. Then converted index can be utilized for data registration and retrieval. This function allows the data in the same time interval to be easily gathered and manipulated in bulk mode. This is useful for making judgements using all the data that are close in time space.
- Utilization of multiple BBs: Allowing utilization of multiple BBs at the same time enables, for example, separation between BBs with different quantization intervals and the construction of a flexible network system configuration, like the one shown in Figure 8.4 without modification of the basic system concept. In this figure, the on-site BB works as a raw data logging database that is required for control and the remote BB works as a database for sum-up data that is required for system working condition surveillance. This configuration is extendible for surveillance of multiple plants with more complicated network configuration.

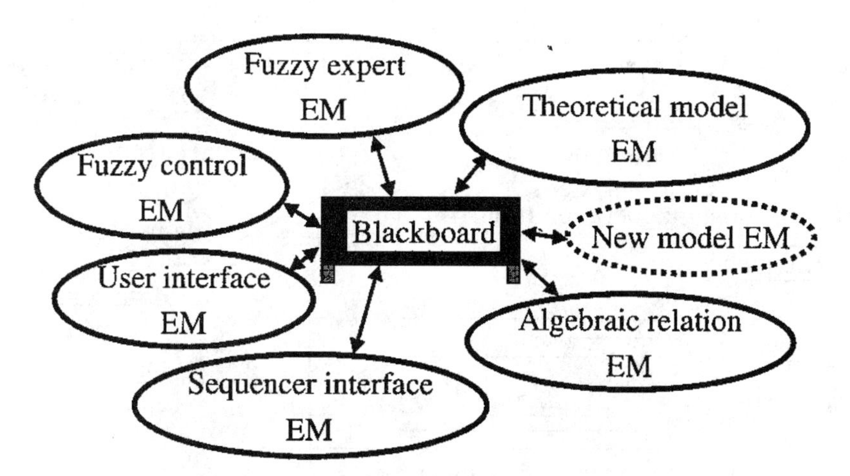


**Figure 8.2** Typical control system based on blackboard concept. Blackboard concept allows incremental system upgrade, and uniform and networked interface to database (blackboard). EM: expert module.

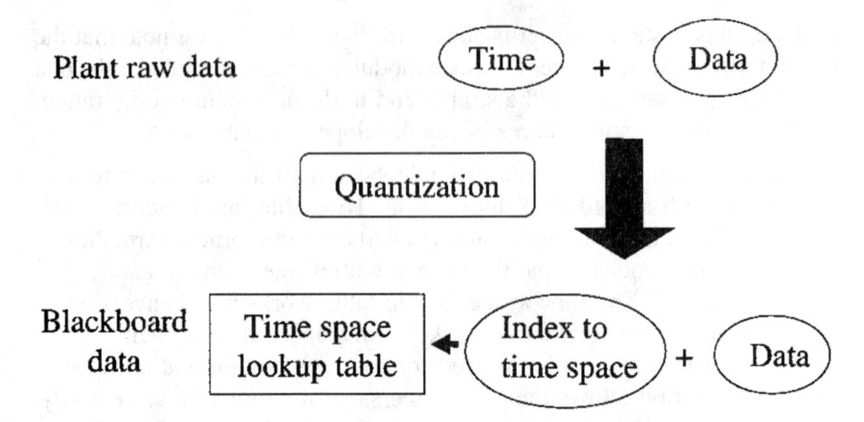


**Figure 8.3** Quantization of plant data in blackboard. Raw time information is converted to the index to the time space. Conversion between real time and index is managed by time space lookup table.

- Network transparency: BB and EMs configure typical client-server relationship. Implementation of a data communication mechanism between them using standard networking technology allows the creation of LAN and WAN mixed systems like the one shown in Figure 8.4. This system uses Windows NT as its base network operating system, and the socket interface on TCP/IP as its data communication protocol. We can utilize the remote access utilities of the operating system to establish TCP/IP connections through

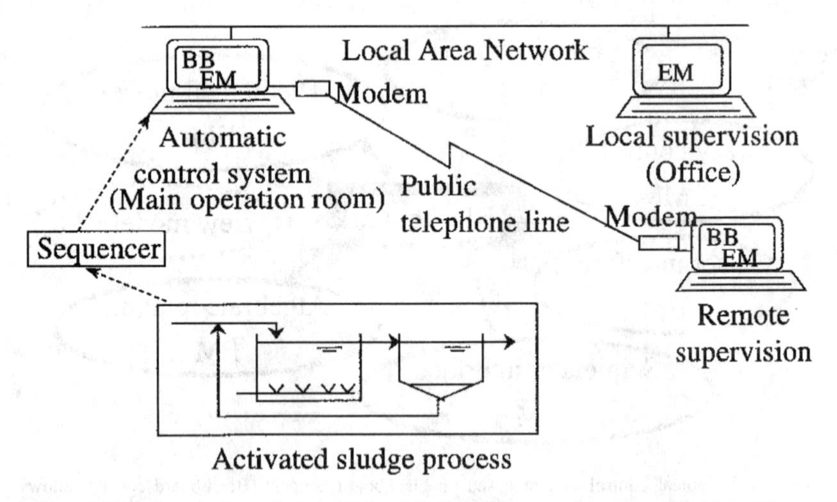


**Figure 8.4** System configuration example of the intelligent control system. Multiple BBs are used as the local database for control and the remote database for supervision.

telephone lines or the Internet. Once the connection is established, for any EM there is no difference between the communication with local BB and remote BB.

## SHARED DATA TYPES ON BLACKBOARD

Shared data on blackboard are required to be able to express a broad range of information that can be used for various model representations. On the other hand these data should not be too much specific to specific model representation. Expression of data in a more generic way permits sharing between different model representations and a longer lifetime for the database. Table 8.1 is the list of data types we adopted for the blackboard database. Some data types that will need explanations are described below;

### FUZZY NUMBER

Fuzzy number data (Figure 8.5) are composed of a central float number, a range value for vagueness and a center membership value for ambiguity. The fuzzy number can be used to describe data such as; effluent quality threshold is about 10 mg/L with the range value of 2, and estimated water quality value with the maximum membership 0.8 based on yesterday's measurement. The fuzzy number is always described by a triangular membership function with the specified maximum membership value.

### JUDGEMENT DATA

Judgement data (Figure 8.6) are expressed by the confidence values of affirmative judgement and negative judgement from a proposition. The fuzzy

TABLE 8.1. The Data Types Utilized in the Developed Intelligent Control System.

| Type of data | Sample of data |
|---|---|
| Real value | Reactor volume, reactor load |
| Integer value | Batch numbers of treatments per day |
| Trend data | Trend of DO measurement |
| String data | Explanation of judgment |
| Fuzzy number | Estimated NH4-N value based on previous measurement |
| Fuzzy membership | Foaming conditions |
| Judgment data | Appropriateness of aeration |

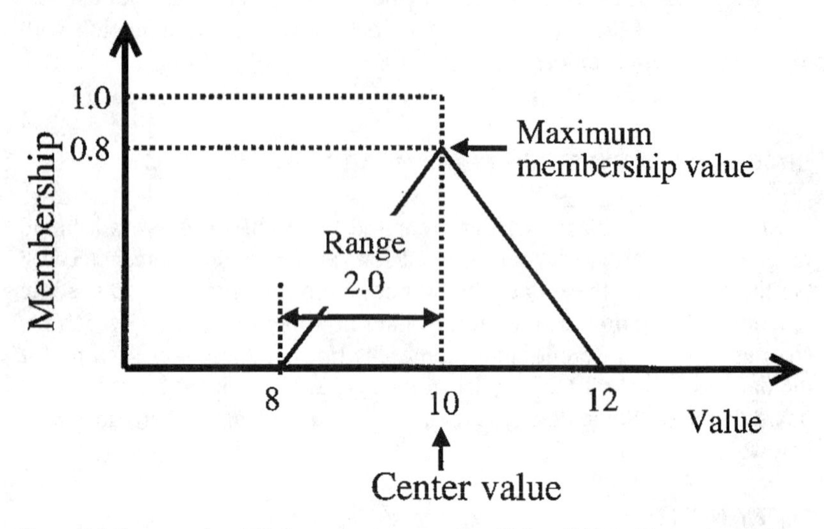


**Figure 8.5** Fuzzy number. This figure shows an example of ''about'' 10. In this case the maximum membership value is 0.8 and vagueness is 2.0.

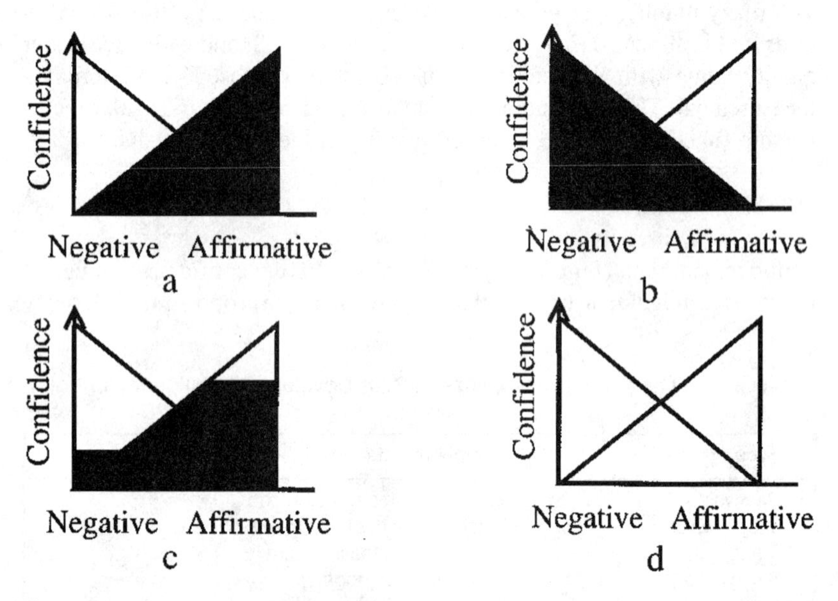


**Figure 8.6** Judgement data. (a) very affirmative, (b) very negative, (c) relatively affirmative, (d) no information.

expert EM described below mainly generates these data. If affirmative confidence is 1.0 and negative confidence is 0.0, this means that the proposition is completely affirmed, Oppositely, if affirmative confidence is 0.0 and negative confidence is 1.0, this means that the proposition is completely negated. If these confidences are more than 0.0 and less than 1.0, this data means fuzzified judgement of the proposition. If both the affirmative and negative confidence are zero or equal, it means that the proposition cannot be judged because there is insufficient data to discriminate positive or negative judgements. This membership representation of judgement is similar to the concept of measures of belief and disbelief that was adopted for the MYCIN project (Shorliffe and Buchanan 1975).

### FUZZY MEMBERSHIP

Fuzzy membership data is composed of a number of discourse and a train of membership values. This expression can be used for generic membership data representation and for various fuzzy modeling. For example with this data we can save observations such as ''Foaming is very severe.'' Judgement data is a special case of membership data.

### TREND DATA

Trend data is bounded by a fixed interval of the time space lookup table in BB. Then, each data is appended with sampling interval, trend data point number and train of float values. This expression of continuous trend data as a collection of bounded trend data allows easy retrieval of trend data that belong to a specific time interval. This also maintains the uniformity of data manipulation interface with other data types.

## IMPLEMENTED EXPERT MODULES

Below is a description of the major EMs implemented. Most EMs activate themselves at previously determined, fixed intervals, read necessary data from BB, and write reusable data to BB. That is, in many cases each EM works spontaneously.

### FUZZY EXPERT EM

Experienced operators make an overall judgement based on all the knowledge applicable to a given situation. Because the certainty of data and knowledge ranges from high to low, some weighting mechanism is necessary to make an overall judgement. The developed fuzzy expert EM has the capability of raw data fuzzification as shown in Figure 8.7. In this figure the raw nitrogen concentration value is converted to membership value using the membership function for ''Nitrogen concentration is low.'' Then, this membership value is converted to judgement data; the positive confidence value is set to the membership value and the negative confidence value is set to 1−membership value. In this conversion, the membership function changes the raw data to positive and negative confidence under the assumption that the summation of both confidences is 1. If we follow the definition of Shorliffe and Buchanan (Shorliffe and Buchanan 1975), these confidence values are equivalent to the measure of belief (MB) and disbelief (MD). They defined the certainty factor (CF) as MB—MD. When the certainty factor is 1, it means we are very positive about the judgement. Oppositely, when the certainty factor is −1, it means we are very negative about the judgement. If the membership value is 0.5, both confidence values will be 0.5 and the certainty factor will be 0.0. This situation means we cannot decide whether the judgment is positive or negative because there are equal confidence values for both positive judgement and negative judgement.

It is also possible to write a syllogistic rule to make a new judgement based on other multiple judgements. This rule knowledge can describe how much its judgement is dependent on other judgements using weight values. How these judgements should be combined (AND or OR operation) can also be described. Figure 8.8 schematically breaks down a judgement concerning sufficient aeration conditions using three separate data sources.

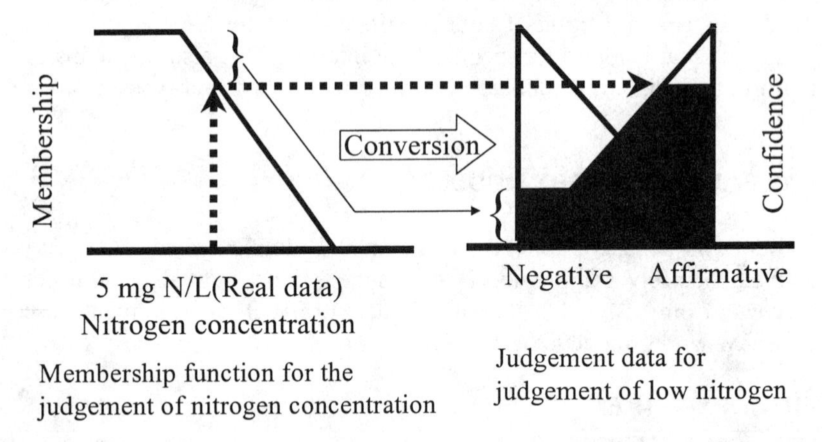


**Figure 8.7** Example of fuzzification of data in fuzzy expert EM. This example shows fuzzification of nitrogen concentration data to low concentration judgement. Real data is converted to the fuzzy membership using a membership function, then translated to judgement data. Membership value is assigned to the positive value of judgement and (1-membership) value is assigned to the negative value.

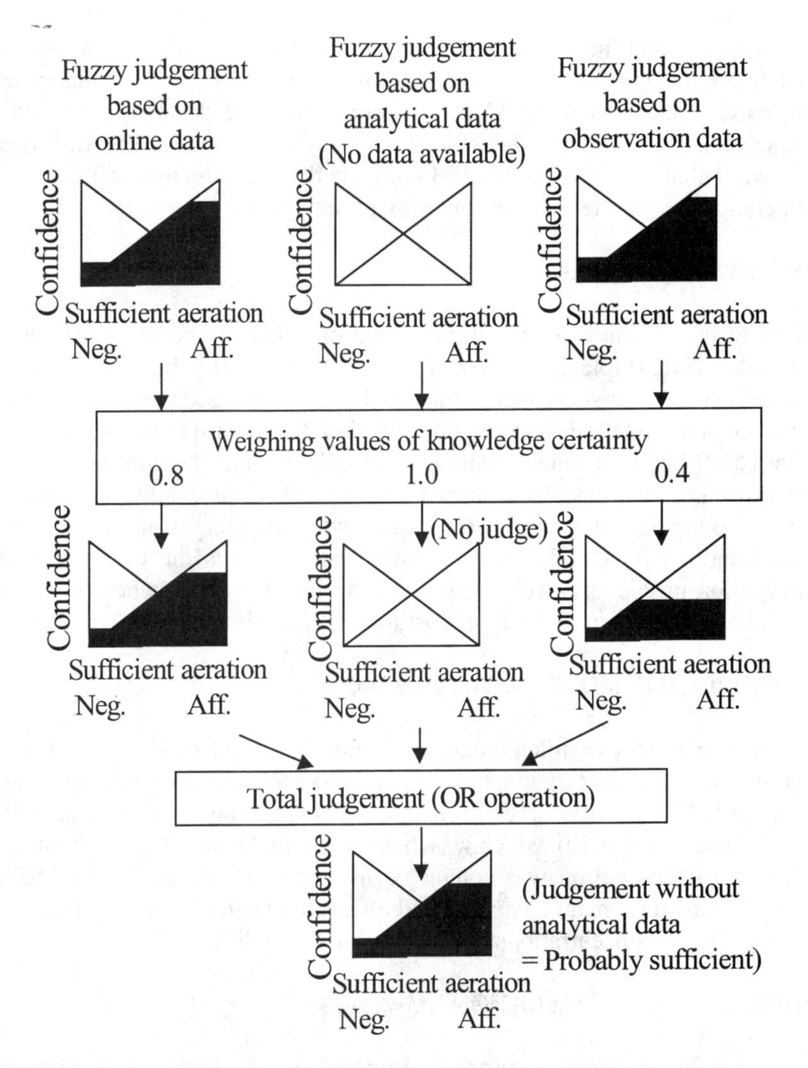


**Figure 8.8** Example of judgement by fuzzy expert EM (aeration situation judgement).

This example shows that, even when a dependent judgement with high weight (analytical data source weight = 1.0) is not available, the system can give a plausible judgement on the basis of qualitative judgements with low certainly (weights less than 1.0). At the same time, even if the judgement based on the observation data is very positive, when its weight value is low, the confidence of syllogistic judgement is lowered. Fuzzification of the situation judgement mechanism makes it easy to incorporate judgements based on collective thinking. In this case the OR operation is selected.

This means that the maximum positive and negative confidence value of each judgement is selected as the confidence value for the new judgement using this rule. When the AND operation is selected the minimum value is selected for the new judgement. The inference engine for this EM uses forward chaining, that is, this EM converts data to judgements. Inference operation is executed at previously fixed intervals.

## FUZZY CONTROL EM

The fuzzy control EM generates control values based on judgements by other EMs (typically, a fuzzy expert EM). Control value generation is based on the mini-max method and defuzzification is carried out by the centroid method. In the developed control system the rule matching conditions (left-hand, or antecedent sides) usually are not the combination of multiple propositions. Because the fuzzy expert EM can conduct the combination operation in a more sophisticated way, the fuzzy control EM does not handle this step. In this way this system avoids the combinatorial explosion problems in fuzzy controller design that occur when we want to take into account various propositions in the rule matching conditions.

## THEORETICAL MODEL (IAWQ NO. 1) EM

A theoretical dynamic model based on the IAWQ No. 1 model (Henze et al. 1987) for the target activated sludge process was constructed using the GPS-X™ system of Hydromantis Inc. according to the plant flow sheet (see Figure 8.20), which was then transported to the PC environment. Numerical integration was conducted using the ACSL system of MGA Inc. The model is manually calibrated offline to match observed respiration rates, sludge concentrations, and water quality indices.

## RESPIROGRAM EVALUATION EM

A newly developed respirometer is installed in the target plant to obtain kinetic information on the activated sludge. This respirometer measures respiration rate by the oxygen concentration decline in the gaseous phase that is equilibrated with the liquid phase oxygen concentration. Gaseous phase measurement allows contamination free evaluation of respiration rate, because the sensing device does not come in contact with the activated sludge. The configuration of the online respirometer is shown in Figure 8.9. It consists of a reaction vessel, an oxygen gas sensor, an air pump, and chemical injection pumps for standard ammonium solution and ATU. A reaction vessel, oxygen gas sensor holder and an air pump make up a closed-loop during measurement to enable respiration rate evaluation. With

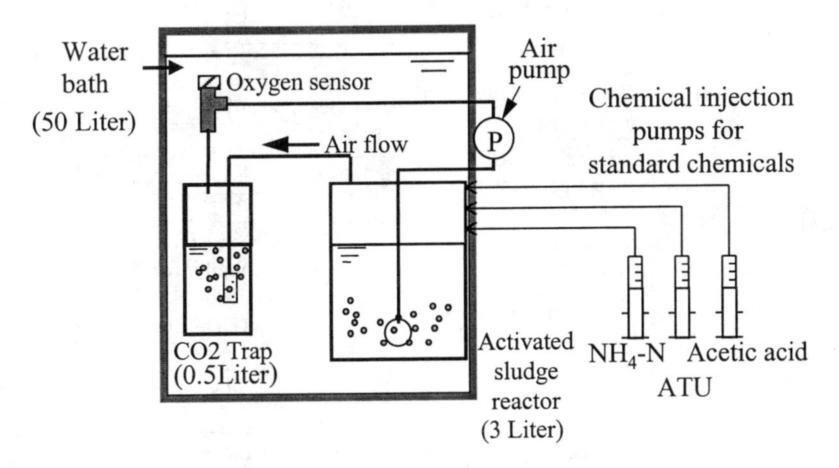


**Figure 8.9** Configuration of online respirometer.

the change of setup information this respirometer can measure various indices relating to respiration rate. In this paper this respirometer is setup to measure influent TOD (Total Oxygen Demand), nitrification activity, ATU-TOD (TOD under ATU injection), activity of heterotrophs, and endogeneous respiration rate. Figure 8.10 is the actual sequence of respiration rate evaluation. In another study, this respirometer was used for nitrification inhibition evaluation based on dose-response relationship with different setup information (Masui et al. 1997).

Figure 8.11 is the user interface of the respirogram evaluation EM. This EM analyzes the raw output from the respirometer, extracts indices and registers them to BB. The first peak in the respirogram is for the TOD, the second is for nitrification activity, the third is for ATU-TOD and the fourth is for carbonaceous activity.

## APPLICATION EXAMPLES OF THE INTELLIGENT CONTROL SYSTEM TO WASTEWATER TREATMENT SYSTEM CONTROL

### HIGH-RATE ACTIVATED SLUDGE PROCESS

As an example, the developed system is applied to the control of a high-rate activated sludge system. In suburban areas of Japan human waste and excess sludge from household-septic-tank are collected by vacuum cars, and are treated by a specially designed activated sludge process. Figure 8.12 is the typical high-rate, activated process to which

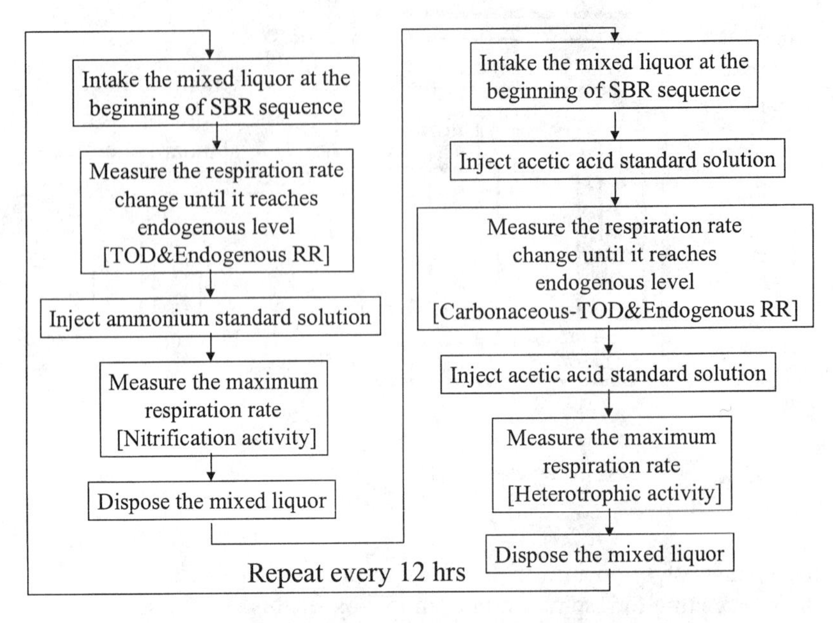


**Figure 8.10** Sequence of respiration rate evaluation.

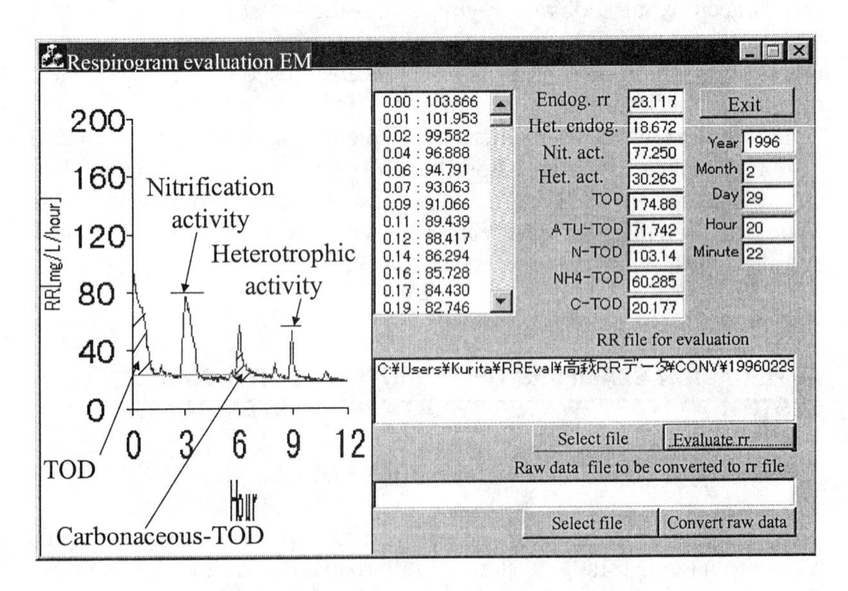


**Figure 8.11** User interface of respirogram evaluation EM.

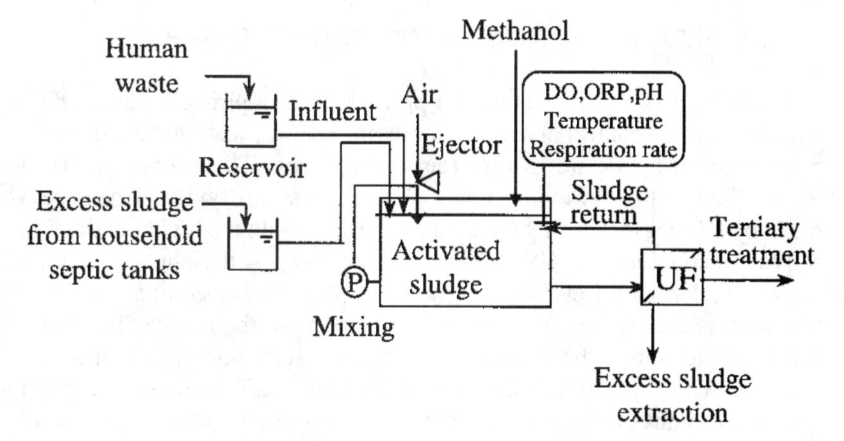


**Figure 8.12** Schematic flowsheet of high-rate activated sludge process. This process is a sequential batch reactor. Maximum MLSS concentration is 20,000 mg/L achieved by ultrafiltration. Nitrogen load is about 0.3 kg N/$m^3$. Influent ammonium concentration of more than 1000 mgN/L is reduced to less than 10 mgN/L in the reactor.

the developed control system is applied. This process is a sequential-batch activated sludge process for BOD and nitrogen removal. Its main reactor volume is 400 $m^3$. A one-batch sequence consists of influent load, aeration, anoxic mixing with methanol injection, and effluent withdrawal. Effluent sludge is separated by ultrafiltration (UF), the condensed sludge is partially returned to the main reactor and the rest is extracted for sludge treatment. In this way the MLSS in the main reactor can be maintained up to 20,000 mg/L without SS leakage to the effluent. The ammonium nitrogen load is about 0.3 kgN/$m^3$. With this process 99% removal of ammonium nitrogen is typically achieved, that is, more than 1000 mg/L of influent ammonium nitrogen is reduced to less than 10 mg/L of total nitrogen by one reactor. In this system DO, ORP, temperature, and pH is continuously measured. DO value is used to detect the completion of the aeration stage by peak emergence. Online respirometer (Figure 8.9) is installed for this experiment and used to measure influent TOD, nitrification activity, ATU-TOD, heterotrophic activity, and endogenous respiration rate as is described in the previous section. These values are recorded twice a day. Aeration and mixing (anoxic) phase control is manipulated by the operation of an air ejector that is provided on the mixing line of the reactor. The aeration strength is controlled by the flow rate of the mixing line. This system is equipped with a sequencer and a PC-based system monitor. In the usual operation, the operator must manually set the loading conditions and tune the aeration intensity, methanol injection dose, and sludge extraction volume to maintain the optimal control condition.

## CASE 1. AERATION AND METHANOL CONTROL EXAMPLE

In the high sludge concentration process, it is empirically known that, with appropriate control of aeration strength, a denitrification reaction takes place even during the nitrification (aeration) process. This phenomenon can be ascribed to the extremely low DO concentration (typically near zero mg/L by DO sensor) during the aeration period. With the high nitrification activity maintained by UF, the system can keep sufficient nitrification activity under such low DO environment. With this operating condition this process can efficiently utilize influent organic carbon as electron donor for denitrification without dividing the main reactor into anoxic and oxic portions. This operation condition poses some difficulty for the plant operator to control the aeration intensity because they cannot get useful information from the DO values, which are always near zero during the whole nitrification process. In place of DO values, they rely on ORP values for the fine-tuning of the aeration intensity. Figure 8.13 shows the typical time course of different nitrogen species concentrations. The ammonium concentration has a peak when the influent is poured into the reactor. After aeration begins, the ammonium gradually decreases and the sum of nitrite and nitrate increases. At the end of the nitrification period the ammonium falls almost zero, and at this time the sum of nitrite and

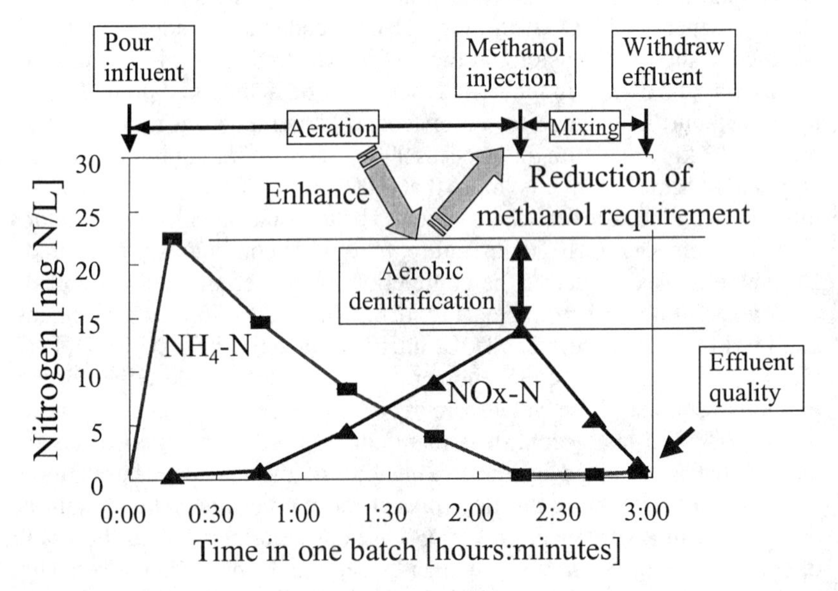


**Figure 8.13** Time course of nitrogen concentration in one batch of the SBR high rate activated sludge system.

nitrate concentration reaches a peak. But the sum of nitrite and nitrate concentration does not reach the concentration of the ammonium peak, and this is the result of the denitrification reaction during the aeration stage. We need to inject methanol that is required to reduce the peak concentration of nitrite and nitrate. Enhancement of this denitrification reaction during the aeration stage decreases the concentration of nitrite and nitrate nitrogen at the end of the aeration stage, and reduces the required volume of methanol for denitrification under anoxic condition. Experienced operators empirically optimize this aeration strength by observing effluent qualities, DO and ORP trends, and other observations such as foaming.

Figure 8.14 shows the schematic configuration of the developed control system. Online data, such as DO and ORP are recorded to the BB by the sequencer interface EM. The user interface EM gathers offline analytical and observed data and records them to the BB, prompting its operator for manual data input. Empirical rules for aeration and methanol situation judgement are implemented by the fuzzy expert EM and this EM judges whether aeration intensity and methanol injection are excessive, appropriate, or deficient based on data on the BB. These judgment data are recorded to the BB again. Rules include judgements based on analytical data such as $NH_4$-N and $NO_3$-N, empirical ORP value interpretation, and empirical knowledge that relies on, for example, foaming observation, and even the plant smell intensity. With these rules, the expert system can make an overall decision based on available data. Even if analytical data are not available, the fuzzy expert EM can make a plausible judgement on aeration and methanol, just like the operators do in manual daily operation. The fuzzy control EM gets the judgement from the BB and determines set-points for aeration and methanol. If the judgement is very confident, the control action will be very aggressive. On the other hand, if it is not so confident, the control action will be very sluggish. These control set points are applied to real plant through the sequencer interface EM. When an operator does not wish to apply these set-points to the real plant automatically, operators can skip the last step by using the system's open-loop advisory mode. In this mode, the system only displays the suggested control set points on its user interface, and operator makes the final judgement on the suggested values and decides whether they are applied to the real plant or not. In this case the system works as a typical operator assistance system.

Figure 8.15 shows the user interface of fuzzy expert EM. Registered rules are displayed in a tree format according to the dependency relation of each rule. In Figure 8.15 some of the rules are hidden leaving the higher level judgement for overall view of the rules. The left bar in the small box shows the affirmative confidence and the right bar shows the negative confidence. In this interface, for quick review of each rule's condition,

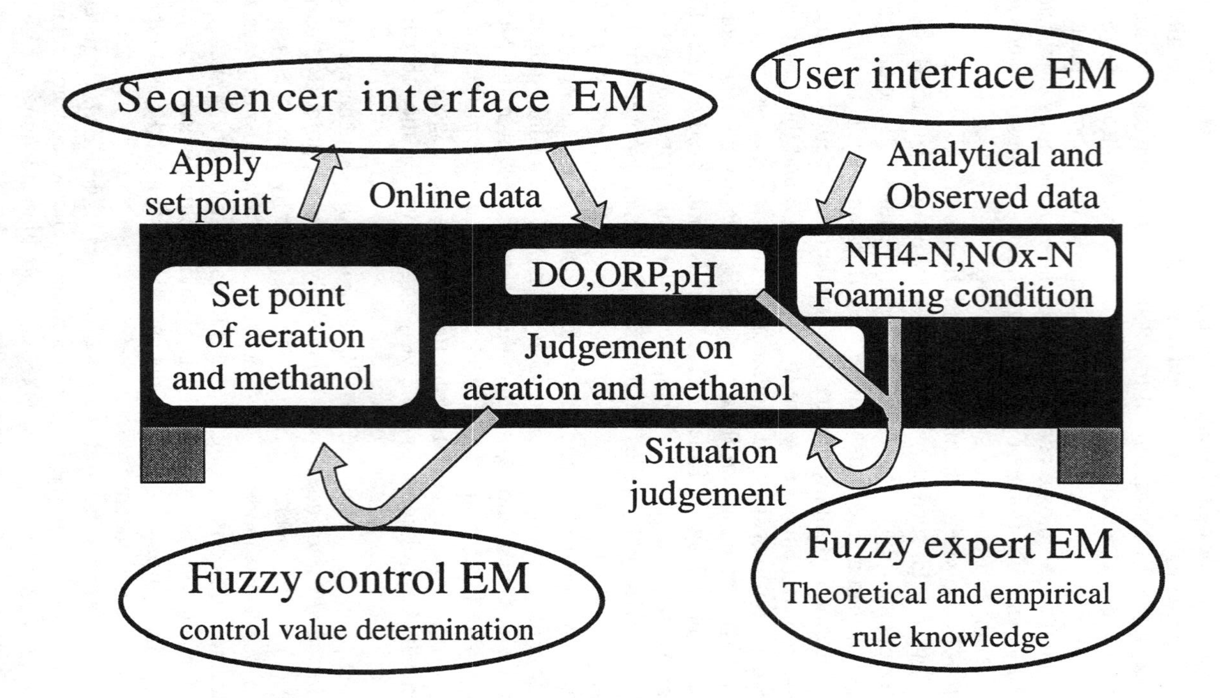


**Figure 8.14** Intelligent control system setup for aeration and methanol control.

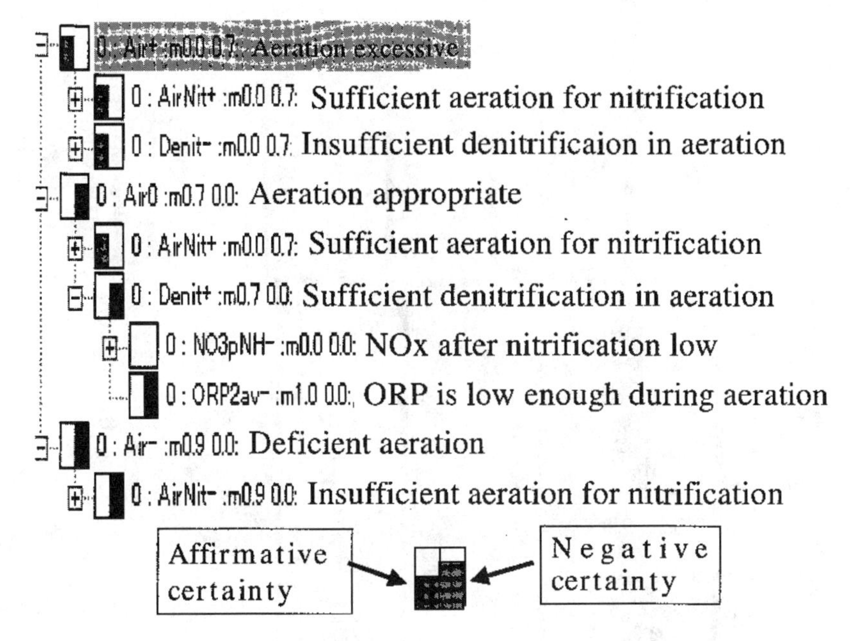


**Figure 8.15** Judgement of the situation by the fuzzy production system. (Aeration is excessive.)

the bar size is defined by the absolute value of the certainty factor that was described in the previous section according to Shorliffe's definition. If the certainty factor is positive, the bar is put on the positive certainty (left), and if the certainty factor is negative the bar is put on the negative certainty (right). Very small or no bar means that there is no enough data to discriminate between positive and negative judgement. In the example of this figure, propositions of ''Deficient aeration'' and ''Aeration appropriate'' are clearly negated, and proposition of ''Aeration excessive'' is affirmatively judged. This situation is just like the case where an operator gets the conclusion after considering all factors. Similar rules are implemented for the judgement of methanol injection conditions.

The fuzzy control EM (Figure 8.16) gets the judgement from the BB and determines set points for aeration and methanol. In this example three very simple rules are implemented in this fuzzy controller. That is, ''If aeration is deficient, increase the aeration'', ''If aeration is appropriate keep the current aeration,'' and ''If aeration is excessive, decrease the aeration.'' As is described in the previous section, this simplification of fuzzy control is made possible by the combination with the fuzzy expert EM. Three judgements from the fuzzy expert EM are converted to the membership values of three conditions (left-hand side) and finally converted to the set point value. If the judgement is very confident, the control

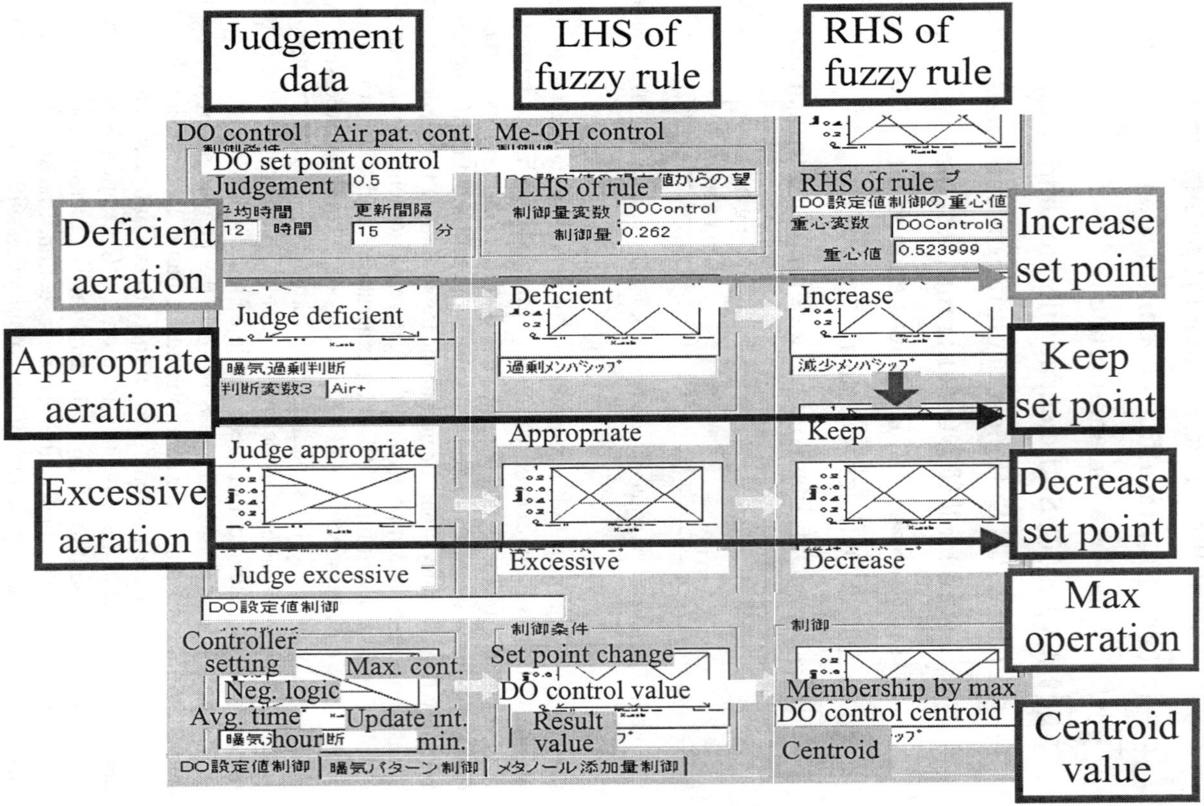

**Figure 8.16** DO set point control by fuzzy controller.

action will be very aggressive. On the other hand, if the judgement is not so confident, the control action will be very sluggish. Similar rules are set for the control of methanol injection volume. These control set points are set to the plant through the sequencer interface EM, if the operator wishes to do so without his intervention.

With this setup the control system successfully controlled the aeration intensity without upsetting effluent quality and minimized the methanol injection volume from 0.87 kg/m$^3$ to 0.5 kg/m$^3$ (Figure 8.17). Please note that the knowledge about the maximum allowance of effluent nitrogen concentration (less than 10 mg N/L) is incorporated into the fuzzy expert knowledge EM. This knowledge admits the effluent $NO_x$-N to increase within the allowance and consequently succeeded to minimize the methanol injection volume.

## CASE 2. INFLUENT LOAD CONTROL BASED ON NITRIFICATION ACTIVITY FORECAST

The influent source for the target system, includes excess sludge from household septic tanks. This is a primary source of toxic contaminants (Figure 8.18). Sterilizers from hospitals and herbicides from farm site are examples of such contaminants. If nitrification activity is continuously monitored, this can be a good index of toxicity, because nitrifiers are known to be highly sensitive to various toxic materials (Hockenbury and Grady 1977). At the same time, if we can get the current nitrification activity, we can estimate the future activity of nitrification under proposed operation conditions, and then we can make a plan for loading based on theoretical considerations. This kind of approach is potentially very useful,

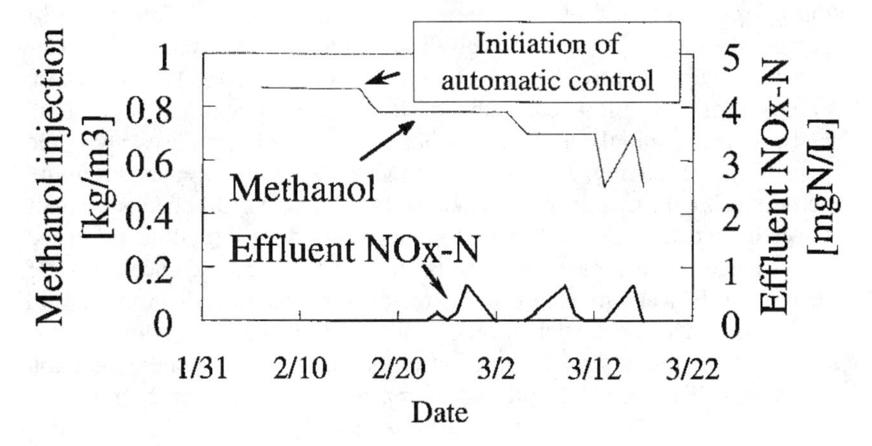


**Figure 8.17** Example result of aeration and methanol control by intelligent control system.

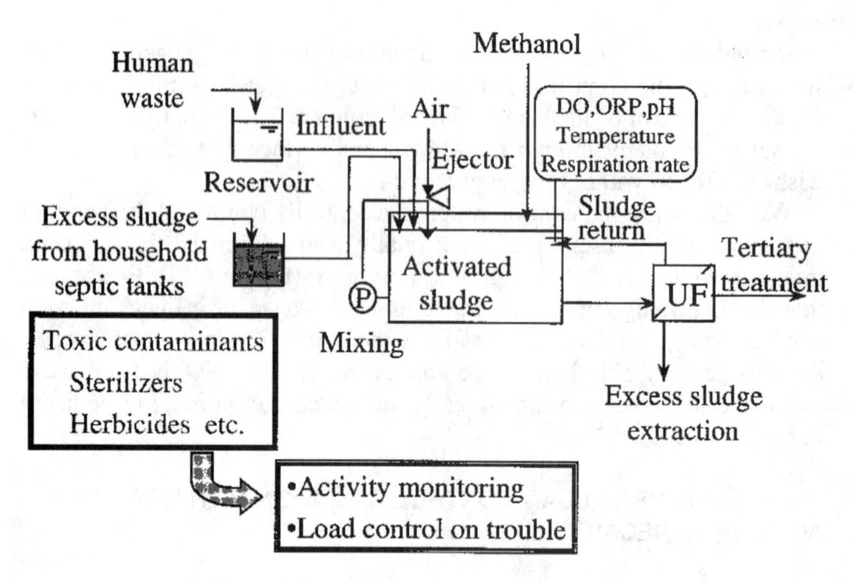


**Figure 8.18** Activity monitoring and load control.

when the activity of activated sludge is the rate limiting factor in the treatment system such as system start up period and recovery period from activity loss by toxicity. Figure 8.19 is the proposed configuration of the intelligent control system for this purpose. In this set up, operator gives the constraints relating system load planning through the user interface EM. Such constraints include, for example, ''On the next Sunday the plant loading should be stopped for pump maintenance'' or ''By the end of this year we need to fill up the influent reservoir to keep the enough loading during the holidays,'' etc. At the same time the sequencer interface EM gathers the current reservoir levels. The respirogram evaluation EM analyzes the current nitrification activity and records it to the BB. These values form the initial condition of the future estimation. With these constraints and initial conditions, the load planning EM proposes a possible plan of future loading, and the theoretical model EM estimates the future activity using these conditions. Then the load planning EM checks whether the estimated activity is sufficient for the proposed loading. If the proposed plan exceeds the estimated activity, the plan is modified and checked using theoretical EM again. This process requires an iteration of proposal and verification. Once a sufficient plan is obtained, the operator confirms this plan, and the today's plan is applied to the plant as the today's operation conditions. This planning process is repeated everyday for operator assistance.

Figure 8.20 is the GPS-X system flow diagram that defines the target

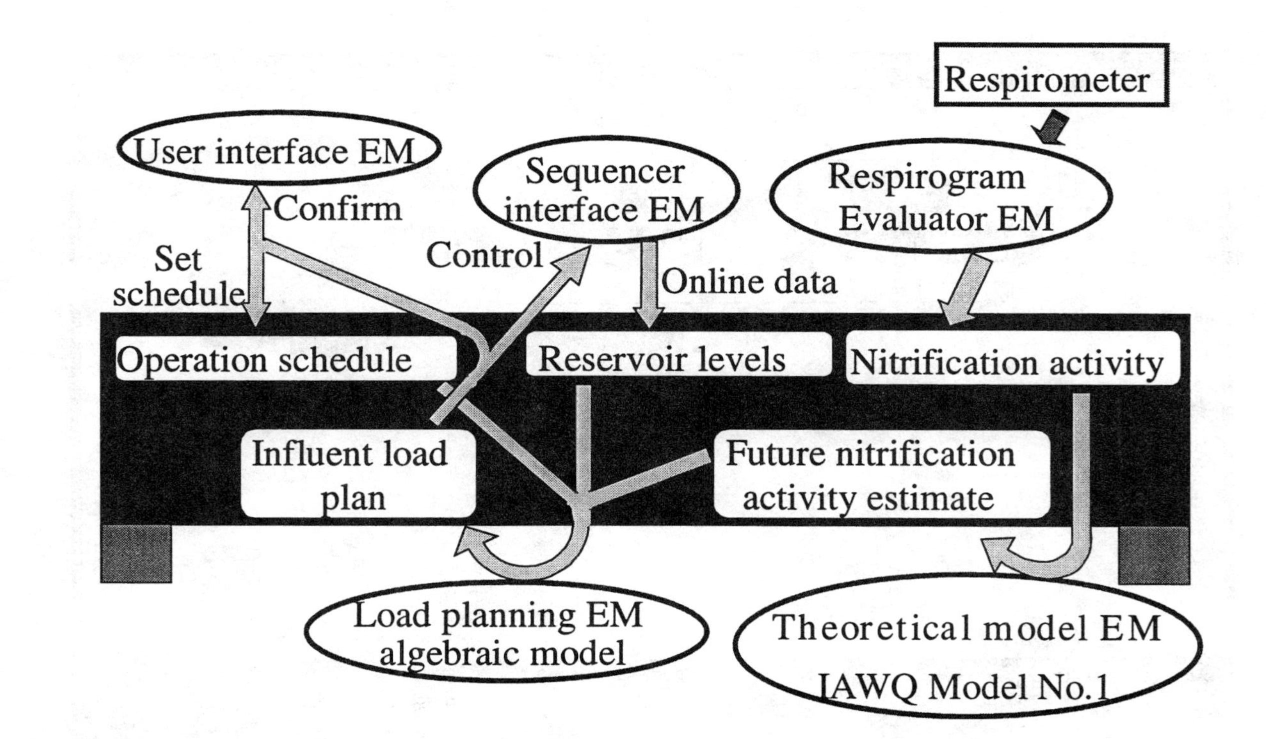


**Figure 8.19** System setup for influent load control based on nitrification activity forecast.

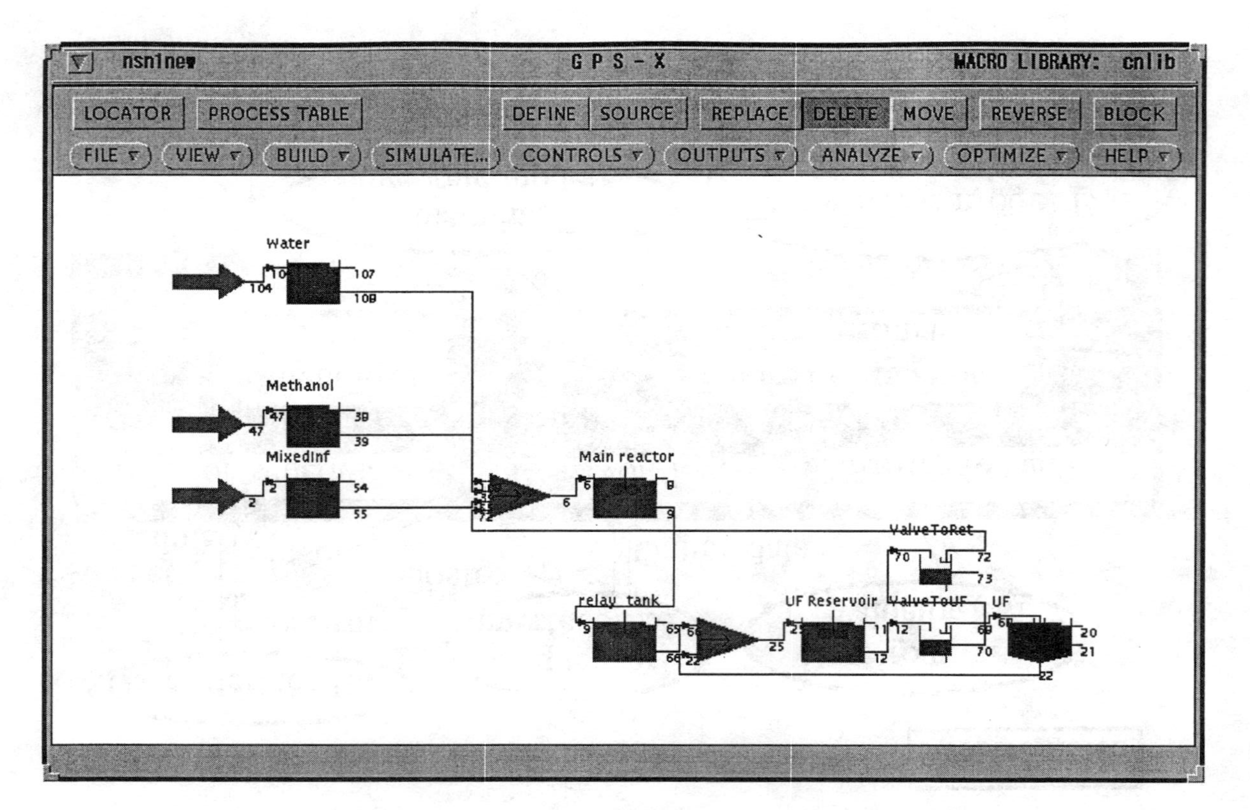


**Figure 8.20** Dynamic model for high-rate SBR process using the GPS-X system.

high rate activated sludge process. In this model, two types of influent are mixed together to obtain the plant influent. Dilution water and methanol are defined as additional influents. Beside the main reactor, a relay tank and UF influent reservoir are defined as biological reactors, because in the real plant we know that even in these tanks non-negligible biological reactions are taking place. Valves are set up to control the return sludge volume and the flow to the UF unit. In this model the UF unit is modeled as an idealized settler with 100% SS recovery efficiency. With the GPS-X system we can build a sophisticated SBR process model with many process controls in a short time.

Figure 8.21 is a simulation example using the constructed model. Model parameters, such as the growth rate of autotrophs and the decay rate of heterotrophs, have been calibrated using the actual time course of nitrogen concentration, the sludge concentration, the maximum respiration, and the endogenous respiration rate. With appropriate calibration, this model can mimic the dynamics in the real plant. By comparing Figures 8.21 and 8.13 it can be seen that denitrification phenomenon during the aeration period can be simulated with the basic IAWQ No. 1 model.

Figure 8.22 is obtained from the online evaluation of nitrification activity and represents typical activity deterioration caused by influent toxicity. This graph also includes the change of effluent ammonium concentration.

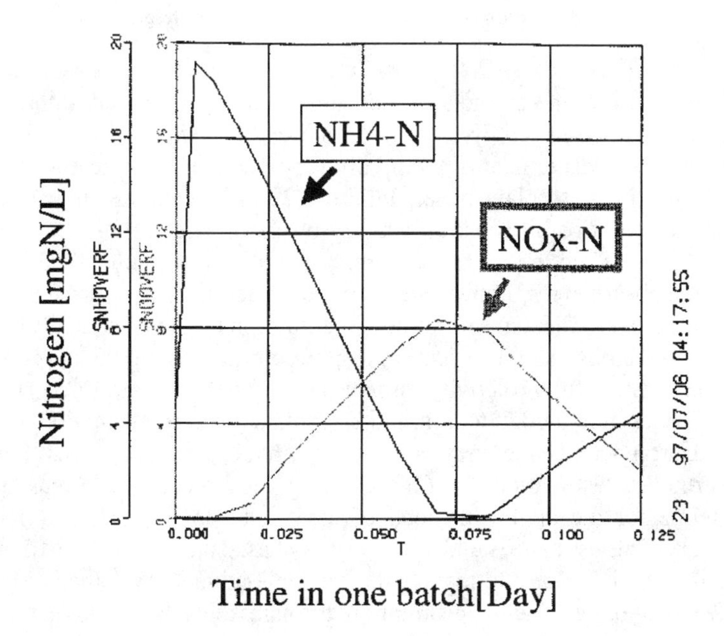


**Figure 8.21** One batch simulation of SBR high rate activated sludge process. Model was built by GPS-X system and numerical integration was conducted by ACSL system.

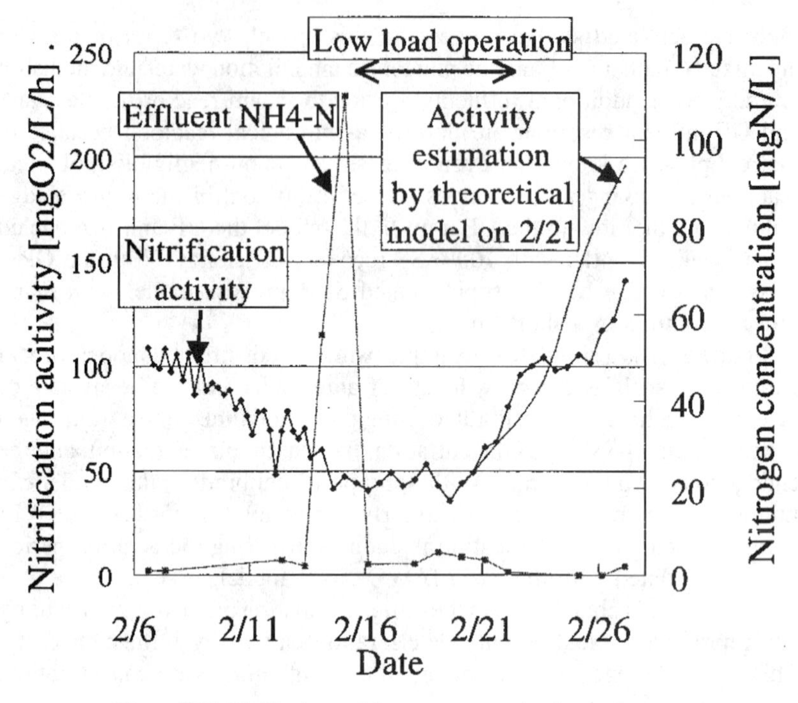


**Figure 8.22** Nitrification activity measurement by the respirometer.

A sudden rise in the ammonium concentration was observed with a one-week delay after an activity deterioration trend became obvious. It is clear that activity measurement is potentially very useful as an indicator because it is a precursor to plant upset. Figure 8.22 also includes the estimated future recovery of nitrification activity by the theoretical model on Feb. 21st. As described in the last paragraph this model is manually calibrated to match observed respiration rates, sludge concentrations, and water qualities. Moreover, the concentration of autotrophs is automatically tuned based on the observed nitrification respiration rate to give the appropriate initial condition for the activity estimation. Unfortunately the IAWQ model No. 1 does not model toxicity, therefore, a precise estimation of the nitrification activity recovery is impossible under inhibitory environment. Nevertheless, estimation like Figure 8.22 gives a theoretically reasonable guideline for the applicable load of today and an optimistic period of recovery. This estimate is updated every day based on real-time nitrification activity, and is utilized to make a reasonable loading plan. Other EMs can utilize this basic kinetic information through the BB to judge process conditions, and to determine other control set points.

Figure 8.23 is the user interface for the load planning. In this interface the operator can change the proposed loading plan from the load planning EM. If the operator wants to stop the loading of the plant during Saturday and Sunday for maintenance purpose, he changes the proposed loading during these two days to zero with the computer's mouse input device. This operation means putting new constraints on the load planning EM, and if he wants to refresh the plan based on his inputs he pushes the refresh button on the user interface. This trigers the load planning EM to start a new plan based on the new constraints. The load planning EM uses various heuristics to generate the initial loading plan proposal. These heuristics include expected inflow volume, desirable reservoir levels, rough estimate of the activity increase and decrease, and rules for desirable load changing patterns, etc. Then, the proposed plan is verified by the estimated activity produced by the thoretical model EM. The proposed plan is tuned based on verification strategies. The example of Figure 8.23 shows the case where the operator sets the load suspension for two days, and the load planning EM modified the proposal based on the new constraints. Please note that the loading before the suspension is increased to avoid reservoir overflow. The loading after the suspension is suppressed considers the estimated activity loss under no loading conditions. With the estimated recovery of nitrification activity the sludge loading is gradually increased after this event.

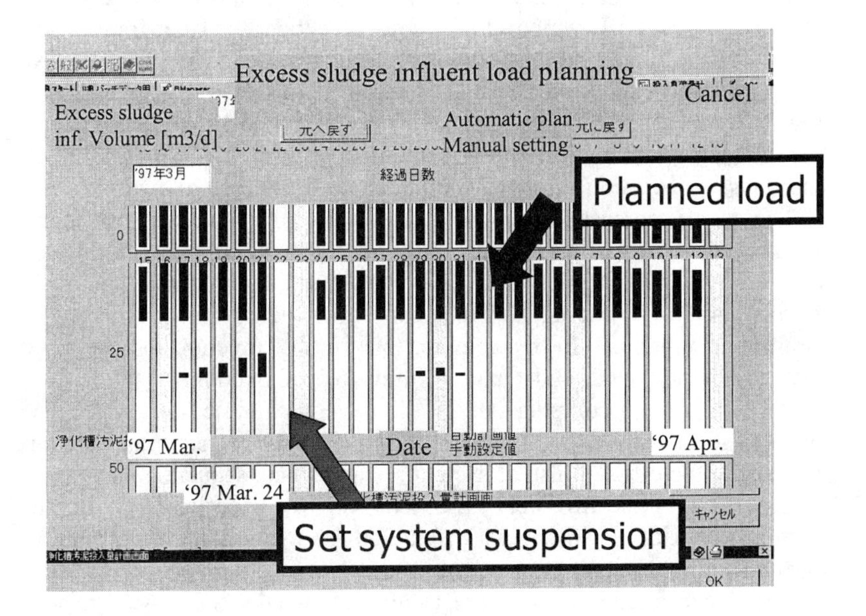


**Figure 8.23** User interface for load planning.

## DISCUSSIONS

It seems that there are various approaches to the development of an intelligent control system. Some approaches focus on the incorporation of humane vagueness and ambiguity to control systems by fuzzy techniques (De Silva 1995). Recent approaches seem to focus on learning, that is, adaptive, capability using neural nets or genetic algorithms (Medsker 1995) in a hybrid architecture with fuzzy control and fuzzy expert methodologies, etc. In the developed system we adopt an even more broad definition for an intelligent control system, that is, a system that can incorporate any modeling method that helps to represent human intelligent activities. The developed blackboard architecture gives a simple but flexible framework for this purpose. We intentionally restrict the data set on the BB to the very basic representations and avoid assigning explicit meanings to the data. Each EM gives meaning to the data by their own modeling objective. We cannot tell what kinds of model representations might be added to the system in the future and some model representations may be removed; therefore data representation that is specific to one modeling method should be carefully avoided. Fuzzy representations of the data, such as fuzzy numbers or judgement data are adopted as basic data types, since they are generic extensions of usually crisp data. As a future project, incorporation of data relationship can be an effective extension of the current system. Object relation is an another kind of static data that has a relatively long lifetime. Hearsay-III, which is a famous blackboard system framework, uses AP3 as a database manipulation language that is embedded in INTERLISP, and can use a flexible relational data structure (Erman et al. 1988). Recent object-oriented databases may be able to provide a promising base framework that allows the flexible data structures, reuse of old data, and data modification through inheritance with the system evolution.

The developed system does not have a scheduling function like many blackboard systems (Erman et al. 1988). The scheduling function decides which EM should be activated at a certain situation. This function is usually used to select an EM that can solve the current problem best (arbitration function). In my opinion, this kind of function restricts how each EM is activated, and reduces the degree of freedom of the system making the system complicated. On the developed system, each EM decides by itself when to work without arbitration. The scheduling function can ensure the efficient use of system resources and sequential activation of EMs when some EMs rely on other EMs. We experimentally implemented this kind of interlocking function by using a message passing mechanism between EMs, just like modern agent-oriented systems. This way of local communication assists the organization of closely related EMs without spoiling system extendibility. The communication between user interface of load planning and load planning EM is one example. The conflict resolution problem, that can occur when multiple EMs try to modify the same datum at the same time, requires other considerations. In the current implement, we avoid this problem by enforcing the rule that one datum is always modified by one EM.

Currently, the developed system relies mainly on the fuzzy expert EM to judge situations and this modeling method provides a flexible way of representing knowledge in rules. But the weights of knowledge must be set manually by the designer, and it is difficult in practice to set an appropriate weight to meet the mental image of the knowledge. We should note that these weight values also affect the aggressiveness or sluggishness of the fuzzy controller EM that is working with the fuzzy expert EM. This is a classical problem of fuzzy controllers, and here this problem appears in a more sophisticated way. Various methods of systematic weight evaluation are available (Maeda and Murakami 1986), but they are complicated and difficult to apply when there are a large number of rules. Some kind of automatic learning method for weight values must be studied for large rule-base construction.

## CONCLUSION

- A framework of an intelligent control system based on a blackboard concept was applied to the controller development for wastewater treatment systems. This approach allows utilization of multiple modeling methodologies to construct an overall model for wastewater treatment systems.
- An automatic control system for a high-rate, activated sludge system was successfully implemented using this system framework.

## REFERENCES

Box G E P and Jenkins G M (1976). *Time series analysis, forecasting and control.* Prentice Hall, New Jersey.

De Silva C W (1995). *Intelligent control: fuzzy logic applications,* De Silva, C W. CRC Press, Florida.

Erman L D, London P E and Fickas S F (1988), *The design and an example use of Hearsay-III,* In: *Blackboard systems,* Engelmore, R S, Addson-Wesley, 281–295.

Henze M, Grady C P L, Gujer W, Marais G v R and Matsuo T (1987). *Activated sludge model No. 1.* IAWPRC Scientific and Technical Report No. 1. IAWPRC, London.

Hertz J, Krough A and Palmer R G (1993). *Introduction to the theory of neural computation.* Addison-Wesley Publishing Company.

Hockenbury M S, Grady Jr. C P L (1977). Inhibition of nitrification—Effects of selected organic compounds. *J. Wat. Poll. Fed., 49,* 768–777.

Hydromantis Inc. (1997). GPS-X version 2.3 User's guide. Hydromantis Inc.

Lesser V R, Erman L D (1988). A retrospective view of the Hearsay-II architecture, In: *Blackboard systems,* Engelmore, R S (Ed.). Addison-Wesley, pp. 87–121.

Maeda H and Murakami S (1986). A fuzzy decision making method for multi-objective problems with a preference structure representation by fuzzy connectives, Vol. 23, No. 5. *Transaction of the society of instrument and control engineers* (In Japanese).

Masui, T., T. Ohtsuki, K. Kawazoe, A. Watanabe and T. Fukase (1997). Online respirometer for nitrification inhibition monitoring, Submitted for presentation in biannual conference of IAWQ 1997.

Medsker L R (1995). *Hybrid intelligent systems.* Kluwer Academic Publishers, Massachusetts.

Meyer B (1994). *Object-oriented software construction.* Second edition. Prentice Hall.

MGA Software (1995). ACSL reference manual Edition 11.1. MGA Software.

Patry G G, and Chapmann D T (1989). *Dynamic modelling and expert systems in wastewater engineering.* Lewis Publishers, Inc., Chelsea, MI.

Quinlan J R (1993). *C4.5: Programs for machine learning.* Morgan Kaufmann Publishers, Inc.

Shorliffe E H, and Buchanan B G (1975). A model of inexact reasoning in medicine. *Mathematical Biosciences 23,* 351–379.

Siler W and Tucher D (1989). Patterns of inductive reasoning in a parallel expert system. *International Journal of Man-Machine Studies 30,* 113–120.

Weijers S R, Engelen G L, Preisig H A and Schagen K V (1997). Evaluation of model predictive control of nitrogen removal with a carrousel type wastewater treatment plant model using different control goals, Instrumentation, control and automation of water and wastewater treatment and transport systems. *Preprint book,* 401–408.

CHAPTER 9

# Dynamics and Control of the High Purity Oxygen (HPO) Activated Sludge Process

## INTRODUCTION

THE high purity oxygen (HPO) activated sludge process is very popular in the United States for large scale applications. The process became popular in the 1970's after the passage of the 1972 Amendments to the Clean Water Act. Many large US cites found themselves in a difficult position. They had to build secondary treatment plants and their developed cities had no room for new, large treatment plants.

The Union Carbide Corporation developed the modern HPO process and marketed under the name Unox™ (McWhirter and Vahldieck, 1978). At first the process was frequently called the "pure oxygen" activated sludge process, but this is somewhat of a misnomer, since the oxygen used is never pure. The oxygen feed to the process generally ranges from 90 to 98%. Reactor head-space purity generally ranges from 40 to 80%. For this reason the process is now usually referred to as the high purity oxygen activated sludge process (HPO-ASP). The process continues to be marketed under trade names, but the patents have expired and no royalty is required to build the process. More than one company markets the technology and sells process and design knowledge with experience and equipment to potential customers.

The results of early pilot work suggested that the HPO process had several advantages over conventional activated sludge processes. Among the claimed advantages were energy savings, better settling sludge, less land area requirements compared to the conventional process, and a cleaner

Michael K. Stenstrom, Civil and Environmental Engineering Department, University of California, Los Angeles, 4173 Engineering I, Los Angeles, CA 90024-1593.

process, due to the covered reactors. The combination of a clean, low odor process with land savings made the process very attractive to large cities. Many large American cities use the process, including the City and County of Los Angeles, San Francisco, Sacramento, Detroit, Boston among others. Today, many of these advantages are questioned by design engineers, but most agree that the process requires less land area compared to conventional processes. It is becoming common practice, at least in California, to cover aeration basins at conventional plants, which is no longer a singular advantage of the HPO process. The process's reputation for land savings continues to be an important reason for selecting the process.

The HPO process requires a source of high purity oxygen. Most often, a pure oxygen plant is constructed with the treatment plant. To effectively utilize the oxygen, a plant is usually composed of 4 to 6 reactors in series (often called stages). Large plants may have several parallel sets of reactors, called "trains." Large plants usually generate oxygen cryogenically while smaller plants tend to use pressure swing oxygen generators. Very few plants are located near sources of high purity oxygen. Controlling and operating an oxygen generation facility is somewhat different than operating an activated sludge plant or other equipment typically associated with wastewater treatment. This presents challenges for treatment plant operators and oxygen generation plants are often operated by contractors.

The combination of covered aeration tanks-in-series with high purity oxygen, and an oxygen generation system, present interesting and challenging control problems. Not only must mixed-liquor dissolved oxygen (DO) concentration be controlled, but gas space purity must also be controlled. The cost of operating with excessive DO concentration is two-fold: it cost more to transfer the oxygen; the second cost is producing the oxygen, which does not exist in a conventional air plant. Because the HPO process uses reactors in series, there is a time delay or "transportation lag" associated with control oxygen feed. Other control problems also exist. HPO-AS processes must be instrumented to detect explosive gases (hydrocarbons) that may enter the process through a spill. It is possible to shut down an entire process if a gasoline spill is not detected before it is pumped into the reactors.

This chapter presents several of the prevailing control stratégies facing HPO plant operators. Both conventional and more modern, knowledge-based strategies are presented.

## PROCESS DESCRIPTION

Figure 9.1 shows a schematic of an HPO process. Usually 4 to 6 reactors are operated in series. Typical reactor sizes are 10 to 19 m square. Reactor

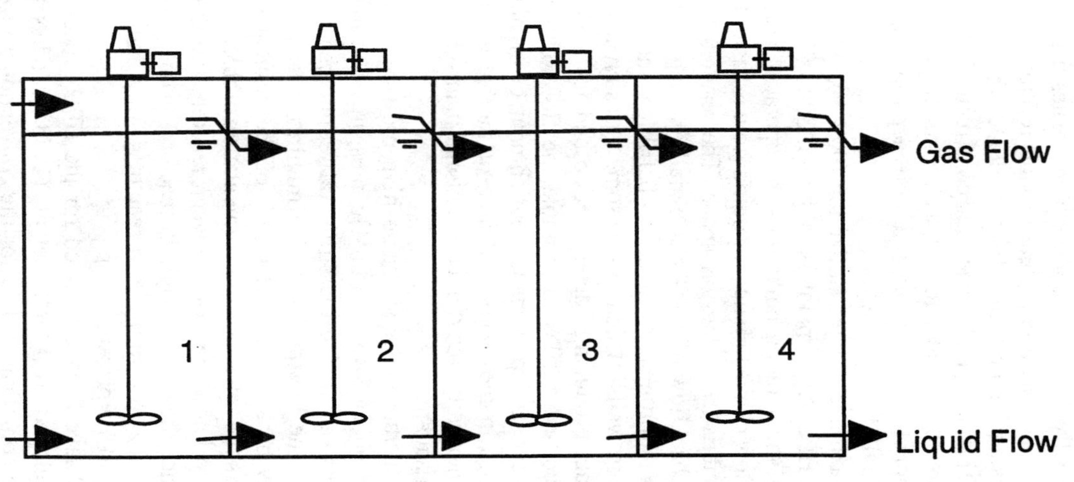

Figure 9.1 Typical HPO activated sludge process.

depth ranges from 3 to 4 m for designs using mechanical surface aerators to as much as 9 m for designs using submerged turbines or surface aerators with draft tubes. Most surface aerators are low speed as opposed to high speed. This is desirable in order to use only one aerator per reactor. Gas head space is usually 1 to 2 m high, which creates a reservoir of HPO gas. The reservoir acts as a complete mixing reactor and its affects on process dynamics and control must be considered when designing plants. Many of the HPO plants were designed and constructed before the popularization of fine-pore aeration technology; as a result no plants, to the knowledge of the author, use submerged fine pore aerators. Large plants may have 10 to 20 parallel trains.

In order to maximize oxygen utilization (90% oxygen utilization is a frequent goal of HPO-ASP operators, although 80 to 90% is more commonly obtained), the aeration tanks must be operated as tanks-in-series. Obtaining plug-flow characteristics for liquid flow is easier than obtaining plug flow characteristics in the gas space. The reactors have baffles or walls to prevent backflow of HPO gas. Small openings are provided for gas passage. One unfortunate result of this baffling is the creating of scum and foam retaining walls, which has in some cases created problems by enriching for scum producing bacteria (Kido and Jenkins). Newer HPO designs have modified baffles to promote the flow of foam and scum through the process, and prevent the enrichment of organisms such as nocardia. Openings in the submerged area of the baffle for liquid flow are smaller than normally found in conventional plants. Openings may be only 1 m to 2 m square.

The reactors operate under slight positive pressure. The first reactor may be operated under 8 to 12 mm Hg pressure. The low pressure is required in order to minimize the structural requirements of the reactor walls and cover. The low pressure also minimizes leakage, which in a typical plant may be as large as 2 to 4% of the influent gas flow. The very low pressures present special instrumentation problems. For example, it is very useful to know the flow between reactors and the exit or vent gas flow rate. Since the total pressure in reactor 1 is only 8 to 12 mm Hg, which decreases to 2 to 4 mm Hg in reactor 4, flow measuring devices must exhibit almost no pressure drop.

Many models have been developed for the HPO process. One of the earliest was developed by Mueller et al. (1973). Models for the HPO process are more complicated than for the conventional process because gas-phase mass balances and gas transfer terms must be included. The gas phase must contain balances on oxygen, nitrogen and carbon dioxide (water vapor is usually considered to be in equilibrium, and is calculated from saturated vapor pressure). Liquid phase mass balances include those required for biological activity as well as for the dissolved gases. The pH of the mixed-liquor must be accurately calculated in order to insure proper gas phase carbon dioxide partial pressure. Incorrect nitrogen or carbon dioxide partial pressures result in incorrect oxygen pressures which cause errors in mass transfer rates. McWhirter and Vahldieck (1978) and Linden (1979) presented similar models to Mueller et al.'s model; all use Monod-style (Lawrence and McCarty, 1970) kinetics and single material balances on substrate and microorganism concentration. Stenstrom et al. (1989) presented a similar model and used it to determine the oxygen transfer capacity of a full scale HPO-ASP.

More recently structured models have been used to describe the HPO-ASP. These models divide the influent substrate and biomass into different pools. Influent substrates are typically divided into soluble and particulate, and sometimes into soluble, particulate and slowly-degrading particulate substrates. Biomass is divided into active and inert portions. These models often include nitrification. Examples of structured models include Busby and Andrews (1975), Stenstrom and Andrews (1979), the IAWQ Model No. 1 (Dold et al., 1980) and Clifft and Andrews (1981). Stenstrom (1990) and Tzeng (1992) were the first to apply structured models to the HPO-ASP. They calibrated the model using full scale plant data and showed that the model was useful for sizing aerators in designing treatment plants. Yuan (1994) and Yuan et al. (1993, 1994) have applied the IAWQ model to the HPO-ASP and have shown that different decay rates are observed in the HPO and conventional processes.

Structured biomass models are more accurate in describing plug flow processes and processes under non-steady state conditions. The single mass/substrate Monod model cannot simulate substrate storage and increased endogenous respiration under high and low loading conditions, respectively. Models used for understanding process dynamics and control must be capable of describing non-steady state conditions if they are to be useful.

## CONVENTIONAL CONTROL STRATEGIES

Conventional control strategies for the HPO-AS process primary address high purity oxygen gas feed control. This is important for several reasons: the biological process must receive sufficient oxygen for operation; also the oxygen generation plant has limited "turn-up" and "turn-down" capability. During periods of high loading increased oxygen feed is required. During low loading much less feed is required. The oxygen feed rate to precisely match the oxygen uptake rates, as predicted by a structured model, may vary by 2 to 1 (Tzeng, 1992). The ratio is somewhat reduced from the ratio of high to low chemical oxygen demand (COD) loading because of the creation of stored substrate mass in the biomass; however, the process has inadequate capacity to assimilate or store excess oxygen caused by the changes in load. Therefore large amounts of oxygen will be wasted unless some form of control is used to modulate the oxygen feed from the oxygen generation plant.

## REACTOR 4 OXYGEN PURITY AND TOTAL SYSTEM PRESSURE CONTROL SYSTEMS

Figure 9.2 shows two typical control loops for modulating oxygen rate. The most straight forward approach is to use reactor 4 oxygen partial pressure to modulate oxygen feed to reactor 1. It is very easy to measure oxygen partial pressure. Meters that use fuel cells as sensors are inexpensive and reliable. They usually have less maintenance requirements that DO probes, because their probe is exposed to a gas, as opposed to a liquid with potential for biomass fouling. The turn up and turn down capacity of a cryo plant can be used to modulate oxygen flow rate.

The problem with this approach is the time lag associated with sensing a disturbance in the gas headspace. The gas retention time (reactor head space volume divided by gas flow rate) varies with gas flow rate. The gas flow rate between reactors decreases as the gas flows through the process. The gas flow rate typically decreases by 80% from the influent to reactor 4 vent. Typical gas retention time may increase from 0.4 hours in reactor 1 to 1.4 hours in reactor 4 at high plant loading, and from 0.4 to 4.6 hours at low loading. The total retention time may be 3 to 8 hours depending upon plant load. A feedback controller must overcome this variable lag to regulate oxygen flow based upon reactor 4 oxygen purity. If process disturbances are rapid, they may occur too quickly to be sensed in reactor 4. The net result of this variable time lag is that few full scale plants operate using this control strategy. Figure 9.3 shows a simulation of a near-optimally tuned proportional-integral (PI) feedback controller which is sensing reactor 4 partial pressure and manipulating oxygen feed rate. The treatment plant is treating a diurnally varying wastewater COD concentration and flow rate. The control system is performing very well, and upon initial examination appears to be adequate to control full-scale treatment plants. Later we will see that this control system is too finely tuned and fails when a more severe disturbance, such as wet weather flow, occurs.

The second control system shown in Figure 9.2 is based upon sensing reactor 1 pressure. As oxygen is consumed due to biological activity, carbon dioxide is liberated; however, less carbon dioxide is liberated in part because of its much greater solubility in water. The net result is a decrease in total system pressure as oxygen is consumed. Because all four

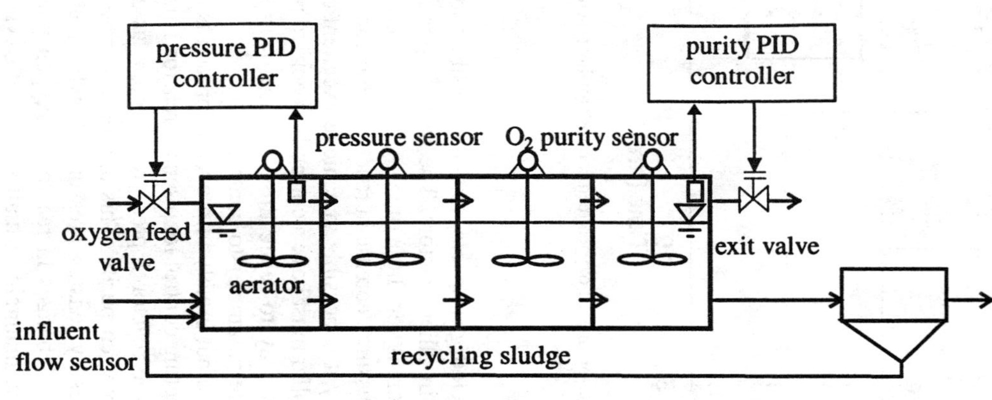


**Figure 9.2** Two conventional oxygen feed control strategies.

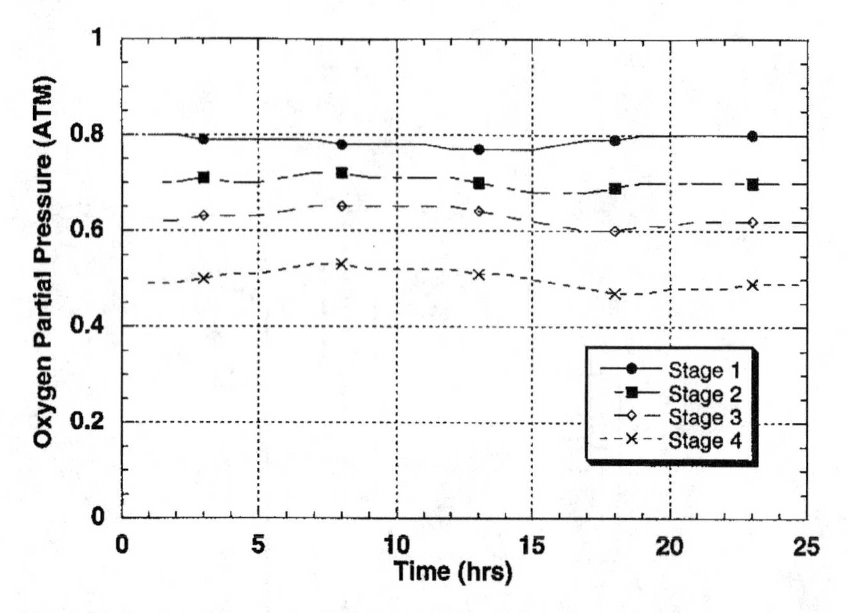


**Figure 9.3** Results of a near-optimal PI oxygen feed controller using reactor 4 purity as a sensed variable.

reactors have connected gas spaces, oxygen consumption in any reactor affects the pressure in other reactors. The most important aspect of this phenomena is that there is very little time lag between oxygen consumption in one reactor and pressure decrease in reactor 1; the controller sees almost no time lag. In theory it is possible to achieve excellent control using this approach. The controller must have a correctly specified pressure set point, and there is no easy way to relate this set point with reactor 4 purity, which is the desirable parameter to control.

A slightly modified but greatly improved approach is to use a two loop controller. The fast acting control loop senses reactor 1 pressure. A slower acting loop adjusts the reactor 1 pressure set point as a function of the desirable reactor 4 oxygen purity. This approach can respond to quickly changing disturbances by sensing pressure. Slowly changing phenomena are sensed by average changes in reactor 4 oxygen purity. This control system can, in theory, maintain virtually constant oxygen purity in each reactor for commonly found diurnal variations in influent COD and flow rate. Unfortunately it suffers from two problems: large changes in conditions, such as wet weather flow, and unreliable pressure signals.

The pressure in reactor 1 is a problematic control signal. As indicated earlier, the pressure in reactor 1 of a typical HPO-ASP is only 8 to 12 mm Hg above ambient pressure. Sensing this signal is complicated by conditions in the reactors. An increase in hydraulic flow rate can produce increased pressure in the reactor. This is because flow through the reactors is regulated by hydraulic structures such as weirs; the water level in the reactor must increase with increasing flow rate to increase flow over the weir or other hydraulic structure. This small increase in pressure is interpreted by the control system as a decrease in oxygen demand, when it most like will produce an increase in oxygen demand.

Sensing pressure is also complicated by the effect of the aerators. Surface aerators may surge and create standing waves in the reactor. This can result in noisy pressure signals which prevent proper controller operation.

Different treatment plants have had varying degrees of success using reactor 1 pressure control systems. Some plants have great success and others quickly turn off the control system in favor of manual operation. The differences leading to varying degrees of success probably relate to the plant's hydraulics. A cause and effect relationship between plant hydraulics and reactor 1 pressure, to the best knowledge of the author, have not been developed. There are probably many detailed, empirical factors which affect the relationship. In plants where reactor 1 pressure controls do not work well, various types of stilling wells to create less noisy signals have been investigated, with limited success.

A novel scheme proposed by Clifft (1991) uses a blower to vent gas from reactor 4. The blower takes suction from reactor 4 headspace. This allows the pressure in stage 4 to be reduced, because elevated pressure is not needed to force gas from the reactor. The reduced gas pressure in reactor 4 reduces all reactor pressure which will decrease leakage and save on HPO costs. A second advantage is that gas flow rate can be measured on the pressure side of the power, and standard devices (e.g., orifice plates, flow tubes, etc.) can be used.

## REACTOR DISSOLVED OXYGEN CONTROL

It is also desirable to control reactor DO concentration. HPO plants typically operate at higher DO concentrations than conventional air plants. The DO concentration set points in HPO plants are often 6 mg/L, which is much higher than the 0.5 to 2.0 mg/L found at air plants. At first this may seem wasteful, but one must realize that the increased oxygen purity results in much greater DO saturation concentrations. At 80% oxygen purity, which is typical for reactor 1, the DO saturation at 20°C is approximately 36 mg/L.

Controlling DO at HPO plants is more difficult than for air plants. There are a variety of reasons. The first relates to the oxygen partial pressure in each reactor. If it is not maintained relatively constant by the previously cited control systems, the driving force for oxygen transfer will vary. The control system has to respond to two disturbances: changing oxygen uptake rate due to changing plant load, and changing driving force due to changing plant load. The impact of the two disturbances tend to be correlated. Increasing plant load decreases DO concentration and decreases oxygen purity due to greater demand. The decrease oxygen purity decreases driving force, further decreasing DO concentration. Therefore, controlling oxygen purity in each reactor is the first step in providing DO control. In is also more important to control DO at an HPO plant. High DO not only creates more aerator power draw, it also consumes more HPO gas.

Figures 9.4 and 9.5 show the ranges of required aerator powers of a hypothetical treatment plant, corresponding to a large west coast HPO-ASP (Tzeng, 1992). Figure 9.4 shows the range when vent purity is manually set at a constant value while Figure 9.5 shows the range when optimally controlling gas purity. The four small graphs show the ranges for different loading conditions, such as an average day or a maximum day. Aerator transfer rate ($\alpha K_L a$, the mass transfer coefficient in process water, $hr^{-1}$) is shown for various loading conditions at various reactor 4 oxygen purity. Upon visual inspection it is easy to see the dramatically reduced rage of $\alpha K_L a$ required to maintain the set point DO. Surface aerators normally do not have sufficient turn up and turn down capability for the uncontrolled gas purity case.

To control DO the power to the aerator must be manipulated. Recall that most HPO plants use surface aerators. Aerator power and mass transfer can manipulated by changing aerator speed or blade submergence, if liquid-level sensitive propellers are used. Blade tip extensions are another way to change aerator power but the change cannot be made in real time. Variable frequency drives can effectively change motor RPM which will reduce or increase aerator power. Reducing liquid level using movable weirs is also possible, but more cumbersome. Some liquid-level sensitive propellers have experienced surging under some conditions.

Another complicating aspect of DO control at HPO plants is DO probe maintenance. The probes must be inserted into covered reactors. Removing a probe requires that the orifice in the reactor be sealed. The probes are not visible to the operators and tend to be harder to maintain.

## REACTOR SIZES

Reactor size can influent process dynamics and process controllability. Tzeng (1992) also investigated the relative reactor sizes and head space volume. He found that the relative reactor sizes (there is no requirement that all four reactors have the same size) had little effect on the natural

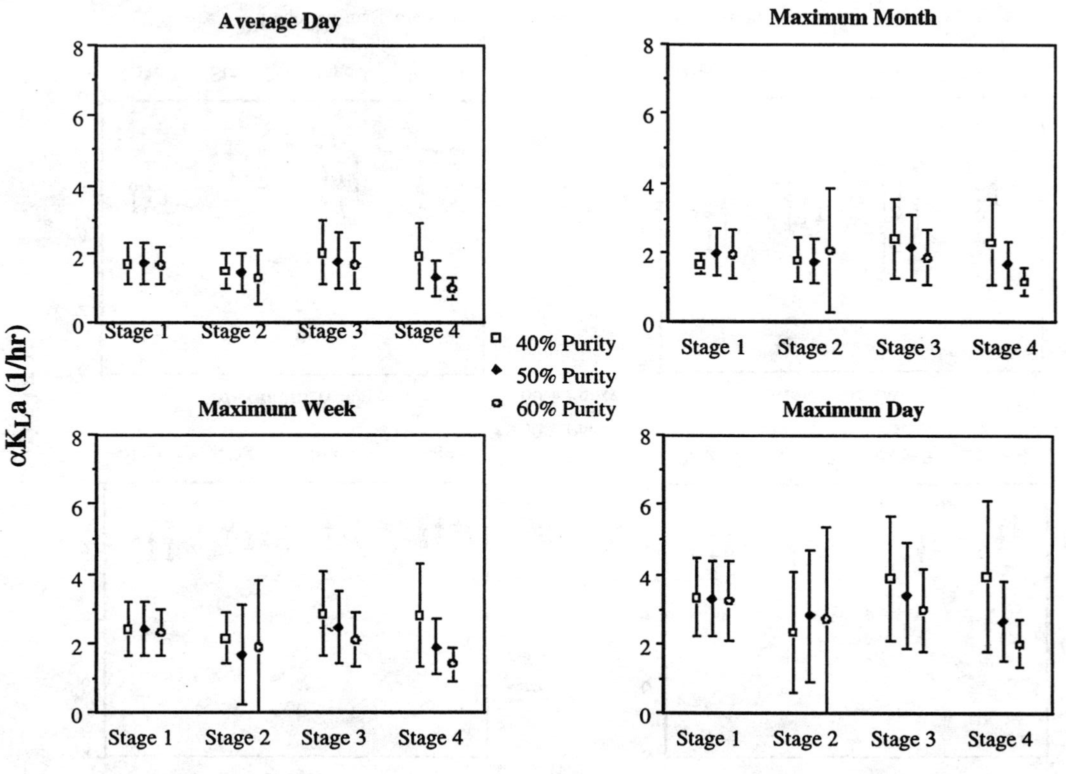


**Figure 9.4** Required mass transfer coefficients to control DO without gas purity control.

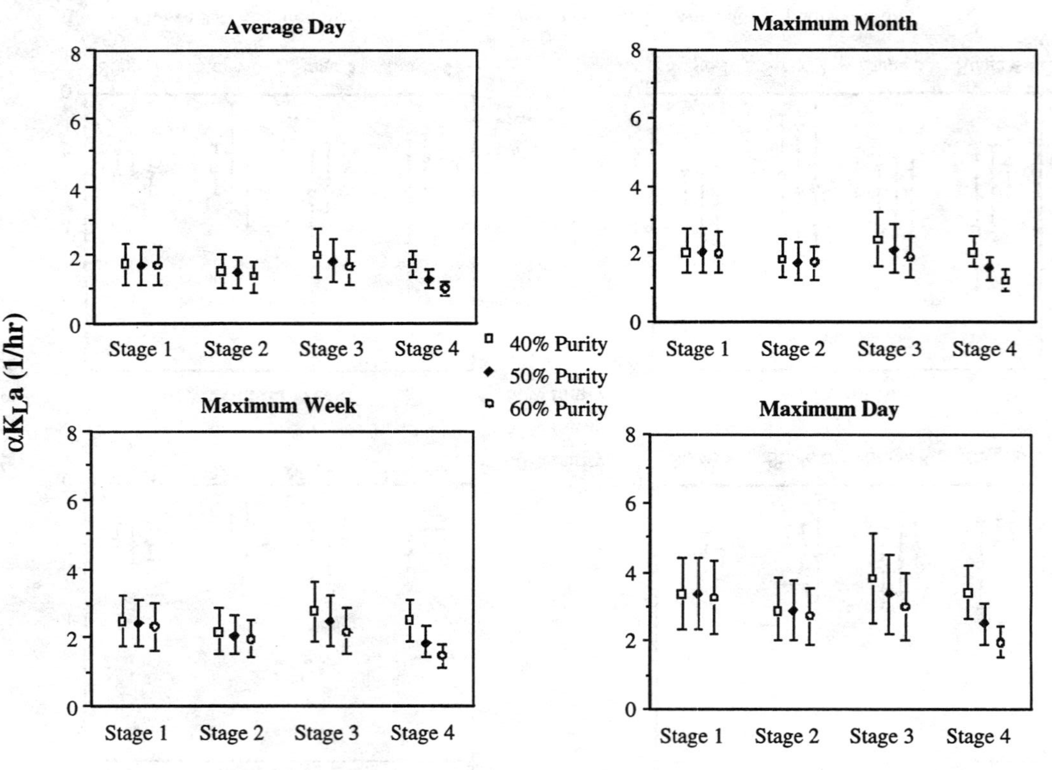


**Figure 9.5** Required mass transfer coefficients to control DO with gas purity control.

or uncontrolled DO and oxygen purity variability. Head space volume did have a large effect. It was possible to reduce the uncontrolled variability in head space purity by 36% by increasing head space volume by a factor of three. This is about 10% increase in total reactor volume. The reduced variability would facilitate control.

## KNOWLEDGE BASED CONTROL STRATEGIES

The previous section has discussed and described conventional control strategies that can use simple signals for control actions. There is much more process knowledge available from an HPO-APS that can be used for improved process control. For example, the oxygen purity profile over reactors 1 to 4 change with plant loading. It should be possible to use information from this profile to better control the process. Yuan (1994), Yin and Stenstrom (1994) and Yin (1995) have investigated knowledge-based control strategies.

Fuzzy logic is a powerful tool that can be used to improve control techniques. Fuzzy logic is useful when fundamental information about a process is not known, but empirical knowledge is known. Yin and Stenstrom (1994) simulated a fuzzy logic controller for an HPO-ASP and compared its performance to conventional control strategies. The fuzzy logic controller was used in a feed-forward as well as feedback mode.

Figure 9.6 shows the fuzzy logic controller block diagram. A key aspect of this controller is that it can use information before it has affected the process. In this case, disturbances in wastewater flow rate can be sensed and control actions taken before a process upset occurs. Table 9.1 shows five control strategies that were built upon the fuzzy logic controller and the previously described process variables used in the conventional control strategies.

The various inputs and feed back variables are manipulated in the fuzzy logic algorithm to determine the desired control actions. The fuzzy logic algorithm might reason as follows to control oxygen partial pressure:

> ''If the error is negative very large then the oxygen feed valve opening is very small; if the error is positive very large then the oxygen valve opening is very large.''

The fuzzy logic controller in this instance has only seven rules. More information is provided in Yin and Stenstrom (1994).

The five strategies were simulated in a manner shown previously and the results for normal weather (e.g., only the normal diurnal variations in process inputs) are shown in Figure 9.7, which shows the simulation for a 48 hour period. Strategy 5 in Table 1 most resembles a conventional

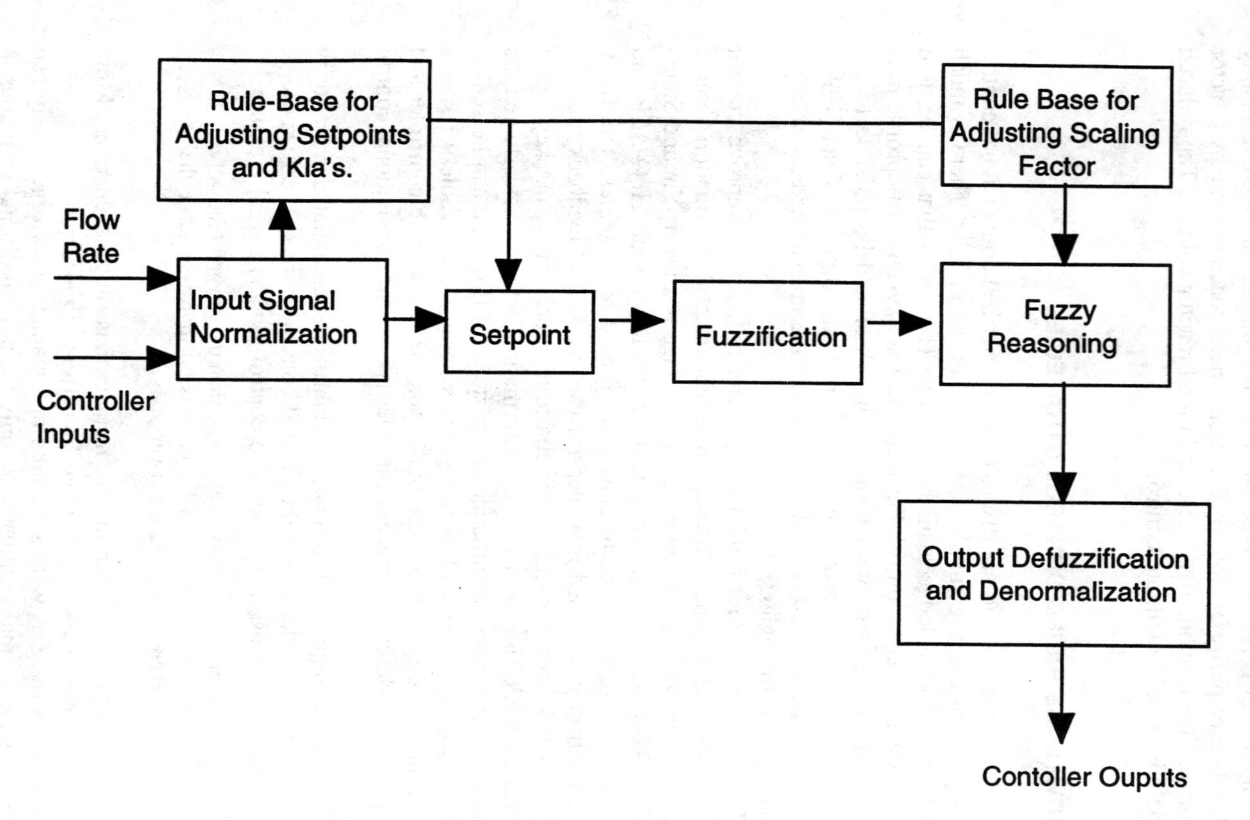


**Figure 9.6** Fuzzy logic control structure.

TABLE 9.1. Fuzzy Logic Control Strategies.

| Strategy No. (1) | Strategy Description (2) | Controller Inputs (3) | Controller Outputs (4) |
|---|---|---|---|
| 1 | set point (SP) control for stage 4 DO combined with $K_L$ a regulation for each stage based on influent flow measurement | stage 4 DO concentration; influent flow rate. | $O_2$ feed valve opening; $K_La$'s of each stage. |
| 2 | SP control for stage 1 total pressure with SP adjusted based upon influent flow rate, SP control for stage 4 $O_2$ purity and with SP is adjusted based upon stage 4 DO error | stage 4 $O_2$ purity; stage 4 DO; influent flow rate. | $O_2$ feed valve opening, motor speed of exit vent device. |
| 3 | stage 1 total pressure SP control with SP adjusted based upon influent flow rate, adjusting vent O2 purity SP based upon influent flow rate, and maintaining SP by regulating vent flow rate | stage 1 total pressure; stage 4 $O_2$ purity; influent flow rate. | $O_2$ feed valve opening; motor speed of exit vent device. |
| 4 | stage 1 pressure SP control with SP adjusted based upon influent flow rate, and regulating vent flow rate based upon stage 4 DO error | stage 1 total pressure; stage 4 DO; influent flow rate. | $O_2$ feed valve opening; motor speed of exit vent |
| 5* | stage 1 pressure SP control based upon stage 1 total pressure, and stage 4 $O_2$ purity SP control by regulating vent flow rate | stage 1 total pressure; stage 4 $O_2$ purity. | $O_2$ feed valve opening; motor speed of exit vent |

*This strategy is most similar to conventional strategies and is provided for comparison.

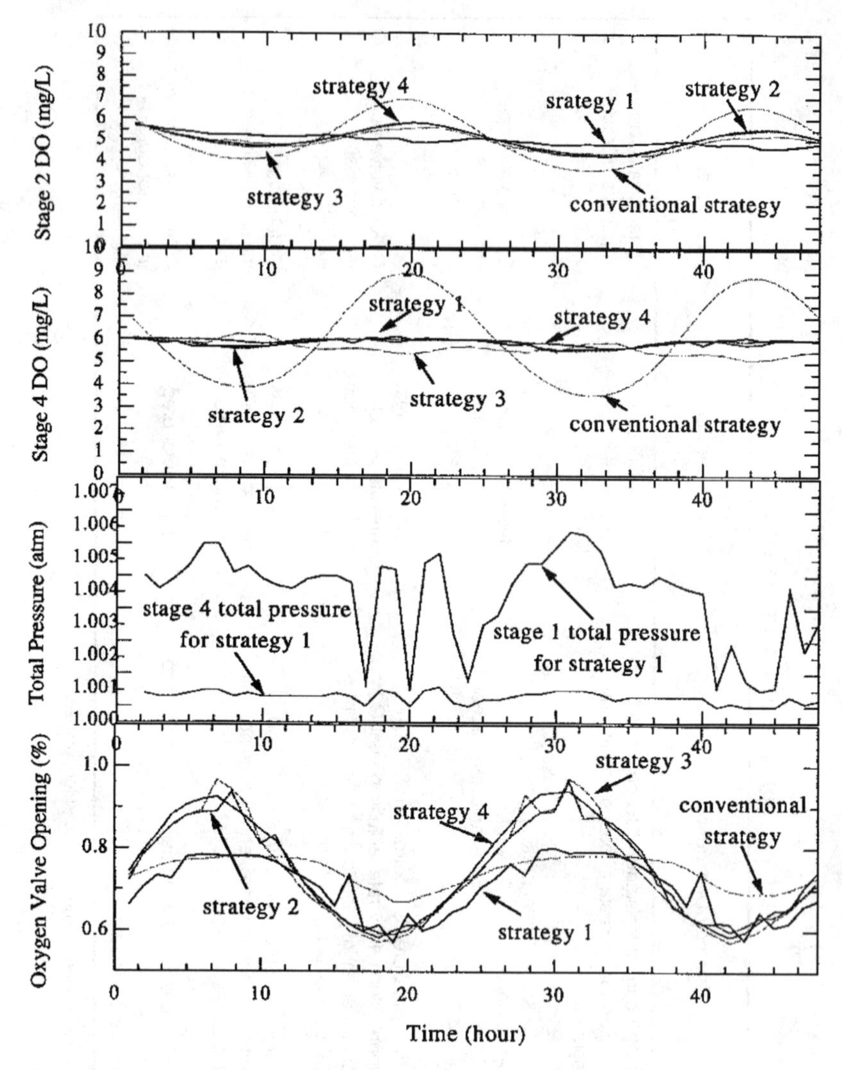


**Figure 9.7** DO concentration and oxygen feed valve position for the five normal weather simulations.

strategy. The top portion of Figure 9.7 shows the Reactor 2 DO concentration profile. The set point is 6 mg/L. All 5 strategies do reasonably well in maintaining the proper DO concentration. The DO concentrations for Reactor 4 are shown in the second panel from the top. In this case there is considerable deviation in DO for the conventional strategy.

The third and bottom panels of Figure 9.7 show reactor 1 total pressure and the oxygen feed valve opening. The value represents the oxygen generation plant capacity. It is desirable to maintain as constant operation as possible, while still meeting the oxygen demands of the process. Strategy 1 shows superior performance because it maintains the proper DO concentration while turning the oxygen plant up or down as little as possible. One can intuitively understand this improvement by considering the feed forward aspects of the controller. The fuzzy logic controllers are able to sense the process input (flow rate) and anticipate the required oxygen feed control action. These strategies do not need to turn up or turn down the oxygen plant to return to the set point.

Figure 9.8 shows another advantage of the fuzzy logic controller. In this simulation, a large flow is applied to the process, simulating wet weather storm flow. A conventional PI or PID controller will have some difficulty adjusting to the rapid change. By applying scaling factors to the fuzzy controller and triggering the rules when certain, predetermined large flow rates are observed, the controller can rapidly adjust to new conditions. Similar rules can be developed for dry weather conditions.

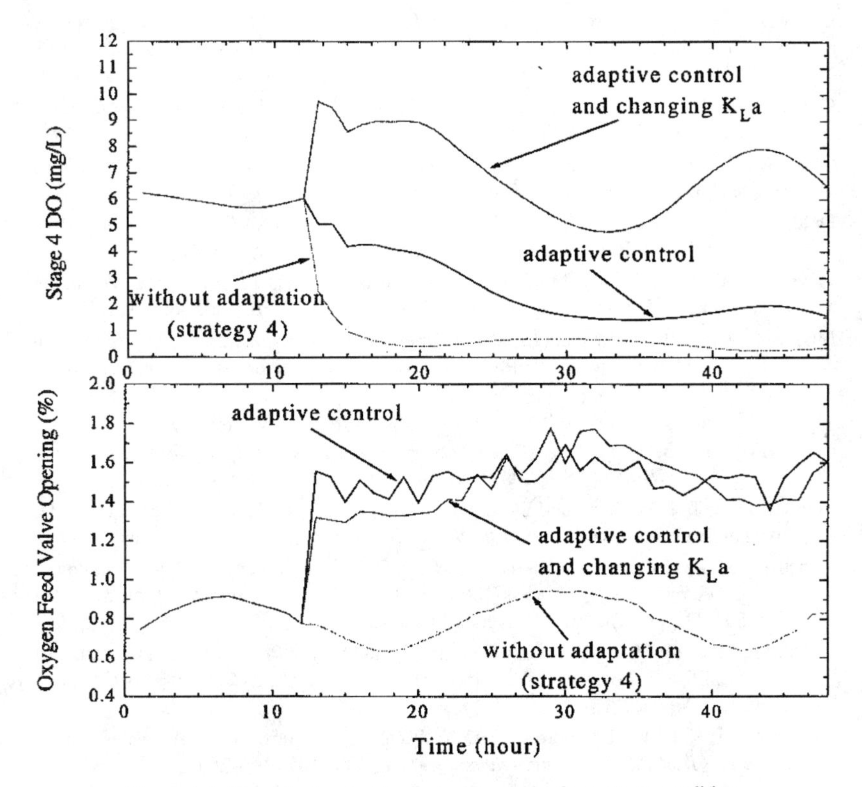


**Figure 9.8** Fuzzy logic adaptive controller results for a storm condition.

## CONCLUSIONS

This chapter has briefly reviewed conventional and knowledge-based control systems for the high purity oxygen activated sludge process. Several problems unique to this process have been discussed and simulated. Conventional, feedback strategies have been simulated, which, under ideal conditions adequately control the process. More powerful and adaptive fuzzy logic controllers have also been simulated. These controllers achieve better results for the same disturbances than the conventional counter parts. The fuzzy logic controllers require no information that is not routinely collected at a modern treatment plant. They only require computer based controllers to implement the fuzzy rules. Although not explicitly discussed in the chapter, other types of knowledge based control systems are also possible.

## REFERENCES

Busby, J. B. and J. F. Andrews (1975). ''Dynamic Modeling and Control Strategies for the Activated Sludge Process,'' *Journal of the Water Pollution Control Federation,* Vol. 59, 957.

Clifft, R. C. (1991). ''Gas Phase Control for Oxygen Activated Sludge,'' *J. of the Envi. Eng. Division, ASCE,* 118(3), 390–401.

Clifft, R. C. and J. F. Andrews (1981). ''Gas-Liquid Interactions in Oxygen Activated Sludge,'' *J. of the Envi. Eng. Division,* ASCE, Vol. 112, No. 1, 61–77

Dold, P. L., G. A. Ekama, G. A. and G.v.R. Marais (1980). ''A General Model for the Activated Sludge Process,'' *Prog. Wat. Technol.,* Vol. 12, No. 6, 47.

Lawrence, A. W. and P. L. McCarty (1970). ''Unified Basis for Biological Treatment Design and Operation,'' *J. of the Sant. Eng. Division, ASCE,* Vol. 96, No. 3, 757.

Linden, R. K. S. (1979). ''Model for Minimizing Energy Requirements in the Pure Oxygen Activated Sludge Process,'' Ph.D. dissertation, University of California, Davis, Davis, CA.

McWhirter, J. R. and Vahldieck (1978). Oxygenation Systems Mass Transfer Design Considerations. In *The Use of High-Purity Oxygen in the Activated Sludge Process,* 1, 235–260, J. R. McWhirter, Ed., CRC Press, Inc., West Palm Beach, Florida.

Mueller, J. A., T. J. Mulligan and D. M. Di Toro (1973). ''Gas Transfer Kinetics of a Pure Oxygen Transfer System,'' *J. of the Envi. Eng. Division, ASCE,* Vol. 9, No. 3, 269.

Stenstrom, M. K. (1976). ''A Dynamic Model and Computer Compatible Control Strategies for the Activated Sludge Process,'' A Dissertation Submitted in Partial Fulfillment for the Ph.D. Degree, Clemson University, Clemson, SC.

Stenstrom, M. K. (1990). ''Westpoint Treatment Plant Oxygen Process Modeling,'' University of California, Los Angeles, Engineering Report Number UCLA ENG 90-17, March 1990, Los Angeles, CA.

Stenstrom, M. K. and J. F. Andrews (1979). ''Real-Time Control of the Activated Sludge Process,'' *J. of the Envi. Eng. Division, ASCE,* Vol. 105, 245–260.

Stenstrom, M. K., W. H. Kido, R. F. Shanks, and M. Mulkerin (1989). ''Estimating Oxygen

Transfer Capacity of a Full Scale Pure Oxygen Activated Sludge Plant,'' *Journal of the Water Pollution Control Federation,* Vol. 61, 208–220.

Tzeng, C-J. (1992). ''Advanced Dynamic Modeling of the High Purity Oxygen Activated Sludge Process,'' A Dissertation Submitted in Partial Fulfillment for the Ph.D. Degree, Univ. of California, Los Angeles, CA.

Yin, M. T. (1995). ''Developing a Decision Support System for Operation and Control of the High Purity Oxygen Activated Sludge Process,'' A Dissertation Submitted in Partial Fulfillment for the Ph.D. Degree, Univ. of California, Los Angeles, CA.

Yin, M. T. and Stenstrom, M. K. (1994). Application of Fuzzy Logic Algorithm to Gas Phase Control for High Purity Activated Sludge Process, *J. of the Envir. Eng. Division, ASCE,* Vol. 122, No. 6, 484–492.

Yin, M. T., Yuan, W. and Stenstrom, M. K. (1994). A Decision Support System for Wastewater Treatment Plant Operations, *Proceedings of First ASCE Congress on Computing in Civil Engineering,* June 20–22, Washington, D.C.

Yin, M. T., W. Yuan, M. K. Stenstrom and D. Okrent (1994). ''A Simulator for the High-purity Oxygen Activated Sludge Process,'' *Proceedings of the 67th Annual Water Environment Federation Conference and Exposition,* Chicago, IL, October 15–18, 1994, #AC942706, Vol. I., 203–211.

Yuan, W., D. Okrent and M. K. Stenstrom (1993). ''Model Calibration for the High-Purity Oxygen Activated Sludge Process—Algorithm Development and Evaluation,'' *Water Science and Technology,* Vol. 28, No 11–12, pp. 163–171.

Yuan, W. (1994). ''Dynamic Modeling and Expert Systems for the Activated Sludge Process,'' A Dissertation Submitted in Partial Fulfillment for the Ph.D. Degree, Univ. of California, Los Angeles, CA.

Yuan, W., M. T. Yin, M. K. Stenstrom and D. Okrent (1994). ''Modeling the Oxygen Activated Sludge Process Using the IAWQ Activated Sludge Model No. 1,'' *Proceedings of the 67th Annual Water Environment Federation Conference and Exposition,* Chicago, IL, October 15–18, 1994, #AC946603, Vol. I. pp. 325–335.

# Index


activated sludge process, 103
adaptive data filter , 221, 223
Advanced Continuous Simulation Language (ACSL), 284
aerobic respiration, 100
agents, 163
analog computer, 91
artificial neural network (ANN), 11, 153
automatic calibration, 222, 259
automatic control, 181
backward chaining, 140
biochemical oxygen demand (BOD), 109, 125
blackboard models, 286
block language, 89
BOD meters, 109
causal relationships, 242
certainty, 143
certainty factors, 143, 294
Clean Water Act, 315
combined sewer overflows, 175
communications, 92
computer simulation, 3, 10
constraints, 6
control, 124, 167
control design, 238
control objectives, 124, 283
control strategies, 316, 319
control system design, 4, 281
coupling (of controllers), 210
death-regeneration, 102
decay rates, 115
design, 4
deterministic models, 12
direct methods, 111
dissolved oxygen, 100
dissolved oxygen control, 323
distributed control systems, 134
drainage basins, 176
drainage systems, 175
dynamic control elements, 176
dynamic model, 3, 5, 9, 283
dynamic parameter estimator, 221, 231
empirical model, 11, 282, 284
endogenous metabolism, 101
endogenous respiration rate, 103
Ethernet, 95
expert systems, 139
fault detection, 222, 260, 273
feedback controller, 245, 320
feedforward control, 242, 253
fieldbus, 93
fine-pore aeration, 318
forecasting, 223, 261
FORTRAN, 195
forward chaining, 140
fuzzy algorithm, 147
fuzzy control, 296, 303
fuzzy expert system, 293
fuzzy logic, 327
fuzzy membership, 293
fuzzy models, 9, 286
fuzzy number, 291
fuzzy relation, 147

fuzzy rules, 145
fuzzy sets, 144
fuzzy variable, 146
gain scheduling, 87
generalization, 137
genetic algorithms, 157
GPS-X, 112, 284
high purity oxygen (HPO) activated sludge process, 315
hydraulic models, 186, 208
if-then rules, 7
inflow forecasts, 197
inhibitory environment, 310
instrumentation, 176
instrumentation, control & automation (ICA), 4
integrated computer control system, 219
intelligent control, 281, 297
intelligent system, 135
interactions, 4
internal model, 197
iterative process, 13
kinetics, 283
knowledge based control, 133, 327
linguistic model, 285
linguistic value, 146
loading plan, 311
local area network, 94
manipulated variables, 242
manometric method, 108
mass balances, 318
mathematical model, 3
Matlab, 239
mechanistic models, 10, 150, 152
methods, 141
model based control, 219
model calibration, 208
model classification, 7, 10
model development, 8
model optimization, 129
model predictive control, 196
modeling, 7, 149
modeling respiration, 110
Monod equation, 282
mutation, 159
nitrification, 112, 272
nitrification activity, 305
non-linear behavior, 88
numerical techniques, 186
object-oriented, 136, 288
object-oriented design, 138
onion model, 133
open systems interconnection (OSI), 92
operational objectives, 124
operator interface, 256
optimal control algorithms, 190
optimization, 119, 156, 261
over fitting, 227
oxidation-reduction potential (ORP), 301
oxygen demand, 11
oxygen measurement, 105
oxygen uptake rate, 89
physical models, 8
PID controller, 247
pollution control plant, 203
process, 4
process optimization, 223
programmable logic controller (PLC), 180, 91
pyramid model, 134
rainfall forecasting, 190, 206
readily biodegradable substrate, 111
real time control, 176
regulator station, 193
remote terminal unit, 180
representation, 135
reproduction, 159
RespEval, 221, 228
respiration rate, 99, 124
respirogram, 112, 296
respirometry, 99
rule based system, 183
rules, 139
runoff model, 207
sampling time, 85
scheduling operations, 312
scientific method, 13, 14
self tuning control, 88
sequencing, 90
settler model, 236
sewer networks, 189
signal tracking, 227
slowly biodegradable matter, 115
sludge blanket threshold concentration, 245
sludge settling, 271
specialization, 137
static control elements, 176
steady state models, 9, 107
stochastic models, 12, 152
supervisory control and data acquisition (SCADA), 134, 220
supervisory remote control, 178

time delays, 87
time scale, 13
total oxygen demand, 297
toxicity meters, 109
ultrafiltration (UF), 299
uncertainty, 13
volumetric method, 108
wastewater collection, 175
yield coefficient, 117
zeta potential, 236